Advances in Plant Reproductive Ecology and Conservation Biology

Advances in Plant Reproductive Ecology and Conservation Biology

Editors

Brenda Molano-Flores
James Cohen

MDPI

Basel • Beijing • Wuhan • Barcelona • Belgrade • Novi Sad • Cluj • Manchester

Editors
Brenda Molano-Flores
University of Illinois
Urbana-Champaign
Champaign, IL
USA

James Cohen
Weber State University
Ogden, UT
USA

Editorial Office
MDPI AG
Grosspeteranlage 5
4052 Basel, Switzerland

This is a reprint of articles from the Special Issue published online in the open access journal *Plants* (ISSN 2223-7747) (available at: https://www.mdpi.com/journal/plants/special_issues/plant_reproductive_ecology_conservation_biology).

For citation purposes, cite each article independently as indicated on the article page online and as indicated below:

Lastname, A.A.; Lastname, B.B. Article Title. *Journal Name* **Year**, *Volume Number*, Page Range.

ISBN 978-3-7258-1957-7 (Hbk)
ISBN 978-3-7258-1958-4 (PDF)
doi.org/10.3390/books978-3-7258-1958-4

Contents

About the Editors

Brenda Molano-Flores

Brenda Molano-Flores is a Principal Research Scientist with the Illinois Natural History Survey, Prairie Research Institute at the University of Illinois Urbana-Champaign, IL, USA. For almost three decades, she has conducted research in the areas of plant reproductive biology, conservation biology, and plant ecology.

James Cohen

James Cohen is an Assistant Professor and Director of the Mary Carver Hall Herbarium (WSCO) at Weber State University in Ogden, UT, USA. Throughout his career, he has focused on various aspects of plant systematics, ranging from phylogenetics and conservation biology to floristics.

plants

Editorial

Introduction to the Special Issue of *Plants* on "Advances in Plant Reproductive Ecology and Conservation Biology"

Brenda Molano-Flores [1,*] and James I. Cohen [2,*]

1 Illinois Natural History Survey, Prairie Research Institute, University of Illinois Urbana-Champaign, 1816 South Oak Street, Champaign, IL 61820, USA
2 Department of Botany and Plant Ecology, Weber State University, 1415 Edvalson St., Dept. 2504, Ogden, UT 84408, USA
* Correspondence: molano1@illinois.edu (B.M.-F.); jamescohen@weber.edu (J.I.C.)

Plant reproductive ecology explores aspects of the biology and ecology of plants ranging from breeding systems, plant–pollinator interactions, seed germination, floral traits, and much more. Plant conservation biology is an interdisciplinary field encompassing plant reproductive biology, population genetics, systematics, modeling, management, and policy, among others. This Special Issue on "Advances in Plant Reproductive Ecology and Conservation Biology" focuses on three main areas of research: population genetics, breeding systems, and ecology of common and rare plants. Each paper provides insights into new discoveries associated with these themes based on a wide variety of methods, with conservation biology being a thread that ties all of these studies together. Lastly, this Special Issue brings research voices from five of the seven continents, and from the USA which, in 2023, celebrated 50 years of the Endangered Species Act of 1973 [1]. We have received papers from California, Tennessee, and Florida, hotspots of biodiversity in North America. Below we provide a synopsis of each paper.

Eight studies in the Special Issue use population genetic methods and demonstrate that these approaches remain useful for exploring conservation biology. Each tried to gain a better understanding of rare species using a distinct approach. Tsaballa et al. [2] and Mansour et al. [3] investigated economically important species that have been impacted by human development. Tsaballa et al. [2] employed molecular barcoding to uncover the identity of wild species of orchids in Salep, a powder used in beverages and food for medicinal purposes that is culturally important in the eastern Mediterranean, and this has resulted in the overharvesting of orchids. The authors recognized that Salep, in Greece, is composed of species from four genera, with the greatest percentage from *Dactylorhiza* Neck. ex Nevski (Orchidaceae), an unexpected result given that species of *Orchis* L. (Orchidaceae) produce superior Salep. Barcoding is useful in this case as it provides evidence that overharvesting has resulted in a shift in the species used for Salep production. Mansour et al. [3] also examined a species, *Commiphora gileadensis* (L.) C.Chr. (Burseraceae) used by humans for perfume and medicine. Using microsatellites, the researchers found that even though the number of populations has declined, genetic diversity can still be recognized, and this diversity is not geographically structured, possibly due to human activity in Saudi Arabia.

Morris et al. [4], using microsatellite loci, and Cantley et al. [5], Hanko et al. [6], Cohen and Turgman-Cohen [7], and Guilliams and Hasenstab-Lehman [8], employing single nucleotide polymorphisms (SNPs) generated via reduced representation sequencing methods [9,10], elucidated patterns of genetic diversity for rare plant species. Chamorro-Premuzic et al. [11] studied *Dalea foliosa* (Gray) Barneby (Fabaceae), a species restricted to populations in Alabama, Illinois, and Tennessee, USA. These authors found that the majority of the genetic diversity was present in the center of geographic diversity, Tennessee, and nearby populations tended to be genetically similar in this area; however, populations in

Citation: Molano-Flores, B.; Cohen, J.I. Introduction to the Special Issue of *Plants* on "Advances in Plant Reproductive Ecology and Conservation Biology". *Plants* **2024**, *13*, 605. https://doi.org/10.3390/plants13050605

Received: 15 February 2024
Accepted: 20 February 2024
Published: 23 February 2024

Alabama and Illinois were quite homogenous, suggesting that genetic diversity decreases toward the edges of the geographic range of the species. These authors also used ecological niche modeling to suggest that at present the niche for the species may be wider than observed currently, but that by 2070, this will have shrunk considerably, which can result in an issue for long-term persistence for the species. Cantley et al. [5] examined species of *Solanum* L. (Solanaceae) in the Australian Monsoon Tropics. These authors compared dioecious and cosexual species and found that while dioecious species had greater genetic diversity compared to cosexual ones, all populations of the five studied species had high rates of inbreeding. Hanko et al. [6] studied a rare species, *Scutellaria floridana* Chapm. (Lamiaceae), that frequently reproduces sexually and asexually. The researchers identified low overall genetic diversity across the species, which was, in part, the result of clonal reproduction; however, some populations harbored higher levels of genetic diversity providing the species with population structure. Cohen and Turgman-Cohen [7] investigated the Great Lakes endemic, *Iris lacustris* Nutt. (Iridaceae), a species that previously had been demonstrated to have limited genetic diversity across its geographic range [12]. Using SNPs, these authors not only recognized genetic diversity and population structure across the northern Great Lakes, but also hypothesized a pattern of western to eastern migration for the species. Guilliams and Hasenstab-Lehman [8] examined a rare species of *Eriodictyon* Benth. (Namaceae), *E. capitatum* Eastw., restricted to western Santa Barbara County in southern California. Most of the studied populations were quite genetically homogenous, which may be due to the clonal growth of the species. A phylogenomic analysis of *Eriodictyon*, also based on SNPs, recovered *E. capitatum* as monophyletic and as sister to *E. altissimum* P.V. Wells. Both species bear narrow leaves, an uncommon feature in the genus.

Among these eight studies focusing on population genetics, two common threads are woven. First, the breeding system and human influence are important in the current population structure of the species. This ranges from the orchid species used in Salep to the role of climate change in the shifts in weather in Australia to the modifications in the fire regime in south Florida [2,5,6]. Second, the hypothesized patterns of genetic diversity were not often recovered. Greater than anticipated genetic variation was found in *S. floridana*, *I. lacustris*, and *E. capitatum*, but only minimal differences were identified between dioecious and cosexual species of *Solanum*, *C. gileadensis*, *D. foliosa*, and for species used in Salep [2,3,5–7]. Collectively, these studies demonstrate the important role that field and laboratory studies play in ensuring a comprehensive understanding of population biology.

Five studies in the Special Issue examined aspects of plant breeding systems to illuminate conservation biology of diverse taxa. Pimienta and Koptur [13] examined the sphingophilous *Guettarda scabra* (L.) Vent. (Rubiaceae) and found that although the species has nocturnal flowers, the flowers remain open into the early morning, which allows for the plant to play host to a larger arthropod community. Ramalde et al. [14] explored dioecy in *Diospyros sericea* A.DC. (Ebenaceae), and these authors not only recognized vestigial sexual organs in the unisexual flowers but also identified some plants as being sexually leaky, with plants that develop staminate flowers also producing some fruits. In multiple species of *Iris*, Lozada-Gobilard et al. [15] investigated the role of visual floral cues as honest signals. The researchers found evidence of honest signaling, across multiple species of *Iris*, in a garden setting, but the effect was population-specific in natural environments, with abiotic factors possibly playing a role. In the rare *Macbridea alba* Chapm. (Lamiaceae), Johnson et al. [16] surveyed seven populations of the species. While seed production varied across the populations, with two having the majority of the seed output, overall seed production was low. The authors attributed this to multiple factors, including a small number of floral visitors and seed herbivory, a factor that had not been previously identified. Across three species of *Sorbus* L. (Rosaceae) endemic to eastern Slovakia, Kolarčik et al. [17] recognized tetraploid and triploid species and found that while seed production was uncommon, rare fertilization events were necessary and sufficient to retain the long-term viability of the small populations of these endemic species.

These studies on breeding systems collectively point to the need to understand the myriad manners in which plants reproduce in order to develop successful conservation biology projects. Additionally, and possibly more importantly, these studies demonstrate that it is crucial to take a closer look at supposedly understood biological phenomena. For example, Ramaldes et al. [14] recognized fruit development on staminate plants of the dioecious *D. sericea;* Pimienta and Koptur [13] identified new floral visitors for *G. scabra* given that the plants were open during dawn, not just at night; Johnson et al. [16] recognized seed herbivory as a potential limiting factor in the success of *M. alba;* and Kolarčik et al. [17] found the critical role rare reproductive events play in maintaining species. These studies on plant breeding systems should serve as a reminder to botanists and conservation biologists of the important role of careful field observations.

Lastly, three studies in the Special Issue focused on several aspects of the ecology of three rare species. Wadl et al. [18] provide a comprehensive review of the research accomplished with the endangered *Pityopsis ruthii* (Small) Small (Asteraceae), endemic to a small geographic area in Tennessee, USA. A collaborative research team has worked for almost three decades to better understand the biology and ecology of this species and factors that could increase its vulnerability to extinction. This paper also highlights the value of partnerships between researchers and state and federal agencies as an integral component for the conservation of the species. Work by Lange et al. [19] focuses on another USA endangered taxon *Sideroxylon reclinatum* Michx. subsp. *austrofloridense* (Whetstone) Kartesz and Gandhi (Sapotaceae). In this paper, the authors used microscopy work to identify morphological differences between the rare cryptic *Sideroxylon reclinatum* subsp. *austrofloridense* and *Sideroxylon reclinatum* subsp. *reclinatum*, the more common subspecies. In addition, the authors developed habitat suitability models (HSMs) for the species. Field verification of the HSMs in search of new populations of that rare cryptic species and new morphological characters to identify the subspecies will assist with identification and protection of critical habitat designation for this rare taxon, even when hybridization is a concern between the two subspecies. Lastly, Heineman et al. [20] focus on the annual USA federally threatened *Acanthomintha ilicifolia* (A. Gray) A. Gray (Lamiaceae). The researchers conducted a common garden study to better understand climate change stressors (e.g., water availability) and adaptability across the range of the species. Focusing on above-ground growth (biomass, height, and width) and reproductive output (flower number, seed number, and seed viability), the study highlights the role of local adaptation in species' responses to climate change. These studies are a reminder of the value of research partnerships and that new and traditional data gathering approaches are key to the field of rare plant conservation.

As the editors of the Special Issue, we hope that as you read the papers associated with this Special Issue, you will find that each one of them advances our knowledge and understanding of plant reproductive ecology in various ways, and that the many collaborative efforts occurring for the study of rare plants are improving the field of plant conservation biology.

Author Contributions: Writing—original draft preparation, B.M.-F. and J.I.C.; writing—review and editing, B.M.-F. and J.I.C.; project administration, B.M.-F. and J.I.C. All authors have read and agreed to the published version of the manuscript.

Acknowledgments: We would like to thank all the authors for including their work as part of the Species Issue. In addition, many thanks to the *Plants* Editorial Office for the support that they provided to us, as without them the Special Issues could not have been possible.

Conflicts of Interest: The authors declare no conflicts of interest.

References

1. Gronewold, N. The International Legacy of the United States Endangered Species Act of 1973. *J. Int. Wildl. Law Policy* **2023**, *26*, 307–332. [CrossRef]
2. Tsaballa, A.; Kelesidis, G.; Krigas, N.; Sarropoulou, V.; Bagatzounis, P.; Grigoriadou, K. Taxonomic Identification and Molecular DNA Barcoding of Collected Wild-Growing Orchids Used Traditionally for Salep Production. *Plants* **2023**, *12*, 3038. [CrossRef] [PubMed]
3. Mansour, H.; Alamer, K.H.; Al-Hasawi, Z.M. Population Genetics, Genetic Structure, and Inbreeding of *Commiphora gileadensis* (L.) C. Chr Inferred from SSR Markers in Some Mountainous Sites of Makkah Province. *Plants* **2023**, *12*, 2506. [CrossRef] [PubMed]
4. Morris, A.B.; Visger, C.J.; Fox, S.J.; Scalf, C.; Fleming, S.; Call, G. Defining Populations and Predicting Future Suitable Niche Space in the Geographically Disjunct, Narrowly Endemic Leafy Prairie-Clover (*Dalea foliosa*; Fabaceae). *Plants* **2024**, *13*, 495. [CrossRef]
5. Cantley, J.T.; Jordon-Thaden, I.E.; Roche, M.D.; Hayes, D.; Kate, S.; Martine, C.T. A Foundational Population Genetics Investigation of the Sexual Systems of *Solanum* (Solanaceae) in the Australian Monsoon Tropics Suggests Dioecious Taxa May Benefit from Increased Genetic Admixture via Obligate Outcrossing. *Plants* **2023**, *12*, 2200. [CrossRef] [PubMed]
6. Hanko, G.R.; Vogel, M.T.; Negrón-Ortiz, V.; Moore, R.C. High Prevalence of Clonal Reproduction and Low Genetic Diversity in *Scutellaria floridana*, a Federally Threatened Florida-Endemic Mint. *Plants* **2023**, *12*, 919. [CrossRef] [PubMed]
7. Cohen, J.I.; Turgman-Cohen, S. The Conservation Genetics of *Iris lacustris* (Dwarf Lake Iris), a Great Lakes Endemic. *Plants* **2023**, *12*, 2557. [CrossRef] [PubMed]
8. Guilliams, C.M.; Hasenstab-Lehman, K.E. Conservation Genetics of the Endangered Lompoc Yerba Santa (*Eriodictyon capitatum* Eastw., Namaceae), including Phylogenomic Insights into the Evolution of *Eriodictyon*. *Plants* **2024**, *13*, 90. [CrossRef] [PubMed]
9. Ott, A.; Liu, S.; Schnable, J.C.; Yeh, C.-T.E.; Wang, K.-S.; Schnable, P.S. tGBS®genotyping-by-sequencing enables reliable genotyping of heterozygous loci. *Nucleic Acids Res.* **2017**, *45*, e178. [CrossRef] [PubMed]
10. Andrews, K.R.; Good, J.M.; Miller, M.R.; Luikart, G.; Hohenlohe, P.A. Harnessing the power of RADseq for ecological and evolutionary genomics. *Nat. Rev. Genet.* **2016**, *17*, 81–92. [CrossRef] [PubMed]
11. Chamorro-Premuzic, T.; Akhtar, R.; Winsborough, D.; Sherman, R.A. The datafication of talent: How technology is advancing the science of human potential at work. *Curr. Opin. Behav. Sci.* **2017**, *18*, 13–16. [CrossRef]
12. Hannan, G.L.; Orick, M.W. Isozyme diversity in *Iris cristata* and the threatened glacial endemic *I. lacustris* (Iridaceae). *Am. J. Bot.* **2000**, *87*, 293–301. [CrossRef] [PubMed]
13. Pimienta, M.C.; Koptur, S. More than Moths: Flower Visitors of a Night-Blooming Plant in South Florida Pine Rocklands, USA. *Plants* **2022**, *11*, 2799. [CrossRef] [PubMed]
14. Ramaldes, B.; Santos, R.; Rech, A.R.; Soares, M. Phenology and Floral Biology of *Diospyros sericea* A. DC. (Ebenaceae): Inconstant Males May Be behind an Enigma of Dioecy. *Plants* **2022**, *11*, 2535. [CrossRef] [PubMed]
15. Lozada-Gobilard, S.; Nielsen, N.; Sapir, Y. Flower Size as an Honest Signal in Royal Irises (*Iris* Section Oncocyclus, Iridaceae). *Plants* **2023**, *12*, 2978. [CrossRef] [PubMed]
16. Johnson, S.A.; Coons, J.; Zaya, D.N.; Molano-Flores, B. Assessing the Reproductive Ecology of a Rare Mint, *Macbridea alba*, an Endangered Species Act Protected Species. *Plants* **2023**, *12*, 1485. [CrossRef] [PubMed]
17. Kolarčik, V.; Mirková, M.; Mikoláš, V. Reproduction Modes and Conservation Implications in Three Polyploid *Sorbus* Stenoendemics in Eastern Slovakia (Central Europe). *Plants* **2023**, *12*, 373. [CrossRef] [PubMed]
18. Wadl, P.A.; Dattilo, A.J.; Call, G.; Hadziabdic, D.; Trigiano, R.N. *Pityopsis ruthii*: An Updated Review of Conservation Efforts for an Endangered Plant. *Plants* **2023**, *12*, 2693. [CrossRef] [PubMed]
19. Lange, J.J.; Angelo, C.L.; Revuelta, E.; Possley, J. Population Assessments of Federally Threatened Everglades Bully in Big Cypress National Preserve, Florida, USA, Using Habitat Suitability Modeling and Micromorphology. *Plants* **2023**, *12*, 1430. [CrossRef] [PubMed]
20. Heineman, K.D.; Anderson, S.M.; Davitt, J.M.; Lippitt, L.; Endress, B.A.; Horn, C.M. San Diego Thornmint (*Acanthomintha ilicifolia*) Populations Differ in Growth and Reproductive Responses to Differential Water Availability: Evidence from a Common Garden Experiment. *Plants* **2023**, *12*, 3439. [CrossRef] [PubMed]

plants

MDPI

Article

High Prevalence of Clonal Reproduction and Low Genetic Diversity in *Scutellaria floridana*, a Federally Threatened Florida-Endemic Mint

Gina Renee Hanko [1,†], Maria Therese Vogel [1,‡], Vivian Negrón-Ortiz [1,2,*] and Richard C. Moore [1]

1 Department of Biology, Miami University, Oxford, OH 45056, USA
2 Florida Ecological Services Field Office, U.S. Fish and Wildlife Service, 1601 Balboa Ave., Panama City, FL 32405, USA
* Correspondence: vivian_negronortiz@fws.gov
† Current address: U.S. Forest Service, P.O. Box 19001, Thorne Bay, AK 99919, USA.
‡ Current address: Atlanta Botanical Garden, 1345 Piedmont Avenue NE, Atlanta, GA 30309, USA.

Abstract: The threatened mint Florida skullcap (*Scutellaria floridana*) is endemic to four counties in the Florida panhandle. Because development and habitat modification extirpated several historical occurrences, only 19 remain to date. To inform conservation management and delisting decisions, a comprehensive investigation of the genetic diversity and relatedness, population structure, and clonal diversity was conducted using SNP data generated by ddRAD. Compared with other Lamiaceae, we detected low genetic diversity (H_E = 0.125–0.145), low to moderate evidence of inbreeding (F_{IS} = −0.02–0.555), and moderate divergence (F_{ST} = 0.05–0.15). We identified eight populations with most of the genetic diversity, which should be protected in situ, and four populations with low genetic diversity and high clonality. Clonal reproduction in our circular plots and in 92% of the sites examined was substantial, with average clonal richness of 0.07 and 0.59, respectively. *Scutellaria floridana* appears to have experienced a continued decline in the number of extant populations since its listing under the Endangered Species Act; still, the combination of sexual and asexual reproduction may be advantageous for maintaining the viability of extant populations. However, the species will likely require ongoing monitoring, management, and increased public awareness to ensure its survival and effectively conserve its genetic diversity.

Keywords: plant conservation genomics; *Scutellaria floridana*; clonality; guerrilla strategy; managed lands

Citation: Hanko, G.R.; Vogel, M.T.; Negrón-Ortiz, V.; Moore, R.C. High Prevalence of Clonal Reproduction and Low Genetic Diversity in *Scutellaria floridana*, a Federally Threatened Florida-Endemic Mint. *Plants* **2023**, *12*, 919. https://doi.org/10.3390/plants12040919

Academic Editors: Brenda Molano-Flores and James Cohen

Received: 11 December 2022
Revised: 8 February 2023
Accepted: 13 February 2023
Published: 17 February 2023

1. Introduction

Genetic diversity is a fundamental component of biodiversity and has profound effects on ecological processes, including community structure and the ability of populations to recover from disturbance. Research in the field of conservation genetics has shown that loss of genetic diversity due to inbreeding reduces the ability of a population to persist in unstable environments [1]. Despite its widely recognized importance, the implementation of conservation genetics in management is lagging behind research efforts [2–6]. Information on inbreeding, population divergence, and effective population size has the potential to greatly increase the efficacy of conservation practices such as population viability analysis and adaptive habitat management [7,8].

Understanding genetic factors such as diversity, reproductive strategies, and gene flow in rare plant species is especially crucial in an era of rapidly changing habitats. Due to their limited migration capabilities and often highly specialized adaptations to soil and moisture conditions, rare plants are especially vulnerable to the impacts of climate change and increasing land conversion due to human activities, and their survival will likely depend on their adaptive abilities [9]. Maximizing genetic diversity in rare plants and understanding the genetic impacts of conservation activities such as habitat restoration, assisted migration, and ex situ propagation are vital to their protection [10–12].

One such rare species is the Florida skullcap (*Scutellaria floridana* Chapman, Lamiaceae), a perennial wildflower endemic to four counties in the Florida panhandle (Figures 1 and 2). It grows in fire-dependent habitats such as longleaf pine wet forests and wet meadows and has a strong response to fire, typically flowering from April to December following fires [13]. Development of slash pine plantations has eliminated large areas of suitable habitat, and *S. floridana* was federally listed as threatened in 1992. Ten of the 29 historically documented populations of *S. floridana* have been extirpated due to habitat modification, and the 19 that remain continue to be threatened by urban development, timber farming, and fire suppression [13].

Figure 1. *Scutellaria floridana* plants, flowers, fruits, and habitat. (**A**) vegetative growth, ANF, (**B**) flowers, BRWMA, (**C**) fruits, BRWMA, (**D**) rhizome, BRWMA, and (**E**) typical habitat of *S. floridana*, ANF.

Only one study of genetic diversity in *S. floridana* has been conducted to date. This study used amplified fragment length polymorphism (AFLP) molecular markers to analyze genetic diversity among 197 samples collected from seven populations [14]. The authors reported moderate genetic diversity and low population differentiation. This study provides a basic groundwork for understanding the genetic diversity of the species; however, it is limited by including a small number of populations. In-depth genetic analysis of the remaining populations of *S. floridana* is vital to accurately determine their viability and inform the process of recovery under the Endangered Species Act (ESA).

Surveys have shown increases in the number of stems within several populations on managed lands over the past 10 years, and the majority of the remaining populations have good or excellent estimated viability [13]. However, population trends are poorly understood because plants can spread via underground rhizomes, which makes it difficult to determine how many stems are part of a single ramet rather than whether they represent distinct individuals [13]. The extent of clonal propagation in *S. floridana* represents a significant knowledge gap for the species. In addition, the type of clonal growth strategy, i.e., phalanx (closely spaced ramets with short internodes) or guerrilla (widely spaced ramets with long internodes) [15], is also unknown.

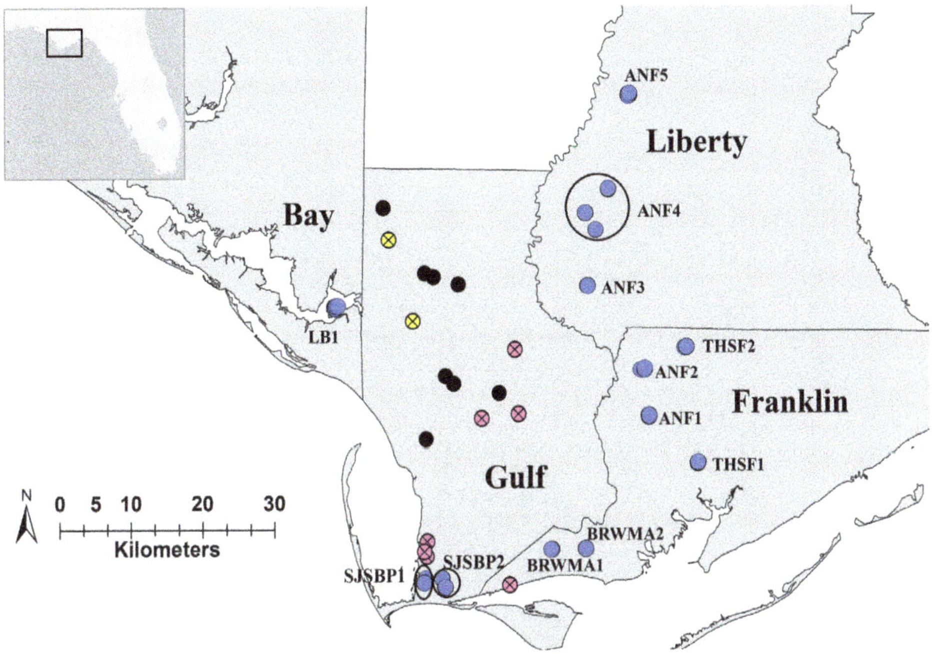

Figure 2. Locations of known occurrences of *S. floridana*. Black circles represent officially extirpated sites, yellow circles represent sites that were inaccessible at the time of our field surveys, pink circles represent sites in which we were unable to locate any individuals during field surveys, and blue circles represent populations that we successfully located and collected samples from for analysis of genetic diversity. Field surveys and sample collection were carried out in March and May of 2021. The Apalachicola River follows the boundaries of Gulf, Franklin, and Liberty counties.

While clonal propagation increases an individual's chance of reproduction, increases population sizes, enhances longevity in disturbance-prone habitats, and allows reproduction when conditions are unfavorable for flowering or seed germination, it also poses issues for rare species and complicates conservation efforts [16,17]. When clonal propagation becomes the primary reproductive strategy of a population, it can impact genetic variation and effective population size, or the number of reproductively viable individuals present in a population [18,19]. Since the stems that comprise a ramet are genetically identical, they are essentially a single individual for genetic purposes. As a result, a population that experiences high levels of clonal propagation and is composed of only a few genetically distinct individuals, or genets, would have an effective population size that is smaller than the number of stems observed. Additionally, it is often impossible for biologists to identify stems as genets or ramets in the field, leading to overestimations of population size and viability [17].

This study has three objectives to help evaluate the current status of *S. floridana*. These are to (1) investigate changes in land use that resulted in population extirpation over the last 20 years, (2) assess the genetic diversity of several populations across the species' range, and (3) determine the prevalence of clonality in the species. The conclusions drawn from this study will contribute to the recovery actions of *S. floridana* as specified in its recovery plan [20] and will address the current knowledge gap regarding the prevalence of clonality, which was identified as a significant problem impeding the ability to accurately assess the abundance of *S. floridana* [13]. Our assessment of genetic diversity will allow us to better evaluate the viability of several remaining populations, and our findings regarding the prevalence of clonality will allow us to generate more accurate estimates of effective population size.

2. Results

We were able to locate individuals of *S. floridana* in 12 of the 17 accessible populations (Figure 2). Five populations did not have any above-ground specimens of *S. floridana* after prolonged surveying throughout the entirety of the area. Three of these are located along power easements; one is located in a private reserve; and one is located in an empty lot in central Gulf County. While most of these locations contained at least one of the species commonly associated with *S. floridana*, such as *Aristrida stricta* (Wiregrass) and *Sphagnum strictum* (Pale bug-moss), they all lacked an overstory of *Pinus palustris* (Longleaf Pine) and were located less than 100 m from major roadways. We were also unable to locate plants in one population that is located in a small area of St. Joseph Bay State Buffer Preserve (SJSBP, Port St Joe, Gulf Co., USA) to the north of the main preserve. Several of the sites within the population were inaccessible due to fencing, but those that we could access were flooded by approximately 30–50 cm of water with an extremely thick understory of *Cyrilla racemiflora* (Titi) and little to no herb layer. Two of the populations that we attempted to visit were located in active rangeland or timber harvest operations in central Gulf County and were inaccessible. Of the 12 populations that contained actively growing individuals, Box-R Wildlife Management Area population 1 (BRWMA1, Apalachicola, Franklin Co., USA) was the smallest, containing seven small patches (<10 stems/patch and <5 m^2), while larger populations contained hundreds of stems. At the time of collection, we observed flowering individuals in several sites in the Apalachicola National Forest (ANF, Sumatra, Liberty Co., USA), one site in Tate's Hell State Forest (THSF, Franklin Co.), one site in the BRWMA, and the only site in the Lathrop Bayou Habitat Management Area (LB, Bay Co., USA).

2.1. Spatial Analysis

Between 2001 and 2019, the approximately 309,000 km^2 area that represents the entirety of *S. floridana*'s range experienced significant increases in medium- and high-intensity development (21% and 14%, respectively). Deciduous and evergreen forests decreased by 33% and 14%, respectively, while mixed forest, shrubland, grassland, and herbaceous wetlands increased by 18%, 19%, 39%, and 22%, respectively. Approximately 22% of *S. floridana*'s range falls within the borders of public lands, including state forests, national forests, habitat management areas, and preserves. These areas experienced no significant change (>2%) in land cover during this time frame, suggesting that changes were restricted to private lands (Table S1).

Areas within 500 m of the 12 *S. floridana* populations that we confirmed to be extant during our surveys are composed almost entirely of woody wetlands and evergreen forest, which account for over 88% of land cover. Notably, these areas also contain no mixed forest, pasture, or cultivated crops, and low, medium, and high intensity development and deciduous forest each account for <0.5% of land cover. There was minimal change in land cover in these areas between 2001 and 2019 (Figure 3).

Areas within 500 m of the five populations where we failed to find *S. floridana* contained more woody wetlands and less evergreen forest than the areas containing extant populations, with woody wetlands accounting for over 70% of land cover and evergreen forest accounting for 8.5% of land cover. They also contained more developed space than evergreen forest and, unlike the areas that contained extant populations, included high intensity development, pasture, and cultivated crops.

2.2. Population Structure and Diversity

The final dataset that we used for downstream analysis included 10,223 loci and 28,210 variable sites that were each present in at least 60% of individuals. The observed heterozygosity ranged from 0.04 to 0.14, with an average value of 0.11 (Table 1). The observed heterozygosity was substantially less than expected for BRWMA1, ANF5, SJSBP1, and THSF1 (ΔH = 0.099, 0.080, 0.055, and 0.050, respectively; Table 1). These populations also displayed high inbreeding coefficients (F_{IS} = 0.56, 0.50, 0.27, and 0.30, respectively).

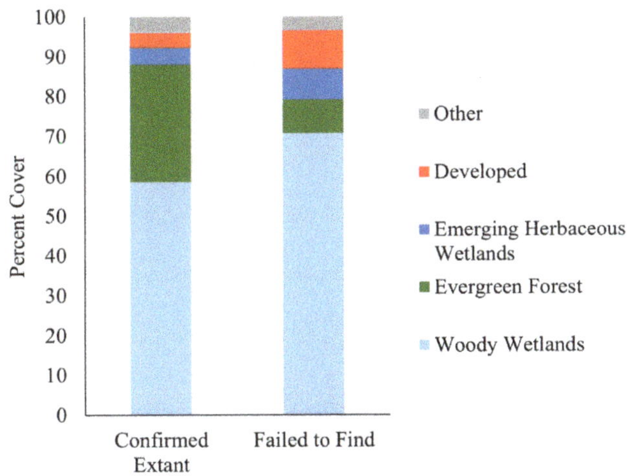

Figure 3. Percent cover of dominant land cover categories within 500 m of *S. floridana* populations that were confirmed to be extant and populations in which we were unable to locate any individuals during 2021 field surveys. See Table S1 for more information.

Table 1. Locations, number of sites, and samples collected; expected (H_E) and observed (H_O) heterozygosity; inbreeding coefficient (F_{IS}); relative evidence of population divergence (based on F_{ST} and DAPC); evidence of unique ancestry; and relative genetic value of twelve populations of *S. floridana*. Samples from LB were collected by Ms. Amy Jenkins (FNAI) and mailed to Miami University, OH, in April 2020. Genetic value determined as described in Materials and Methods.

County	Population	Sites	*N*		H_E	H_O	F_{IS}	Population Divergence	Unique Ancestry	Genetic Value
Bay	LB1	1	23		0.142	0.125	0.113	Moderate	Yes	Moderate
Franklin	ANF1	1	19		0.125	0.106	0.137	Moderate	Yes	Moderate
Franklin	ANF2	1	14		0.128	0.127	0.024	Very High	Yes	High
Franklin	BRWMA1	1	10		0.139	0.04	0.555	Moderate	No	Low
Franklin	BRWMA2	1	11		0.136	0.139	−0.01	Very High	Yes	High
Franklin	THSF1	1	14		0.131	0.081	0.297	Moderate	No	Low
Franklin	THSF2	2	18		0.131	0.13	0.025	High	Yes	High
Gulf	SJSBP1	2	7		0.145	0.090	0.268	Moderate	No	Low
Gulf	SJSBP2	1	12		0.145	0.136	0.060	Moderate	Yes	Moderate
Liberty	ANF3	1	18		0.126	0.132	−0.02	Very High	Yes	High
Liberty	ANF4	3	34		0.138	0.116	0.154	Low	Yes	Moderate
Liberty	ANF5	2	11		0.124	0.044	0.504	Moderate	No	Low
	Total:	17	191	Mean:	0.134	0.106	0.176			

Populations generally displayed moderate divergence from one another (F_{ST} = 0.05–0.15). Nine pairs of populations showed little to no differentiation from one another ($F_{ST} \leq 0.05$), and six pairs of populations showed high divergence from one another (F_{ST} = 0.15–0.25). Overall, LB1, SJSBP1, SJSBP2, THSF1, and THSF2 all showed low divergence from one another as a group, with pairwise F_{ST} falling at or below 0.07 for all pairs. Five populations, ANF1, ANF2, ANF3, BRWMA2, and THSF2, also had high F_{ST} values relative to all other populations and particularly high divergence values ($F_{ST} > 0.20$) relative to ANF5 and BRWMA1 (Figure 4). Within the Apalachicola National Forest, ANF5 showed moderate to high divergence from all other ANF populations despite its geographic proximity (Figure 4). Mantel test detected a significant positive relationship between F_{ST} and geographic distance ($R^2 = 0.29$, $p = 0.02$).

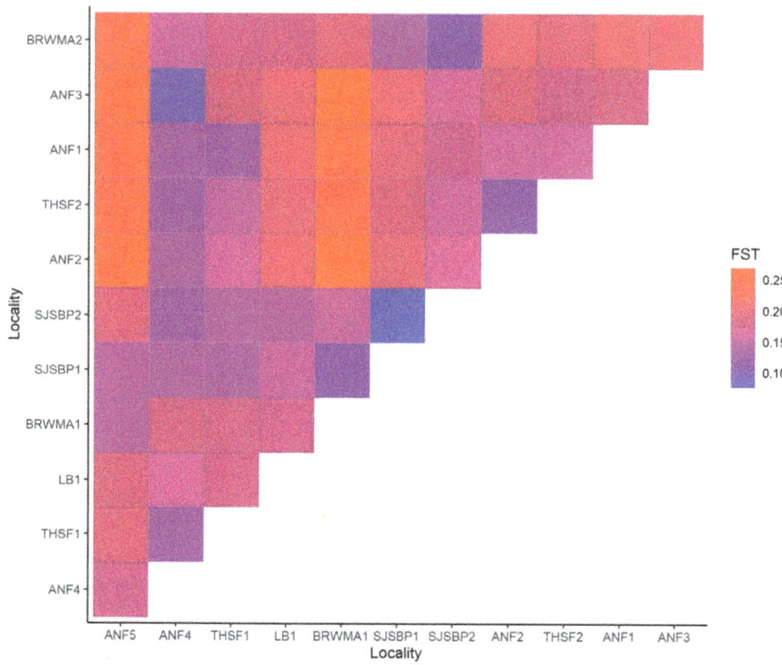

Figure 4. Heatmap visualizing pairwise F_{ST} values for twelve populations across *S. floridana*'s range.

Our analysis of ancestry proportions identified ten ancestral populations (K = 10) as the most likely scenario. Lathrop Bayou, SJSBP2, BRWMA2, ANF1, ANF2, ANF3, ANF4, and THSF2 all contained individuals possessing unique ancestry proportions that were not present in any other population. The four remaining populations (ANF5, SJSBP1, BRWMA1, and THSF1) were almost entirely composed of individuals with highly admixed ancestry estimates and did not display an identifiably unique pattern of ancestry (Figure 5).

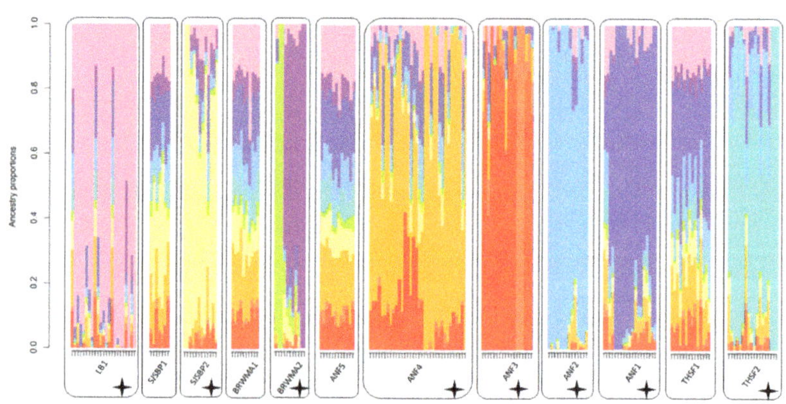

Figure 5. Ancestry proportions of individuals collected from twelve populations across *S. floridana*'s range (K = 10) based on sNMF analysis. Individuals are grouped by population (LB = Lathrop Bayou; SJSBP = St. Joseph Bay State Buffer Preserve; BRWMA = Box-R Wildlife Management Area; ANF = Apalachicola National Forest; and THSF = Tate's Hell State Forest). Populations with unique ancestry are marked with asterisks. Colors indicate different genetic ancestry within individuals.

Cross-validation identified the optimal number of PCs to retain as 20. Of the eight possible discriminant axes, all were retained. The first two axes explained 33.7% and

26.1% of genetic variance, respectively. The DAPC results supported our findings of high admixture between populations and identified ANF2, ANF3, and BRWMA2 as genetically distinct with limited inter-population gene flow (Figure 6). Populations SJSBP2 and THSF2 also appeared to be distinct, and to some extent LB1, ANF1, and ANF4, in a second DAPC (Figure S1) that removes the most divergent populations ANF2, ANF3, and BRWMA2. Therefore, the DAPCs results mirrored, in part, the sNMF analyses (Figures 5, 6 and S1).

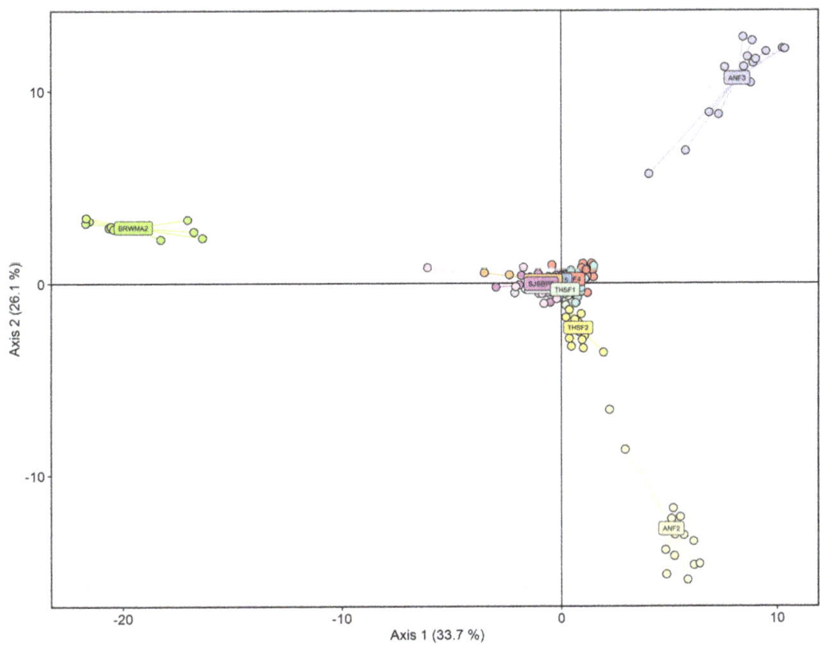

Figure 6. Discriminate analysis of principal components (DAPC) of twelve populations of *S. floridana* using 20 principal components and eight axes. Populations are shown with different colors, circles represent individuals, and labels are placed on the centroid (average) position for each population.

2.3. Clonality

We identified clonal individuals in two 5 m diameter circular plots and identified a pair-wise genetic distance of 0.03 as the threshold below which individuals were considered to be clones (Figure S2). In the plot located in SJSBP, we identified five stems as belonging to one clone and the remaining 23 stems as belonging to another clone (CR = 0.04). The plot located in ANF contained two stems that were genetically distinct individuals and 19 stems belonging to one clone (CR = 0.10). Clones were spread across the entirety of the 5 m plots, and ramets of different genets were intermixed (Figure 7).

An analysis of the clonal diversity in the studied populations points out that 92% are multiclonal. There were one to 16 sampled clones in a single population (Figure 8, Table 2). The average clonal number per population was 8.4 ± 1.4 SE, and the average number of individuals per clone was 1.7 ± 0.23 SE. Clonal richness was the lowest in the LB and THSF1 populations (CR = 0.09 and 0.08, respectively) and the highest in populations ANF 1, 2, 3, 4.3, and SJSBP 2 (CR = 0.82 to 1.0; Table 2). Similar to the circular plots, ramets of the same clone intermingled with other clones (Figure 8).

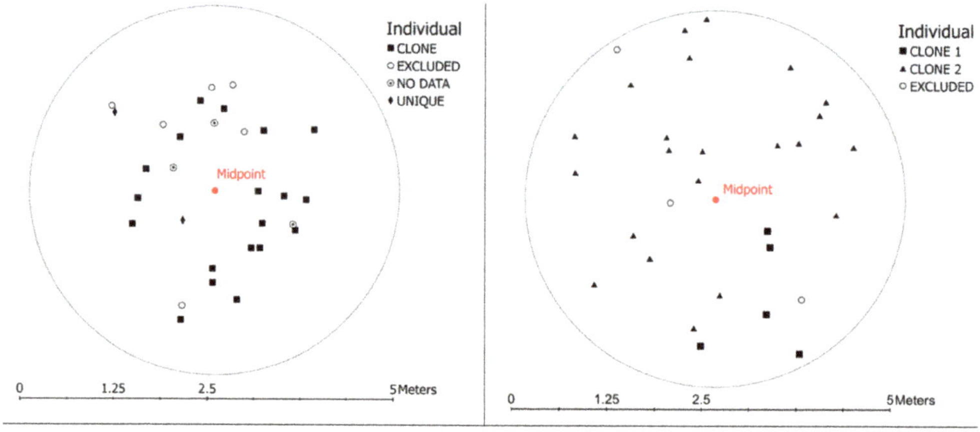

Figure 7. *Scutellaria floridana* clone identification (squares and triangles) and genetically unique individuals (diamonds) in 5m circular plots established in ANF4 (**left**, N = 21) and SJSBP2 (**right**, N = 28). Each plot was 5 m in diameter, with a midpoint selected randomly in an area displaying high stem density. Excluded individuals (open circles) in each plot were missing 45% (ANF4) or 25% (SJSBP2) of SNPS. Individuals with no data (circle with dot) were excluded from analyses.

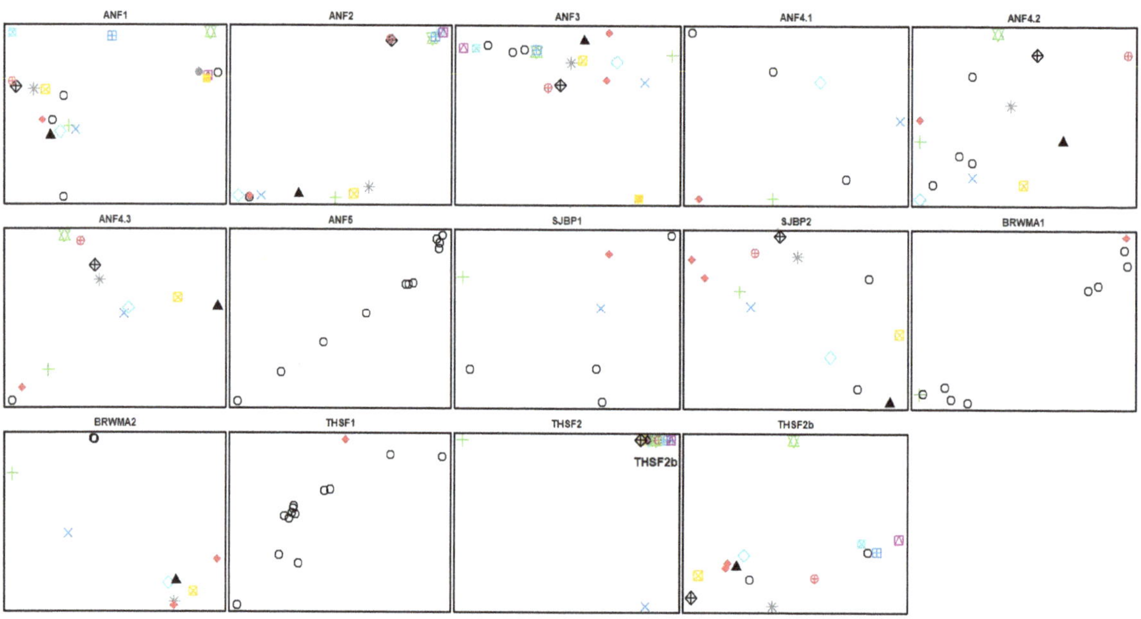

Figure 8. Spatial distribution of *S. floridana* clones and individuals with unique profiles within studied populations (see Figure 2 for locations) based on longitude (x-axis) and latitude (y-axis). Symbols represent each sampled individual. Within each population, ramets belonging to the same clone possess the same symbol. The last panel, THSF2b, shows the spatial distribution of a subset of THSF2 data. ANF4 was divided among three sampling sites because of its broad spatial distribution.

Table 2. Estimation of the number of unique clones (genotypes), ramets, and clonal richness per *S. floridana* population. N = sample size. Fifth column reports the number of ramets per clone (frequency in parentheses). CR = clonal richness, CR = (G − 1)/(N − 1) (49). Clonal analysis for ANF4 was divided among three sampling sites because of its broad spatial distribution. n/a = no gap in the histogram to allow for a threshold designation for clones; for these populations, the minimum pairwise diversity was ≥ 0.04.

Population	N	Clonal Diversity Threshold	Number of Unique Clones	Number of Ramets per Unique Clone	CR
ANF1	19	0.01	16	4, 1 (15)	0.83
ANF2	14	n/a	14	1 (14)	1.00
ANF3	18	0.02	15	3, 2, 1 (13)	0.82
ANF4.1	7	0.01	5	3, 1 (4)	0.67
ANF4.2	15	0.01	11	5, 1 (10)	0.71
ANF4.3	11	n/a	11	1 (11)	1.00
ANF5	11	0.02	1	11(1)	0
BRWMA1	10	0.002	3	8, 1 (2)	0.22
BRWMA2	11	0.02	8	3, 2, 1 (6)	0.70
LB1	23	0.02	3	18, 3, 2	0.09
SJSBP1	7	0.01	4	4, 1 (3)	0.50
SJSBP2	12	0.03	10	2, 2, 1 (8)	0.82
THSF1	14	0.03	2	13, 1	0.08
THSF2	18	0.01	14	3, 2, 2, 1 (11)	0.76
Total, or Average ± standard error	190		117	1.7 ± 0.23	0.59 ± 0.09

3. Discussion

We were able to describe genetic diversity and specifically assess the extent of clonal spread of *S. floridana* using next-generation sequencing, ddRADseq. Our finding that about 62% of the sampled plants were unique genotypes suggests that the *S. floridana* populations consist of a mixture of both asexually and sexually reproducing individuals. This demonstrates that sexual reproduction is occurring in most populations, more than was previously thought based on field observations [13]. Clonal propagation in some populations such as ANF5, LB, and THSF is high (CR ≤ 0.1), but fruit production has been observed over the years during surveys, suggesting the possibility of sexual reproduction.

Scutellaria floridana appears to possess moderately low genetic diversity and high levels of between-population differentiation in contrast to other members of the Lamiaceae of conservation concern [21,22]. Compared with a previous *S. floridana* study based on the AFLP marker [14], we observed lower estimates of expected heterozygosity (H_E = 0.124–0.145 in this study versus H_E = 0.145–0.184 in [14]) and higher levels of between-population differentiation (*S. floridana* F_{ST} = 0.035 in [14]). While differences in methodologies and markers used can make direct comparisons between the two studies difficult, the difference in F_{ST} estimates in particular may be due to the different sampling scheme and greater number of samples and areas included in this study. For example, the two populations that we identified as the most differentiated and least genetically diverse, BRWMA1 and ANF5, were not included in the previous work. This likely led to an underestimation of population differentiation and an overestimation of genetic diversity in the previous study.

Several populations displayed relatively large inbreeding coefficients and low observed heterozygosity compared with other Lamiaceae species and other *S. floridana* populations (Table 1) [21,22]. This could be due to low genet abundance, limited gene flow with other populations as a result of fragmentation, and decreased sexual reproduction, with clonal propagation dominating as a reproductive strategy [23–25]. Population BRWMA1, which possessed the lowest observed heterozygosity and highest inbreeding coefficient of all populations, is likely affected by all of these factors, as it displayed high divergence from other populations, a very low abundance of stems, no evidence of flowering during survey periods, and a clumped arrangement of stems across the landscape that were largely clonal

(CR = 0.22, Figure 8). Population ANF5, which possessed similar observed heterozygosity values and inbreeding coefficients, also displayed high divergence and no evidence of flowering during our survey periods but had a much higher abundance of stems spread over a larger area. The sampling of ANF5 was of two subpopulations in close proximity, and the low clonal richness suggested that all sampled stems derived from a single genet (Figure 8). Theoretical models predict for long-lived, strictly clonal populations an increase in heterozygous loci, resulting in lower F_{IS} and F_{ST} [17,26]; however, the presence of even limited amounts of sexual reproduction can lead to increases in homozygosity and higher F_{IS} and F_{ST} [17]. It is possible that sexual reproduction among clonal individuals in these populations, which simulates selfing, could lower observed heterozygosity values and increases F_{IS} estimates.

We identified evidence of substantial clonal reproduction in our circular plots located in SJSBP2 and ANF4 (Figure 7). We also found variation in the number of clones found among the 14 populations and sites of *S. floridana* (Table 2), as well as in the spatial arrangement between ramets (Figure 8). The spatial pattern of clonal growth, where ramets are dispersed over considerable distances and intermixed with other clones across the landscape (Figure 8), can be characterized as the guerrilla type [15,27]. This strategy can increase geitonogamy, i.e., pollination between flowers of the same plant, leading to inbreeding depression. In contrast, the mixing of ramets of different clones can enhance cross-pollination. Therefore, according to our study, cross-pollination could be occurring in the populations in which the circular plots were established, since they possess moderately high genetic diversity relative to other populations and little evidence of inbreeding. It is worth noting that the mating system of *S. floridana* is currently unknown, but the possibility exists that this species exhibits both selfing and outcrossing, as has been documented for *S. angustifolia* complex [28] and *S. indica* in the form of dimorphic cleistogamy [29].

Implications for Conservation and Management

The majority of *S. floridana*'s genetic diversity appears to be mostly encapsulated within a few populations that possess moderate heterozygosity, little or no evidence of inbreeding, unique ancestry, and high clonal richness (Tables 1 and 2). Populations ANF2, ANF3, BRWMA2, and THSF2 fall into the highest category of conservation genetic value based on these criteria, and their continued persistence is vital to maintain the diversity, resiliency, and adaptive potential of *S. floridana* as a species. Populations ANF1, SJSBP2, LB1, and ANF4 fall into an intermediate category of conservation value and represent a significant portion of *S. floridana*'s reproductively viable individuals. Any of these seven populations would likely be suitable sources for ex situ conservation efforts. The remaining four populations, ANF5, BRWMA1, THSF1, and SJSBP1, displayed much lower levels of genetic diversity, moderate evidence of inbreeding, little unique ancestry, and low clonal richness. They were also moderately to highly diverged from most other populations, suggesting a lack of gene flow across the landscape. These populations should be further investigated to identify the cause of inbreeding and determine if management activities can stabilize or improve genetic diversity via the introduction of transplants from populations of high conservation value. Unfortunately, we do not know the level of inbreeding that is acceptable for these populations.

As our survey efforts did not reveal any individuals at the locations of the five previously documented populations on private lands that we were able to access, we suspect that they may be extirpated. Examination of historical survey data revealed that all of these populations had gone over 15 years without any monitoring efforts but were still assumed to be extant. Our findings suggest that monitoring efforts need to be significantly improved for populations of *S. floridana* that are located on private lands. We recommend that these populations be revisited and extensively surveyed to determine their status. If these five populations have been extirpated, the total number of extant populations of *S. floridana* would be reduced to 14, presenting evidence of a continued decline of the species since listing. Moreover, the geographic arrangement of the historically extirpated and suspected extirpated populations represents substantial fragmentation of the species' range and increasing isolation of western populations.

Given how widespread clonal reproduction appears to be, it is likely that it plays a major role as a reproductive strategy. One possible factor driving clonal spread is fire, as clonal growth is common in disturbed habitats, and fire is an important disturbance for maintaining this species' habitat and Florida ecosystems [30–32]. Furthermore, plants in fire-dependent ecosystems have shown positive responses after fire, such as increased resprouting and flowering [30,33–35]. *Scutellaria floridana* is locally abundant in areas managed with fire, such as the ANF and the SJBSBP, where flowering was extensive following recent burns [13]. *Scutellaria floridana* can be considered a resprouting species rather than a seeder, as the latter tends to regenerate solely by post-fire recruitment from a seed bank, and this species does not have a persistent seed bank [14]. Resprouting from below-ground tissue after fire is a key life-history trait of fire-dependent ecosystems, but the extent of *S. floridana* new ramet production and their lateral spread is currently unknown and needs to be investigated.

Our results also suggest that stem counts alone are an imperfect proxy for abundance in populations with low clonal richness, as one individual can be composed of dozens of stems with no clearly delineated shape or arrangement. Based on our circular plots, it appears that on average, ten adjacent stems generally represent 1–2 unique individuals, suggesting that stem counts are usually equivalent to ten times the true population size. This estimate may be even higher in small populations with low clonal richness. Our spatial analysis also reported that the populations on private lands that appear to be extirpated display noticeably different patterns of land cover than those located on public, managed lands. Areas on private lands that historically contained active populations of *S. floridana* contained less evergreen forest, which is generally described as ideal land cover for the species, and more low, medium, and high intensity development than the areas on public lands where populations have persisted [13]. *Scutellaria floridana* has the capability of prolonged vegetative dormancy, allowing it to persist in dense pine plantations; thus, one way to determine whether the species is not extirpated is to reintroduce a fire regime to those areas [13]. This suggests that *S. floridana* is heavily reliant on active, targeted land management with prescribed fire to maintain suitable habitat conditions and allow populations to persist over time.

4. Materials and Methods

4.1. Study Area

Scutellaria floridana is predominantly found in well-established longleaf pine flatwoods with a thin to moderate overstory of longleaf pine, an open understory with little to no shrub layer, and groundcover dominated by wiregrass [13] (Figure 1). This habitat occurs in frequently flooded lowland areas with elevations of approximately 0.5–10 m and sand and fine-sand soils. The annual mean temperature is 20 °C with an average high of 26 °C and a low of 15 °C, and the mean annual precipitation is approximately 147 cm [36]. *Scutellaria floridana* is endemic to the Florida Panhandle and is only documented in Liberty, Franklin, Gulf, and Bay counties (Figure 2). The majority of *S. floridana* populations are located on public lands with regular fire regimes that maintain suitable habitat conditions for growth and flowering.

The Florida Natural Areas Inventory (FNAI), Florida's Natural Heritage Program and state member of the NatureServe network, has previously documented *S. floridana*'s range and identified all known occurrences of the species. Independent populations are defined as occurrences of the species that are at least 1 km away from the next nearest occurrence [37]. Because of this, populations are frequently composed of several spatially fragmented sub-populations, referred to in this paper as "sites." This results in some very large populations composed of many sites and some much smaller populations containing only one site. We collected samples from at least one site in each population where *S. floridana* was found and collected from multiple sites within populations that were spread across a larger geographic area.

Of the 19 populations previously identified as extant, we were unable to access two due to fencing and an active timber harvest operation. All of the extant populations of *S. floridana* that we located in the field are found on managed lands. The majority of these are located along the western edge of ANF in management areas designated as longleaf pine and slash pine adaptive management units that are maintained by prescribed fire [38]. Other populations were located in THSF, BRWMA, SJSBP, and LB (Figure 2). Tate's Hell State Forest is located directly south of ANF in the lower coastal plain along the coast of the Gulf of Mexico and is managed by prescribed fire with a burn frequency of 2–15 years [39]. Box-R Wildlife Management Area is located on the Gulf coast approximately 15 km west of THSF and is managed with selective thinning and prescribed fire with a burn frequency of 1–5 years [40]. St. Joseph Bay State Buffer Preserve is situated on the Gulf coast approximately 10 km west of BRWMA. Relevant areas that contain *S. floridana* are designated as wet prairie and are managed with prescribed fire at a 2–3 year frequency [41]. Lathrop Bayou is located approximately 30 km west of ANF and includes four islands at the eastern end of East Bay. *Scutellaria floridana* is found on the largest of these islands, Raffield Island. The primary management goal of this area has been to restore open-understory pine flatwood conditions using prescribed burns at a frequency of 1–2 years [42].

4.2. Tissue Collection and Storage

We conducted two trips in the spring of 2021 and collected a total of 294 plant tissue samples from 12 populations, representing 17 sites (Table 1). We took samples from individuals separated by at least one meter since the sites were generally patchy. Sampling sites such as ANF5 show a diagonal distribution as plants were found near a wetland (Figure 8). We georeferenced each sample collected using a Bad Elf GNSS Surveyor (Bad Elf West Hartford, CT). Plant collections were permitted under the Florida Department of Agriculture and Consumer Services Division of Plant Industry (#2021-03-002), US Department of Agriculture Forest Service Special Use Permit (WAK04122020), US Department of Interior Fish and Wildlife Service 10(a)(1)(A) permit to VNO, and Florida Fish and Wildlife Conservation Commission (SUO-82149).

For our analysis of genetic diversity, we conducted wide-scale sampling of 10–15 individual stems across each site where *S. floridana* was present, with a minimum distance of approximately 1 m between each sample (Table 1). For our analysis of clonality, we conducted intensive sampling within circular plots of 5 m in diameter, with a randomly selected midpoint placed in an area of high stem density. We established two plots in the ANF and one plot in the SJSBP and collected 30–40 stems per plot. To ensure accurate spatial analysis, we recorded distance (cm) and azimuth to the center of the plot for each sample using a compass and tape measure and manually placed sample locations onto a map using ArcGIS Pro (v. 2.7, ESRI 2011, Redlands, CA: Environmental Systems Research Institute). For each sample, we collected the entirety of the above-ground stem (including stem, leaves, and flowers) to ensure enough tissue for DNA isolation. Samples were stored at 4 °C after collection for up to 5 days, then transferred to −80 °C upon return to Miami University, OH. Voucher-photographed specimens from select populations were deposited in the Miami University (MU) Herbarium.

4.3. DNA Isolation and Sequencing

We extracted total genomic DNA from our plant material samples using a protocol similar to that used by [43]. Briefly, frozen samples were finely ground in liquid N_2 and dissolved in an extraction buffer containing 100mM Tris, pH 8.0, 50 mM EDTA, 500 mM NaCl, and 0.1% W:V PVP 40, followed by 5M potassium acetate precipitation of cellular debris and isopropanol precipitation of genomic DNA. We assessed the quality of the DNA from the samples using gel electrophoresis on a 1.5% agarose gel in Tris-Acetate-EDTA buffer to ensure there was little to no DNA degradation. We estimated the quantity of DNA in our samples using a Qubit 4 fluorometer (Thermo Fisher Scientific, Waltham, MA, USA). Samples that displayed adequate quality and reached a minimum DNA concentration of 20 ng/uL were then sent to Floragenex (Floragenex, Inc., 4640 SW Macadam Ave., Portland, OR), where double-digest restriction site-associated DNA sequencing (ddRAD-Seq) was carried out. To summarize, DNA was first digested using the restriction endonucleases PstI and MseI. Samples were diluted for PCR amplification, and the product was used to construct a ddRAD-Seq library. The library was sequenced at the University of Oregon Genomics and Cell Characterization Core Facility (GC3F) on a NovaSeq 6000 with a SP100 chip, generating 118 bp single-end reads with a mean 27.5× effective coverage per sample. The sequence data were run through the pipeline STACKS (version 2.60) to assemble the short-read sequences from all the samples (via the process radtags program) and to align reads into loci that are genotyped (via the gstacks program) [44,45]. Single nucleotide polymorphism data were exported in VCF version 4.2 file format for downstream data analysis (see below). Three quality cut-off filters were applied, allowing for genotypes to be present in 40%, 60%, or 80% of individuals. The final dataset used maximized the number of variable sites while keeping the proportion of missing data per site at 40% or lower. The 80% presence cutoff dataset was not used as the number of loci was reduced to 227 and the number of variable sites to 102.

4.4. Spatial Analysis

We analyzed changes in habitat over the past 20 years using 2001 and 2019 land cover rasters retrieved from the USGS National Land Cover Database. We used ArcGIS Pro (Esri ArcGIS Pro v. 2.9) to calculate the total area and change in area of each land cover category over the past 20 years across the entirety of *S. floridana*'s range, on managed lands within its range, and within 500 m of each population.

4.5. Genetic Diversity Assessment

Floragenex filtered raw sequence data and identified and assembled loci using the Stacks pipeline [44]. We used a dataset in which each locus was represented by at least 60% of individuals; datasets with less missing data (found in 80% of individuals) resulted in a loss of informative loci. To assess within-population genetic diversity, we calculated heterozygosity and inbreeding coefficients for each population using the R package hierfstat [46]. To assess genetic differentiation between populations, we calculated pairwise F_{ST} for populations using the package StaMPP [47]. To investigate isolation by distance, we ran a Mantel test for a significant relationship between pairwise F_{ST} and geographic distance between populations using the package ade4 [48]. We estimated ancestry coefficients for individuals via an sNMF analysis using the package LEA [49] and performed a discriminate analysis of principal components (DAPC) using the R package hierfstat. We performed cross-validation to determine the optimal number of principal components (PCs) to retain.

The population genetic value was qualitatively assessed based on the level of a population's observed heterozygosity, inbreeding coefficient, population divergence (based on F_{ST} and DAPC analyses), and signature of unique genetic ancestry (based on sNMF analysis) relative to other populations in this study. High to very high genetic value populations had high estimates of observed heterozygosity and low inbreeding coefficients, high or very high estimates of population divergence (based on Fst and DAPC analyses), and signatures of unique ancestry. Moderate genetic value populations have moderate estimates of observed heterozygosity and inbreeding coefficients, moderate estimates of population divergence, and exhibit signatures of unique ancestry. Finally, low genetic value populations have low estimates of observed heterozygosity, high inbreeding coefficients, moderate estimates of population divergence, and lack signatures of unique ancestry.

4.6. Clonal Assignment and Analyses

We identified clonal individuals from two circular plots and from 12 populations sampled across the species range using the package poppr version 2.9.3 [50]. The clonal analysis for ANF4 was divided among three sampling sites because of its broad spatial distribution. We set the threshold to distinguish unique genotypes from clones by generating frequency histograms of genetic distance between samples and identifying the location of a large gap between values (Figure S2). We then constructed UPGMA trees and identified clones as any group of individuals that diverged to the right of the distance value corresponding to this threshold (Figure S2). For the circular plots, we filtered individuals missing more than 25% of SNPs from the plot in SJSBP2 and individuals missing >45% of SNPs from one plot in ANF4, as this method was sensitive to missing data. Because over half of the individuals in our second circular plot located in ANF4 were missing significant amounts of data, we excluded this plot from analysis. Our distance threshold to distinguish individuals was set at 0.03 for both plots. For population clonal analysis, we constrained the ploidy level to diploid before running the analysis in poppr. We identified the number of clones (genets) and the number of individuals per clone (ramets) in each population or site. Clonal richness (CR) was calculated as the number of genotypes (G) relative to the number (N) of samples assessed (CR = (G-1)/(N-1); [51]). The spatial arrangement of the samples assigned to the corresponding clones was visualized using their longitude and latitude coordinates; however, given the classification status of this species, we cannot report the specific coordinates.

5. Conclusions

Overall, *S. floridana* appears to have suffered a continued decline since its listing as threatened under the ESA, and populations seem to be heavily reliant on management activities to remain extant. The 12 populations that we were able to confirm as extant possess low genetic diversity, and four of them present considerable evidence of inbreeding. We recommend thorough surveying of populations located on private lands to determine whether they are still extant and an in-depth investigation of clonal reproduction and management history of the four populations on public lands (BRWMA1, ANF5, THSF1, and SJSBP1) that possess very low genetic diversity and evidence of inbreeding. If the high prevalence of clonality is driving the low genetic diversity in these populations, re-establishing disturbance regimes, especially prescribed fire, could stimulate flowering and potentially sexual reproduction and improve genetic diversity [13]. We identified four populations (ANF2, ANF3, BRWMA2, and THSF2) that possess most of the genetic diversity of *S. floridana* as a species and suggest that they be treated as the highest conservation priority, as their continued existence is vital to preserve the adaptive potential of the species.

Scutellaria floridana will likely require regular monitoring and active conservation efforts, including public outreach, to avoid any further extirpation events. There is a possibility, however, that the *S. floridana* genotypes will persist in a population longer, since the guerrilla strategy tends to reduce the chance of extirpation, lessening the impact of stochastic events on the genetic structure of this species. The combination of sexual and asexual (in the form of clonal growth) reproduction may be advantageous for maintaining the viability of *S. floridana* populations.

Supplementary Materials: The following supporting information can be downloaded at: https://www.mdpi.com/article/10.3390/plants12040919/s1, Figure S1. DAPC with excluded outlying populations (ANF2, ANF 3, and BRWMA2) using 20 principal components and eight axes; Figure S2: Histograms of genetic distances and UPGMA trees for 5 m circular plots established in ANF4 and SJSBP2; Table S1: Percent change in NLCD land cover categories between 2001 and 2019 across *S. floridana's* range, on managed lands within *S. floridana*'s range, within 500 m of populations confirmed to be extant, and within 500 m of populations in which we were unable to locate any individuals.

Author Contributions: Conceptualization, G.R.H., V.N.-O. and R.C.M.; methodology, G.R.H., V.N.-O. and R.C.M.; validation, G.R.H., M.T.V., V.N.-O. and R.C.M.; formal analysis, G.R.H. and M.T.V.; investigation, G.R.H.; resources, V.N.-O. and R.C.M.; data curation, V.N.-O. and R.C.M.; writing—original draft preparation, G.R.H.; writing—review and editing, G.R.H., V.N.-O., R.C.M.; visualization, G.R.H. and V.N.-O.; supervision, V.N.-O. and R.C.M.; project administration, V.N.-O. and R.C.M.; funding acquisition, G.R.H. and R.C.M. All authors have read and agreed to the published version of the manuscript.

Funding: Financial support was provided under grants from the USFWS Coordination and Assistance Program Award F20AC10974-00 and the USFWS Coastal Program Award F21AC03234-00 to RCM and a Willard Sherman Turrell Herbarium (MU) research award to GH.

Data Availability Statement: We uploaded SNP data and vcf files to Dryad.

Acknowledgments: We thank Ann Johnson and Amy Jenkins of Florida Natural Areas Inventory; U.S. Fish and Wildlife Service Panama City Field Office, especially Lydia Ambrose; the staff of St. Joseph Bay State Buffer Preserve, Dylan Shoemaker, Kaylyn Cullen, and Sandra Chafin; Michael R. Jenkins of Florida Department of Agriculture and Consumer Services; and Box-R Wildlife Management Area manager Jerry Pitts for their invaluable assistance in the field. We also thank Angelica Vasilatos, William Gregor, and Wolfgang Graff for their assistance with DNA isolation, David Berg for his intellectual input on analysis and visualization, and Alan Gorchov Negron for help with figures. The findings and conclusions in this article are those of the authors and do not necessarily represent the views of the US Fish and Wildlife Service.

Conflicts of Interest: The authors declare no conflict of interest.

References

1. Hughes, R.; Inouye, B.; Johnson, M.; Underwood, N.; Vellend, M. Ecological consequences of genetic diversity. *Ecol. Lett.* **2008**, *11*, 609–623. [CrossRef]
2. Laikre, L. Genetic diversity is overlooked in international policy implementation. *Conserv. Genet.* **2010**, *11*, 349–354. [CrossRef]
3. Cook, C.N.; Sgró, C.M. Aligning science and policy to achieve evolutionarily enlightened conservation. *Conserv. Biol.* **2016**, *31*, 501–512. [CrossRef] [PubMed]
4. Ralls, K.; Ballou, J.D.; Dudash, M.R.; Eldridge, M.D.B.; Fenster, C.B.; Lacy, R.C.; Sunnucks, P.; Frankham, R. Call for a paradigm shift in the genetic management of fragmented populations. *Conserv. Lett.* **2018**, *11*, e12412. [CrossRef]
5. Holderegger, R.; Balkenhol, N.; Bolliger, J.; Engler, J.O.; Gugerli, F.; Hochkirch, A.; Nowak, C.; Segelbacher, G.; Widmer, A.; Zachos, F.E. Conservation genetics: Linking science with practice. *Mol. Ecol.* **2019**, *28*, 3848–3856. [CrossRef]
6. Willi, Y.; Kristensen, T.N.; Sgro, C.M.; Weeks, A.R.; Orsted, M.; Hoffmann, A.A. Conservation genetics as a management tool: The five best-supported paradigms to assist the management of threatened species. *Proc. Natl. Acad. Sci. USA* **2022**, *119*, e2105076119. [CrossRef]
7. Frankham, R.; Bradshaw, C.J.A.; Brook, B.W. Genetics in conservation management: Revised guidelines for the 50/500 rules, Red List criteria and population viability analyses. *Biol. Conserv.* **2014**, *170*, 56–63. [CrossRef]
8. Shafer, A.B.A.; Wolf, J.B.W.; Alves, P.C.; Bergstrom, L.; Bruford, M.W.; Brannstrom, I.; Colling, G.; Dalen, L.; Meester, L.D.; Ekblom, R.; et al. Genomics and the challenging translation into conservation practice. *Trends Ecol. Evol.* **2015**, *30*, 78–87. [CrossRef]
9. Jump, A.S.; Penuelas, J. Running to stand still: Adaptation and the response of plants to rapid climate change. *Ecol. Lett.* **2005**, *8*, 1010–1020. [CrossRef]
10. Kramer, J.T.; Havens, K. Plant conservation genetics in a changing world. *Trends Plant Sci.* **2009**, *14*, 599–607. [CrossRef]
11. Bragg, J.G.; Cuneo, P.; Sherieff, A.; Rossetto, M. Optimizing the genetic composition of a translocation population: Incorporating constraints and conflicting objectives. *Mol. Ecol. Resour.* **2020**, *20*, 54–65. [CrossRef]
12. St. Clair, A.B.; Dunwiddie, P.W.; Fant, J.B.; Kaye, T.N.; Kramer, A.T. Mixing source populations increases genetic diversity of restored rare plant populations. *Restor. Ecol.* **2020**, *28*, 583–593. [CrossRef]
13. U.S. Fish and Wildlife Service (USFWS). *Scutellaria floridana (Florida skullcap) 5-Year Review: Summary and Evaluation*; USFW: Panama City, FL, USA, 2019; p. 19.
14. Molano-Flores, B.; Coons, J.; Annis, J.; O'Brien, J.; Feist, M.; Koontz, J.; Maruzak, J.; Menglekoch, J. *Seed Ecology and Population Genetics Studies for Pinguicula ionantha (Godfrey's Butterwort) and Scutellaria floridana (Florida skullcap)*; Illinois Natural History Survey: Champaign, IL, USA, 2014; p. 27.
15. Ye, X.-H.; Yu, F.-H.; Dong, M. A trade-off between guerrilla and phalanx growth forms in *Leymus secalinus* under different nutrient supplies. *Ann. Bot.* **2006**, *98*, 187–191. [CrossRef] [PubMed]
16. De Witte, L.C.; Stocklin, J. Longevity of clonal plants: Why it matters and how to measure it. *Ann. Bot.* **2010**, *106*, 859–870. [CrossRef] [PubMed]

17. Tepedino, V.J. Overestimating population sizes of rare clonal plants. *Conserv. Biol.* **2012**, *26*, 945–947. [CrossRef] [PubMed]
18. Balloux, F.; Lehmann, L.; de Meeus, T. The population genetics of clonal and partially clonal diploids. *Genetics* **2003**, *164*, 1635–1644. [CrossRef] [PubMed]
19. Wright, S. Evolution in Mendelian populations. *Genetics* **1930**, *16*, 97–169. [CrossRef] [PubMed]
20. U.S. Fish and Wildlife Service. *Recovery Plan for Four Plants of the Lower Apalachicola Region, Florida: Euphorbia Telephioides (Telephus Spurge), Macbridea Alba (White Birds-in-a-Nest), Pinguicula ionantha (Godfrey's Butterwort), and Scutellaria floridana (Florida skullcap)*; Fish and Wildlife Service: Panama City, FL, USA, 1994; p. 32. Available online: https://ecos.fws.gov/ecp/species/2240 (accessed on 14 March 2022).
21. Kyrkjeeide, M.O.; Westergaard, K.B.; Kleven, O.; Evju, M.; Endrestol, A.; Brandrud, M.K.; Stabbetorp, O. Conserving on the edge: Genetic variation and structure in northern populations of the endangered plant *Dracocephalum ruyschiana* L. (Lamiaceae). *Conserv. Genet.* **2020**, *21*, 707–718. [CrossRef]
22. Zhou, X.; Zhang, Z.; Huang, Y.; Zhao, H.; Wu, J.; Qi, Z.; Wei, Y. Conservation genomics of wild red sage (*Salvia miltiorrhiza*) and its endangered relatives in China: Population structure and interspecific relationships revealed from 2b-RAD data. *Front. Genet.* **2021**, *12*, 688323. [CrossRef]
23. Ellstrand, N.C.; Elam, D.R. Population genetic consequences of small population size: Implications for plant conservation. *Annu. Rev. Ecol. Syst.* **1993**, *24*, 217–242. [CrossRef]
24. Aguilar, R.; Quesada, M.; Ashworth, L.; Herrerias-Diego, Y.; Lobo, J.A. Genetic consequences of habitat fragmentation in plant populations: Susceptible signals in plant traits and methodological approaches. *Mol. Ecol.* **2019**, *17*, 5177–5188. [CrossRef] [PubMed]
25. Vallejo-Marin, M.; Dorken, M.E.; Barrett, S.C. The ecological and evolutionary consequences of clonality for plant mating. *Annu. Rev. Ecol. Evol. S.* **2010**, *41*, 193–213. [CrossRef]
26. Delmotte, F.; Leterme, N.; Gauthier, J.P.; Rispe, C.; Simon, J.C. Genetic architecture of sexual and asexual populations of the aphid *Rhopalosiphum padi* based on allozyme and microsatellite markers. *Mol. Ecol.* **2002**, *11*, 711–723. [CrossRef] [PubMed]
27. Lovet, D.L. Population dynamics and local specialization in a clonal perennial (*Ranunculus repens*) 1. The dynamics of ramets in contrasting habitats. *J. Ecol.* **1981**, *69*, 743–755. [CrossRef]
28. Olmstead, R.G. Biological and historical factors influencing genetic diversity in the *Scutellaria angustifolia* complex (Labiatae). *Evolution* **1990**, *44*, 54–70. [CrossRef] [PubMed]
29. Sun, M. Cleistogamy in *Scutellaria indica* (Labiatae): Effective mating system and population genetic structure. *Mol. Ecol.* **1999**, *88*, 1285–1295. [CrossRef]
30. Maguire, A.J.; Menges, E.S. Post-fire growth strategies of resprouting Florida scrub vegetation. *Fire Ecol.* **2011**, *7*, 12–25. [CrossRef]
31. Brewer, J.S.; Platt, W. Effects of fire season and soil fertility on clonal growth in a pyrophilic forb, *Pityopsis graminifolia* (Asteraceae). *Am. J. Bot.* **1994**, *817*, 805–814. [CrossRef]
32. Hartnett, D.C. Effects of Fire on Clonal Growth and Dynamics of *Pityopsis graminifolia* (Asteraceae). *Am. J. Bot.* **1987**, *74*, 1737–1743. [CrossRef]
33. Slapcinsky, J.L.; Gordon, D.R.; Menges, E.S. Responses of rare plant species to fire in Florida's pyrogenic communities. *Nat. Area. J.* **2010**, *30*, 4–19. [CrossRef]
34. Ames, G.M.; Anderson, S.M.; Wright, J.P. Multiple environmental drivers structure plant traits at the community level in a pyrogenic ecosystem. *Funct. Ecol.* **2016**, *30*, 789–798. [CrossRef]
35. Negrón-Ortiz, V.; Kaeser, M. Timing and patterns of size, reproduction, and seed germination in Florida endemic *Euphorbia telephioides* (Euphorbiaceae): Management and conservation implications. *Nat. Area. J.* **2020**, *40*, 262–272. [CrossRef]
36. NOAA National Centers for Environmental Information (NCEI) U.S. Climate Normals 2020: Annual/Seasonal Climate Normals 2006–2020. Available online: https://is.gd/IsUcIp (accessed on 14 March 2022).
37. NatureServe. *NatureServe Network Biodiversity Location Data*; NatureServe: Arlington, VA, USA, 2022.
38. U.S. Forest Service (USFS). *Forest Plan: National Forests in Florida*; USFS: Washington, DC, USA, 1999.
39. Florida Department of Agriculture and Consumer Services & Florida Forest Service. *Ten-Year Land Management Plan for the Tate's Hell State Forest*; Florida Department of Agriculture and Consumer Services & Florida Forest Service: Tallahassee, FL, USA, 2019.
40. Florida Fish and Wildlife Conservation Commission. *A Management Plan for Box-R Wildlife Management Area*; Florida Fish and Wildlife Conservation Commission: Tallahassee, FL, USA, 2016.
41. Florida Department of Environmental Protection. *St. Joseph Bay State Buffer Preserve Management Plan*; Florida Department of Environmental Protection: Tallahassee, FL, USA, 2016.
42. Bureau of Land Management. *Lathrop Bayou. Final Habitat Management Plan Environmental Assessment*; Bureau of Land Management: Jackson, MS, USA, 2003.
43. Kim, C.S.; Lee, C.H.; Shin, J.S.; Chung, Y.S.; Hyung, N.I. A simple and rapid method for isolation of high-quality DNA from fruit trees and conifers using PVP. *Nucleic Acids Res.* **1997**, *25*, 1085–1086. [CrossRef] [PubMed]
44. Catchen, J.; Hohenlohe, P.; Bassham, S.; Amores, A.; Cresko, W. Stacks: An analysis tool set for population genomics. *Mol. Ecol.* **2013**, *22*, 3124–3140. [CrossRef]
45. Paris, J.R.; Stevens, J.R.; Catchen, J.M. Lost in parameter space: A road map for STACKS. *Methods Ecol. Evol.* **2017**, *8*, 1360–1373. [CrossRef]

46. Goudet, J.; Jombart, T. Hierfstat: Estimation and Tests of Hierarchical F-Statistics. 2022. Available online: https://www.r-project.org (accessed on 1 May 2022).
47. Pembleton, L.W.; Cogan, N.O.I.; Forster, J.W. StAMPP: An R package for calculation of genetic differentiation and structure of mixed-ploidy level populations. *Mol. Ecol. Res.* **2013**, *13*, 946–952. [CrossRef]
48. Dray, S.; Dufour, A. The ade4 package: Implementing the duality diagram for ecologists. *J. Stat. Softw.* **2007**, *22*, 1–20. [CrossRef]
49. Frichot, E.; Francois, O. LEA: An R package for landscape and ecological association studies. *Methods Ecol. Evol.* **2015**, *6*, 925–929. [CrossRef]
50. Kamvar, Z.N.; Tabima, J.F.; Grünwald, N.J. Poppr: An R package for genetic analysis of populations with clonal, partially clonal, and/or sexual reproduction. *PeerJ* **2014**, *2*, e281. [CrossRef]
51. Dorken, M.E.; Eckert, C.G. Severely reduced sexual reproduction in northern populations of a clonal plant, *Decodon verticillatus* (Lythraceae). *J. Ecol.* **2001**, *89*, 339–350. [CrossRef]

plants

MDPI

Article

The Conservation Genetics of *Iris lacustris* (Dwarf Lake Iris), a Great Lakes Endemic

James Isaac Cohen [1,*] and Salomon Turgman-Cohen [2]

[1] Department of Botany and Plant Ecology, Weber State University, 1415 Edvalson St., Dept. 2504, Ogden, UT 84408-2504, USA
[2] E.S. Witchger School of Engineering, Marian University, 3200 Cold Spring Road, Indianapolis, IN 46222-1997, USA; sturgmancohen@marian.edu
* Correspondence: jamescohen@weber.edu

Abstract: *Iris lacustris*, a northern Great Lakes endemic, is a rare species known from 165 occurrences across Lakes Michigan and Huron in the United States and Canada. Due to multiple factors, including habitat loss, lack of seed dispersal, patterns of reproduction, and forest succession, the species is threatened. Early population genetic studies using isozymes and allozymes recovered no to limited genetic variation within the species. To better explore genetic variation across the geographic range of *I. lacustris* and to identify units for conservation, we used tunable Genotyping-by-Sequencing (tGBS) with 171 individuals across 24 populations from Michigan and Wisconsin, and because the species is polyploid, we filtered the single nucleotide polymorphism (SNP) matrices using polyRAD to recognize diploid and tetraploid loci. Based on multiple population genetic approaches, we resolved three to four population clusters that are geographically structured across the range of the species. The species migrated from west to east across its geographic range, and minimal genetic exchange has occurred among populations. Four units for conservation are recognized, but nine adaptive units were identified, providing evidence for local adaptation across the geographic range of the species. Population genetic analyses with all, diploid, and tetraploid loci recovered similar results, which suggests that methods may be robust to variation in ploidy level.

Keywords: genotyping-by-sequencing; *Iris*; Lake Huron; Lake Michigan; polyploidy; polyRAD; rare plants; tGBS

Citation: Cohen, J.I.; Turgman-Cohen, S. The Conservation Genetics of *Iris lacustris* (Dwarf Lake Iris), a Great Lakes Endemic. *Plants* **2023**, *12*, 2557. https://doi.org/10.3390/plants12132557

Academic Editor: Hwan Su Yoon

Received: 5 May 2023
Revised: 26 May 2023
Accepted: 3 July 2023
Published: 5 July 2023

1. Introduction

In 1818, Thomas Nuttall described a new species of crested *Iris* L., *Iris lacustris* Nutt., "on the gravelly shores of calcareous islands of Lake Huron" [1]. Since then, the recognized geographic range of the species has expanded to include the northern regions of Lakes Huron and Michigan in the United States and Canada. Presently, the species is known from 165 occurrences, with more than half in Michigan (89) and the others split between Wisconsin (36) and Ontario (40) [2].

Plants of *I. lacustris* grow less than 15 cm in height [3], and this feature provides the species with its common name, Dwarf Lake Iris. The species bears self-compatible flowers, with purple sepals and purple petals with yellow and white markings, that are visited by various species of bees [4]. Across its geographic range, *I. lacustris* frequently inhabits the understory of coniferous forests along the shore, although a small number of inland populations are known (Figure 1) [2,5,6]. These habitats have thin entisols, and the dominant tree species primarily include *Thuja occidentalis* L., *Abies balsamea* (L.) Miller, and *Picea glauca* (Moench) Voss. The species has become a well-known endemic plant of the Great Lakes and is so characteristic of the region that it was recognized as the state wildflower of Michigan [7].

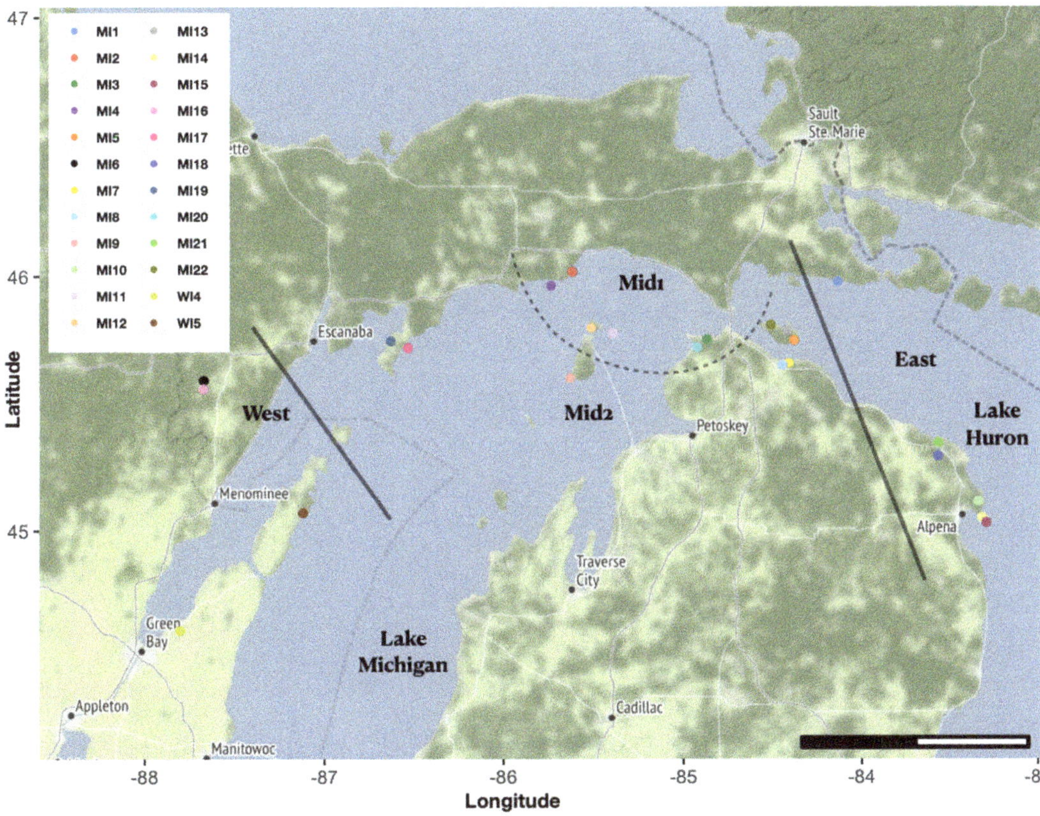

Figure 1. Map of locations sampled in present study. Dark gray entire lines denote division between East, Mid1, Mid2, and West clusters (also recognized as management units). The dashed gray line separates Mid1 and Mid2 populations, and Mid includes both groups of populations together. Light gray lines separate Wisconsin (USA), Michigan (USA), and Ontario (Canada). Scale bar is 100 km, with each section representing 50 km.

In 1988, 170 years after *I. lacustris* was initially described, the species was listed as federally threatened [5]. The small number of populations and individuals is due to multiple factors, including the loss of shoreline habitat, fungal infection of fruits, lack of seed dispersal, and overgrowth of the forest canopy that restricted plant growth, flower production, and sexual reproduction. Plants of the species currently reproduce more by vegetative growth than germination from the myrmecochorous seeds [5]. Despite this low germination rate, seeds can remain viable in the seedbank for at least 15 years [5], a factor that could influence long-term population growth and genetic diversity, although mass germination and recruitment are rare [4].

The ecology of *I. lacustris* has been examined to a greater extent than the genetic diversity of the species. To date, only three studies have explored this topic: Simonich and Morgan [8] examined nine populations in Wisconsin, using 22 allozyme markers, Orick [9] investigated nine populations in Michigan, using 24 isozymes, and Hannan and Orick [10] examined nine populations in Michigan, using 18 isozymes. In two studies, researchers identified genetic homogeneity across the populations; however, Orick [9] found overall heterozygosity to be 3.7%. Hannan and Orick [10] also note gene silencing may have been possible in four loci. In contrast to the genetic diversity recognized in *I. lacustris*, Hannan and Orick [10] found that the sister species, *I. cristata* Aiton [11], which has a wider geographic range across eastern North America, was variable at 11 of 15 loci. These studies suggest that the genetic diversity of this rare species of *Iris* is quite limited. This genetic paucity is intriguing because *I. lacustris* and its sister are both putative tetraploids [10], and polyploid plant species tend to have greater genetic variation than diploid relatives, although selfing tends to be higher in polyploids [12–14]. Importantly, the genetic diversity of the *I. lacustris* may have implications for the ability of the species to respond to the changing environment across its geographic range and for various conservation efforts.

In order to investigate the population and conservation genetics of the species in a comprehensive manner, we examined multiple populations from across Michigan and Wisconsin, and we used tunable Genotyping-by-Sequencing (tGBS [15]), a method of reduced representation sequencing, to identify single nucleotide polymorphisms (SNPs) among the populations. The objectives of the present study are threefold: (1) identify genetic diversity and population structure and substructure across the range of *I. lacustris*, (2) explore patterns of migration, and (3) recognize population clusters for management of this rare species. Given the paucity of genetic diversity identified in previous studies, we hypothesized that there would be limited genetic variation across the species.

2. Results

2.1. DNA Sequencing and Polyploid Filtering

Among the 171 individuals of 24 populations across the geographic range of *I. lacustris* in Michigan and Wisconsin (Figure 1, Table 1), 726,786,603 paired-end reads were sequenced, with a mean of 4,225,503 reads per sample. The consensus sequence included 1,335,996 scaffolds with 196,139,854 bp (N50 = 644,994, L50 = 145). The mean per sample alignment and unique alignment to the consensus sequences are 93.9% and 74.4%, respectively. For the MCR90 dataset, 125 reads were interrogated per SNP across 2,341,730 bases, with 4.8% missing data for the final dataset. For the MCR50 dataset, 31 reads were interrogated per SNP across 23,904,409 bases, with 31.4% missing data for the final dataset. The numbers of SNPs in the diploid and tetraploid datasets identified through analysis in polyRAD are in Table 2.

Table 1. Population and sampling information and assignation of populations to clusters based on results of various population genomic analyses, including recognition of management and adaptive units, based on loci not under and under selection, respectively. Cluster, management unit, and adaptive unit assignation is based on population genetic analyses with fastStructure, discriminant analysis of principal components (DAPC), principal component analyses (PCA), and others described in the text.

Populations Sampled	Number of Individuals Sampled	Four Population Clusters in Analyses	Three Population Clusters in Analyses	Management Units (All Loci)	Management Units (Diploid and Tetraploid Loci)	Adaptive Units
MI1	10	East	East	1	1	1
MI2	3	Mid1	Mid	2	2	2
MI3	8	Mid1	Mid	2	2	3
MI4	7	Mid1	Mid	2	2	3
MI5	7	Mid2	Mid	3	3	4
MI6	14	West	West	4	4	5
MI7	8	Mid2	Mid	2	2	4
MI8	4	Mid2	Mid	3	3	4
MI9	3	Mid2	Mid	2	2	6
MI10	5	East	East	1	3	7
MI11	1	Mid1	Mid	2	2	3
MI12	2	Mid1	Mid	2	2	3
MI13	3	Mid2	Mid	3	3	4
MI14	13	East	East	1	3	7

Table 1. *Cont.*

Populations Sampled	Number of Individuals Sampled	Four Population Clusters in Analyses	Three Population Clusters in Analyses	Management Units (All Loci)	Management Units (Diploid and Tetraploid Loci)	Adaptive Units
MI15	8	East	East	1	1	7
MI16	3	West	West	4	2	5
MI17	10	Mid2	Mid	2	2	6
MI18	10	East	East	1	1	1
MI19	10	Mid2	Mid	2	2	6
MI20	3	Mid1	Mid	2	2	2
MI21	7	East	East	1	3	1
MI22	8	Mid2	Mid	3	3	4
WI4	12	West	West	4	2	8
WI5	12	West	West	4	2	9

Table 2. Information on six SNP (single nucleotide polymorphism) datasets examined including best K (cluster) value under various analyses. Dashes indicate analysis was not performed for dataset. STRUCTURESELECTOR results include MedMedK, MedMeanK, MaxMedK, and MaxMeanK, and, therefore, may have a range of best K values due to different results from these four metrics. DAPC is discriminant analysis of principal components, and for these analyses, best K value is determined via Bayesian Information Criterion. Additional information on identification of loci under selection and best K values in text.

Dataset	SNPs	Loci under Selection	All Loci		Loci under Selection		Loci Not under Selection	
			STRUCTURESELECTOR	DAPC	STRUCTURESELECTOR	DAPC	STRUCTURESELECTOR	DAPC
MCR90	5354	401	6	9	12–14	13	3–4	7
MCR90 diploid loci	2106	29	4–5	7	-	-	-	-
MCR90 tetraploid loci	1382	21	4–5	6	-	-	-	-
MCR50	344,509	65,075	5–7	4	11–13	10	3	1
MCR50 diploid loci	50,134	4311	3–4	2–3	9–10	7	2–3	1
MCR50 tetraploid loci	82,237	6939	3–4	2–3	8	9	3	1

2.2. *Population Genomics*

Across all datasets, observed heterozygosity slightly exceeds expected heterozygosity, and F_{IS} values are, in general, negative (Table 3). Pairwise F_{ST} values vary from 0.1–0.45, and results are similar among datasets (Table 4). Based on various AMOVA results, most of the variation is within samples, followed by between the populations, regardless of the datasets and partitioning of the populations (Supplemental Table S1). Mantel tests for isolation-by-distance analyses identify all datasets as having spatial structure (Supplemental Figure S1) with $p < 0.001$ for analyses of individuals, but only MCR90 datasets had spatial structure for populations ($p < 0.05$).

Results from analyses in fastSTRUCTURE, STRUCTURE, MavericK, and tess3r are similar. Based on the results from STRUCTURESELECTOR, the optimal K values were greater for all loci analyzed together than for either the diploid or tetraploid loci analyzed independently (Table 2, Supplemental Table S2). Similar clusters were recovered with the different datasets (Figure 2, Table 1), with a clear division between three groups—eastern, western, and central populations—and multiple analyses resulted in the central population being divided into two distinct groups at K = 4 and/or 5 (Figure 1, Supplemental Figures S2–S4), especially for all loci in fastSTRUCTURE and multiple datasets with STRUCTURE, MavericK, and tess3r. At K = 4–5, the two Wisconsin populations were often recovered with unique genomic signatures suggestive of admixture, and this is particularly the case with the MCR90 datasets. While the results of conStruct are similar to others, the three distinct groups identified are more opaque, with boundaries between the eastern and western populations overlapping to a larger extent than with the other analyses (Supplemental Figure S5); although, similar patterns can be recognized at K = 4 and 5 for the MCR90 all and diploid loci datasets. Among all methods, the three populations on Bois Blanc Island in Michigan (MI5, MI13, and MI22), in the northwestern geographic range of the species, also include some individuals that show signals of admixture between the eastern and central populations (Figure 2). Graphs of K values for all analyses are included in Supplementary Materials (Figures S6–S16).

Table 3. Observed, expected, and total heterozygosity (H_O, H_S, H_T) and fixation index (F_{IS}) for the three MCR90 datasets for each population. Sample sizes are less than five for MI2, MI8, MI9, MI11, MI12, MI13, MI16, MI20, which could impact calculated statistics.

Population	MCR90 All Loci				MCR90 Diploid Loci				MCR90 Tetraploid Loci			
	H_O	H_S	H_T	F_{IS}	H_O	H_S	H_T	F_{IS}	H_O	H_S	H_T	F_{IS}
MI1	0.0586	0.0516	0.0516	−0.1365	0.064	0.0519	0.0519	−0.2325	0.0548	0.0462	0.0462	−0.1875
MI2	0.0503	0.0411	0.0411	−0.2224	0.0538	0.0417	0.0417	−0.2883	0.0532	0.0431	0.0431	−0.2329
MI3	0.0472	0.0307	0.0307	−0.5394	0.0521	0.0322	0.0322	−0.6191	0.0474	0.0301	0.0301	−0.5768
MI4	0.0581	0.0451	0.0451	−0.2873	0.0651	0.0479	0.0479	−0.3582	0.0628	0.0497	0.0497	−0.2624
MI5	0.0558	0.0532	0.0532	−0.047	0.052	0.0428	0.0428	−0.2161	0.051	0.041	0.041	−0.2444
MI6	0.0957	0.0704	0.0704	−0.3593	0.1043	0.0742	0.0742	−0.4064	0.0933	0.0675	0.0675	−0.3826
MI7	0.054	0.049	0.049	−0.1021	0.0519	0.042	0.042	−0.2372	0.0559	0.0449	0.0449	−0.2455
MI8	0.0655	0.0563	0.0563	−0.1631	0.0677	0.0573	0.0573	−0.1816	0.067	0.0555	0.0555	−0.2074
MI9	0.0594	0.0401	0.0401	−0.4814	0.0586	0.0373	0.0373	−0.5714	0.0673	0.0442	0.0442	−0.5217
MI10	0.0612	0.0477	0.0477	−0.2842	0.0554	0.043	0.043	−0.289	0.0515	0.0383	0.0383	−0.3455
MI11	0.0475	-	-	-	0.0527	-	-	-	0.0499	-	-	-
MI12	0.0522	0.0385	0.0385	−0.3578	0.0592	0.0411	0.0411	−0.4413	0.0551	0.0433	0.0433	−0.2749
MI13	0.0557	0.0447	0.0447	−0.2463	0.0508	0.0409	0.0409	−0.2434	0.0543	0.0401	0.0401	−0.3551
MI14	0.0535	0.0488	0.0488	−0.0981	0.0542	0.0448	0.0448	−0.2105	0.0497	0.0415	0.0415	−0.1989
MI15	0.0573	0.0551	0.0551	−0.0395	0.059	0.0516	0.0516	−0.1434	0.0589	0.0502	0.0502	−0.1724
MI16	0.0961	0.0694	0.0694	−0.384	0.0956	0.0661	0.0661	−0.4464	0.0795	0.0541	0.0541	−0.4686
MI17	0.0671	0.062	0.062	−0.0827	0.0609	0.0488	0.0488	−0.2478	0.0647	0.0524	0.0524	−0.2357
MI18	0.0639	0.0567	0.0567	−0.1277	0.0672	0.0542	0.0542	−0.2401	0.0643	0.0534	0.0534	−0.2054
MI19	0.0651	0.0575	0.0575	−0.1324	0.0645	0.0512	0.0512	−0.2604	0.0591	0.0487	0.0487	−0.215
MI20	0.0467	0.0343	0.0343	−0.3597	0.0481	0.0351	0.0351	−0.3711	0.0516	0.0365	0.0365	−0.4123
MI21	0.0624	0.054	0.054	−0.1557	0.0628	0.0497	0.0497	−0.262	0.0637	0.0512	0.0512	−0.2435
MI22	0.0543	0.0492	0.0492	−0.1035	0.0464	0.0382	0.0382	−0.2135	0.049	0.0397	0.0397	−0.2343
WI4	0.1081	0.0946	0.0946	−0.1424	0.1032	0.0848	0.0848	−0.2179	0.0895	0.0745	0.0745	−0.201
WI5	0.1033	0.085	0.085	−0.2157	0.1015	0.0775	0.0775	−0.3102	0.094	0.0734	0.0734	−0.2814

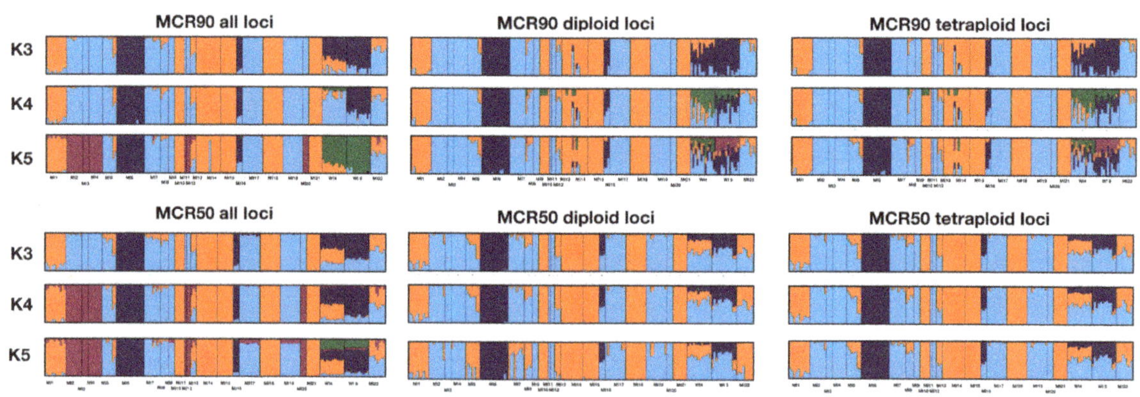

Figure 2. Structure bar graphs from fastSTRUCTURE for the six datasets analyzed in the present study for K = 3–5. Individual ancestry denoted by color.

Table 4. Pairwise F_{ST} values and heatmap for MCR90 all loci (below diagonal) and MCR90 diploid loci (above diagonal). Below the diagonal, red indicates lower values, and blue is for higher values. Above the diagonal, yellow is for lower values, and green is for higher values. Sample sizes are less than five for MI2, MI8, MI9, MI11, MI12, MI13, MI16, MI20, which could impact calculated statistics.

	MI1	MI2	MI3	MI4	MI5	MI6	MI7	MI8	MI9	MI10	MI11	MI12	MI13	MI14	MI15	MI16	MI17	MI18	MI19	MI20	MI21	MI22	WI4	WI5
MI1	-	0.17	0.20	0.21	0.16	0.25	0.18	0.16	0.22	0.14	0.12	0.17	0.13	0.11	0.09	0.26	0.18	0.05	0.20	0.16	0.05	0.15	0.15	0.19
MI2	0.29	-	0.12	0.12	0.16	0.24	0.16	0.18	0.24	0.20	0.07	0.11	0.13	0.16	0.15	0.27	0.16	0.19	0.17	0.00	0.19	0.11	0.16	0.20
MI3	0.36	0.26	-	−0.01	0.19	0.21	0.19	0.20	0.26	0.27	0.00	0.05	0.22	0.21	0.20	0.28	0.13	0.22	0.14	0.14	0.22	0.18	0.13	0.16
MI4	0.37	0.26	0.01	-	0.19	0.24	0.18	0.19	0.20	0.25	−0.08	0.03	0.19	0.22	0.20	0.26	0.14	0.23	0.15	0.10	0.23	0.18	0.16	0.19
MI5	0.26	0.29	0.30	0.31	-	0.20	0.08	0.04	0.20	0.12	0.12	0.18	0.08	0.12	0.12	0.23	0.12	0.17	0.14	0.16	0.16	0.06	0.14	0.15
MI6	0.41	0.39	0.37	0.39	0.31	-	0.23	0.19	0.22	0.24	0.14	0.19	0.20	0.24	0.23	0.13	0.20	0.25	0.20	0.21	0.24	0.22	0.19	0.20
MI7	0.32	0.35	0.37	0.37	0.12	0.35	-	0.08	0.19	0.19	0.11	0.18	0.16	0.15	0.15	0.25	0.12	0.20	0.14	0.14	0.18	0.11	0.15	0.18
MI8	0.31	0.38	0.41	0.40	0.07	0.34	0.14	-	0.18	0.15	0.04	0.14	0.10	0.14	0.12	0.20	0.11	0.17	0.13	0.16	0.16	0.09	0.12	0.15
MI9	0.42	0.43	0.44	0.38	0.32	0.40	0.34	0.39	-	0.29	0.18	0.25	0.28	0.24	0.21	0.25	0.10	0.24	0.12	0.25	0.25	0.24	0.13	0.14
MI10	0.20	0.34	0.44	0.42	0.23	0.38	0.32	0.29	0.45	-	0.19	0.25	0.13	0.03	0.06	0.28	0.20	0.13	0.21	0.21	0.09	0.12	0.14	0.19
MI11	0.30	0.24	0.05	−0.07	0.21	0.31	0.29	0.30	0.38	0.38	-	−0.03	0.12	0.14	0.10	0.13	0.05	0.14	0.06	0.07	0.14	0.12	0.02	0.07
MI12	0.33	0.25	0.07	0.03	0.26	0.35	0.34	0.36	0.41	0.41	0.01	-	0.17	0.19	0.15	0.21	0.13	0.18	0.13	0.14	0.20	0.18	0.09	0.13
MI13	0.21	0.26	0.37	0.33	0.16	0.33	0.27	0.23	0.44	0.22	0.28	0.31	-	0.09	0.09	0.25	0.15	0.14	0.16	0.15	0.14	0.02	0.12	0.15
MI14	0.15	0.31	0.38	0.39	0.24	0.40	0.30	0.28	0.42	0.04	0.33	0.36	0.18	-	0.03	0.26	0.18	0.10	0.19	0.15	0.07	0.09	0.14	0.19
MI15	0.17	0.32	0.38	0.39	0.24	0.39	0.29	0.26	0.41	0.09	0.30	0.34	0.19	0.04	-	0.23	0.16	0.09	0.18	0.13	0.08	0.11	0.12	0.17
MI16	0.43	0.44	0.44	0.43	0.32	0.19	0.37	0.34	0.42	0.43	0.30	0.37	0.39	0.42	0.39	-	0.20	0.25	0.20	0.26	0.26	0.27	0.14	0.14
MI17	0.28	0.25	0.22	0.25	0.17	0.32	0.21	0.22	0.18	0.27	0.13	0.20	0.21	0.27	0.26	0.28	-	0.20	0.07	0.14	0.18	0.13	0.13	0.16
MI18	0.12	0.33	0.38	0.40	0.29	0.41	0.33	0.32	0.42	0.17	0.32	0.35	0.23	0.14	0.15	0.41	0.29	-	0.21	0.18	0.04	0.16	0.15	0.21
MI19	0.33	0.30	0.28	0.31	0.22	0.35	0.25	0.25	0.27	0.32	0.21	0.26	0.26	0.31	0.30	0.34	0.11	0.33	-	0.15	0.20	0.15	0.15	0.17
MI20	0.29	−0.01	0.29	0.25	0.27	0.36	0.33	0.38	0.45	0.35	0.24	0.28	0.29	0.31	0.31	0.42	0.22	0.32	0.28	-	0.18	0.10	0.12	0.16
MI21	0.09	0.34	0.39	0.40	0.26	0.39	0.31	0.30	0.42	0.13	0.32	0.36	0.21	0.09	0.13	0.41	0.27	0.05	0.32	0.33	-	0.14	0.14	0.19
MI22	0.19	0.22	0.28	0.30	0.09	0.33	0.17	0.15	0.34	0.17	0.21	0.26	0.04	0.15	0.17	0.36	0.16	0.22	0.21	0.21	0.18	-	0.15	0.18
WI4	0.27	0.26	0.23	0.27	0.22	0.30	0.28	0.24	0.27	0.22	0.12	0.19	0.20	0.24	0.23	0.25	0.20	0.25	0.25	0.21	0.24	0.22	-	0.15
WI5	0.33	0.32	0.24	0.28	0.26	0.27	0.30	0.27	0.27	0.30	0.16	0.21	0.26	0.32	0.30	0.21	0.23	0.33	0.27	0.28	0.31	0.26	0.24	-

The results of principal components analysis (PCA) and discriminant analyses of principal components (DAPC) are similar to those that explicitly consider a priori population structure. With PCA, three to four clusters were recovered corresponding to the same ones from the population assignation analyses, and this was more evident with the MCR90 datasets compared to the MCR50 ones. In all analyses, three populations—MI6, MI16, and WI5—were recognized as most distinct from the other populations. Across DAPC analyses, individuals from populations tended to cluster together, and this is similar to results from other methods. In general, DAPC analyses recover MI6, MI16, and WI5 as distinct units or as a cluster together, with the results for MCR50 all loci being the only exception. In analyses with this dataset, WI5 was included in a cluster distinct from the other two populations, but with WI4 and populations from Michigan. In some analyses, such as MCR50 and MCR90 diploid loci, the divided cluster of central populations was identified. The number of loci under selection in each dataset is in Table 2.

Patterns of migration inferred from BA3-SNPs suggest that migration is minimal, regardless of the dataset analyzed, and that most individuals are from their original population (Figure 3). While this was certainly the case for all loci for MCR90, analyses with only the diploid loci for three or four population clusters (Table 1) provide evidence of greater rates of migration between adjacent populations (Figure 3). Migration directly between the eastern and western populations was negligible. The relationship among the four population clusters that was most supported by the results of DIYABC-RF and abcranger varies depending on the dataset analyzed. For all, diploid, and tetraploid loci, (West (Mid1 (Mid2, East))), (West (East (Mid1, Mid2))), and (West (Mid1 (Mid2, East))) are recovered as optimal, respectively, and (Mid2 (West (Mid2, East))) and (West (East (Mid1, Mid2))) are identified as close second choices for all and tetraploid datasets, respectively. The one constant among the three optimal trees is that the western population is recognized to have diverged prior to the mid and eastern populations, and this also is the case for one of the near-optimal trees (Supplemental Figure S17).

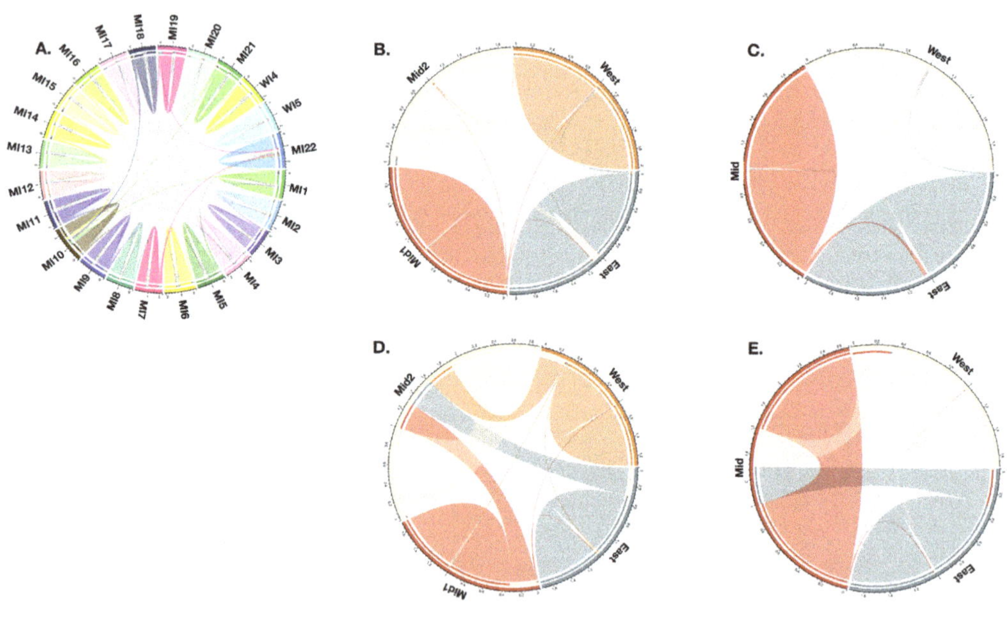

Figure 3. Patterns of migration based on MCR90 all loci (**A–C**) and MCR90 diploid loci (**D**,**E**) as resolved using BA3-SNPs. (**A**) All populations, (**B**,**D**) 4 populations, (**C**,**E**) 3 populations. Outermost circle denotes each population, and inner circle shows origin of migrants from each population. Lines connecting populations demonstrate patterns of migration.

2.3. Conservation Units

Based on the method of Funk et al. [16], evolutionarily significant units (ESUs) were identified using all loci, as described below, and the management units (MUs), which are based on fastSTRUCTURE, PCA, and DAPC analyses with loci not under selection, are quite similar. The largest difference between ESUs and MUs is that the two populations in Wisconsin may or may not be included with the other two western populations, MI6 and MI16, depending on the use of all loci or only diploid or tetraploid loci (Figure 4). The populations on Bois Blanc Island also have mixed ancestry based on these loci. The adaptive units, which are based on fastSTRUCTURE, PCA, and DAPC analyses with loci under selection, provide quite different results. Generally, among analyses, nine adaptive units are recognized, and these are structured based on geography (Figure 4, Supplemental Figure S18, Table 1).

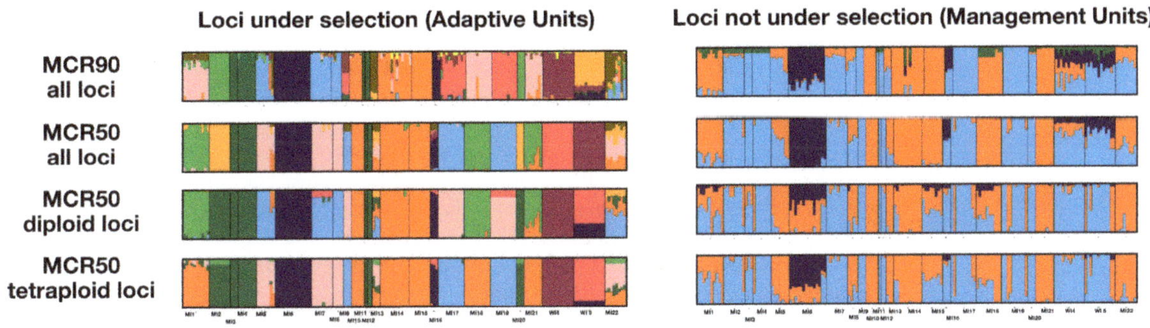

Figure 4. Structure bar graphs for MCR90 all loci and three MCR50 datasets for loci under and not under selection (adaptive units and management units, respectively). Individual ancestry denoted by color. Groups for each listed in Table 1, and best K values noted in Table 2.

The results of the fastSTRUCTURE, PCA, and DAPC are similar, with one exception. Unlike analyses with fastSTRUCTURE and PCA, where individuals of the same population cluster together, with DAPC, some individuals of the same population are members of different clusters. This is likely due to the large number of clusters identified as optimal, which is particularly the case for MCR90 and MCR50 datasets with all loci.

3. Discussion

3.1. Population Structure and Genetic Diversity

Based on the multiple datasets explored using various methodological approaches, three or four different population clusters were frequently recognized for *I. lacustris* across Michigan and Wisconsin. These clusters are structured geographically, with eastern, central, and western groups, and at higher K values, the central group is subdivided into two groups that are also geographically oriented (Figures 1 and 2). In the three prior studies that employed isozymes and allozymes to examine the population genetics of *I. lacustris* [8–10], no to limited genetic diversity was identified in the populations. Each study only investigated the genetic diversity of populations within one state, using markers available at the time, which likely led to the paucity of genetic diversity. In the present study, many more loci were examined, and individuals from across most of the geographic range of the species were analyzed together, which provides a more holistic approach to elucidating the genetic diversity of the species. These results demonstrate that our hypothesis—a lack of genetic diversity among the species—was incorrect.

Across all studied populations, statistically significant isolation-by-distance is noted, and much of the genetic variation occurs within samples and among populations, with little variation within each population. These results are, on some level, unsurprising for a species that is not only clonal but also includes minimal sexual reproduction. Sampling issues, such as small numbers of individuals studied for some populations and potential collection of ramets, could also have contributed to limited within-population genetic diversity. Additionally, almost all populations have negative F_{IS} values, a finding frequently occurring with clonal plants [17]. A similar result was recovered by Edgeloe et al. [18] for another clonal, polyploid species, *Posidonia australis* Hook.f. Despite the clonal growth in these polyploid species, the multiple gene copies may provide sufficient genetic diversity and potential so that rare species, such as *I. lacustris*, do not suffer the negative long-term impacts of vegetative reproduction and inbreeding. The changing climate will certainly be a test as to whether the genetic diversity harbored in each population will be appropriate to adapt to new conditions [19].

Among the identified clusters of populations, there are two notable areas: Bois Blanc Island in the eastern part of the sampled range and the four western populations. In Bois Blanc Island, the populations display mixed ancestry between the eastern and central populations, and these were results recovered with multiple datasets and analyses. This mixed ancestry could occur because of hybridization on the island itself with ancestors from both populations colonizing and interbreeding there. Alternatively, hybridization could have taken place on the mainland of the lower peninsula of Michigan, such as at MI7 or MI8, followed by colonization of the island. While the signature of mixed ancestry identified in the present study may suggest that hybridization is recent, given that the species reproduces clonally, the signature of (older) hybridization could remain for an extended period of time. It is useful to keep in mind that the island and nearby areas on the mainland are some of the more heavily sampled geographic regions in the present study. This greater sampling could hint at a similar pattern in other areas if individuals were sampled to a larger extent. It was not possible to include representatives from Ontario, Canada in the study, and future studies that add these will likely have greater context for the relationship of the central and eastern populations to those even farther east.

The four populations in the western cluster (MI6, MI16, WI4, and WI5) are notable. While these populations form a cluster in most analyses (Figure 2), the two Wisconsin populations (WI4 and 5) differ from those in Michigan, and, in some analyses, from each other. While WI4 and WI5 are geographically close together on the Door Peninsula and tend to cluster together in some analyses, WI4 is sometimes resolved as sharing ancestry with the eastern populations, which is not the case for WI5. This could be due to the retention of ancestral polymorphism or the fact that the establishment of each of these populations differs. However, in analyses that account for both genetic and geographic data (i.e., tess3r and conStruct), both Wisconsin populations are distinct clusters and/or are usually allied with the other western populations. This is particularly the case for the diploid dataset. In another, well-known Great Lakes shoreline endemic, *Cirsium pitcheri* Torr. & A.Gray, a similar pattern was recovered. The populations from the Door Peninsula are also quite distinct from others on Lake Michigan [20], and the northern populations on the peninsula share more alleles with the populations in the Upper Peninsula of Michigan than with some of the populations on the southern part of the peninsula.

MI6 and MI16 are intriguing populations of *I. lacustris* because they are situated inland, and this is not the case for the other sampled populations. While other populations can be found a short distance from the shoreline, these populations are ca. 30 km from the current boundary of Lake Michigan. These two populations are consistently recognized as genetically distinct from the other sampled populations, and these both likely became established during higher water level periods of Glacial Lake Algonquin ca. 12,500 years ago [21,22]. As water levels decreased during the time of Glacial Lake Chippewa and subsequently rose to current levels, these two populations became isolated in suitable habitat (e.g., conifer wetland) that allowed individuals of *I. lacustris* to persist, but without the opportunity to interbreed with other, coastal populations, resulting in their distinct genetic signature (Figure 2).

3.2. Migration and Demography

After deglaciation, *I. lacustris* migrated eastward from the western part of its range. This pattern provides evidence that MI6 and MI16 became established early in the colonization of the species during times of higher water levels and, therefore, are relicts rather than the result of inland dispersal. Additionally, the central and then eastern populations developed via migration across northern Lakes Michigan and Huron, and these populations may have retained some of the ancestral polymorphisms in the more western populations, such as WI4 and WI5. This west-to-east pattern suggests that the populations in Ontario are the most recently established, a hypothesis that can be tested during a future study. The pattern noted here for *I. lacustris* differs from that of *C. pitcheri*, which is hypothesized to have migrated from east to west [20].

Overall, rates of migration, as inferred with BA3-SNPs, among populations are minimal, a result recovered in other species of *Iris* on the Korean Peninsula [23] and a pattern that is not uncommon for narrow endemics [20]. This minimal migration is the case for all 24 populations studied as well as with three and four population clusters inferred (Figure 3). Although the species presently reproduces within populations, migration occurred and may have provided an infusion of new alleles, even if this was not a common occurrence.

In *C. pitcheri*, Fant et al. [20] note that the changes in the water level of the Great Lakes shaped the geographic distribution of this endemic species, with lower water levels allowing for increased connection among populations. Lake level changes could also have impacted the geographic distribution of *I. lacustris*. This is particularly the case for the more inland populations, which could have become established ca. 4500 years ago during the most recent high water levels for the lake. Lower lake levels may have influenced colonization of the islands as well as migration across the northern regions of Lake Michigan and allowed for the exchange of individuals that currently would be more challenging.

An alternative hypothesis for the present geographic distribution of the species also exists. Van Kley and Wujek [6] and Brotske [4] provide evidence that *I. lacustris* can inhabit a diversity of ecosystems and that changes in patterns of disturbance and forest succession following European colonization of the area reduced the suitable habitat for the species (e.g., more forests with more closed canopies). This has resulted in populations primarily being restricted to shorelines where habitat was appropriate. If this is the case, the inland populations, such as MI6, would still represent relicts of a prior time, but this would be due to remnant habitat availability based on adequate disturbance regimes and/or seral stages, not prior establishment during higher water levels of the Great Lakes and subsequent serendipitous survival.

3.3. Subsetting Diploid and Tetraploid Loci

In the present study, polyRAD [24] was used to create datasets of diploid and tetraploid loci, and these were analyzed alongside a dataset of all loci for the MCR90 and MCR50 datasets. In general, analyses of all six datasets produced fairly similar results (Figure 2, Tables 3 and 4). fastSTRUCTURE analyses of MCR90 and MCR50 datasets of all loci resulted in the identification of a cluster of six populations in the central part of the sampled population of *I. lacustris* (MI2, MI3, MI4, MI11, MI12, and MI20) that was not recovered with the diploid or tetraploid datasets, although hints of this cluster can be seen in the MCR90 2N dataset at $K = 5$. This cluster is identified in all of the datasets with loci under selection as either one or two clusters (Figure 2) and with the MCR90 datasets analyzed with STRUCTURE [25] and MavericK [26].

The similar results among the datasets, regardless of ploidy, may provide some evidence that not disentangling diploid and tetraploid loci from all loci may not lead to spurious results using SNP data for population genomics [27]. This statement should be treated with skepticism because it is based only on one, empirical, study. Others who have used polyRAD to subset their datasets and identify diploid loci to use for population genomics [28,29], which is a practice aligned with assumptions of common methods [28], have not explored the use of all loci and/or tetraploid loci in comparison to only ones that segregate as diploids. It would be useful for additional studies on the population genomics of polyploid species to examine data employing all, diploid, and tetraploid (and higher) loci to determine if similar or divergent results are recovered. At the same time, the results presented herein may provide some level of confidence for researchers investigating the population genomics of species of unknown ploidy that use all loci identified via tGBS, and similar reduced-representation methods may not yield incongruent results.

3.4. Conservation Genetics of *I. lacustris*

The evolutionarily significant units (ESUs) were described above with all loci used for population genomic analyses, and the management units (MUs), which were determined using only loci not under selection, are similar, but not identical to the ESUs; however, the differences are minor (Figure 4). Given the similar ESUs and MUs, the management of the populations of *I. lacustris* could be geographically clustered into three to four units. However, the results of the use of the loci under selection to resolve adaptive units (AUs) differ from those of ESUs and MUs (Supplemental Figure S18). The AUs provide evidence of local adaptation, so managing only three or four MUs would not necessarily ensure that all of the genetic diversity of the species is appropriately protected. A total of nine AUs are recognized (Table 1), and while these are also geographically clustered, the AUs are much smaller than are the ESUs and MUs (Figure 4).

This local adaptation is, on some level, unsurprising, because even though the species is generally restricted to the same type of habitat presently (i.e., shorelines), climatic, soil, and vegetation differences occur across the geographic range of the species. Indeed, *I. lacustris* inhabits three of the landscape ecology regions of Michigan and multiple districts and subdistricts within each region [30,31]. Van Kley and Wujek [6] also recognized four soil types, four vegetation types, and pH variation across the species' range. Given that the species primarily reproduces asexually, this can lead to a loss of genetic variation over time as a limited number of successful genotypes dominates each particular climate–soil–vegetation combination. Consequently, the seemingly same type of habitat in a geographically distinct area may result in local adaptations to the specific region and ecosystem and contribute to outbreeding depression, limiting successful offspring from infrequent interpopulation crosses.

4. Materials and Methods

4.1. Plant Material

During the summers of 2019 and 2020, leaf material of 171 individuals of *I. lacustris* was collected from 24 locations in Michigan and Wisconsin (Figure 1) and dried in silica gel. The number of individuals per population ranged from 1 to 12, depending on the suitability of the population for collection. Most individual plants were collected at least 3 m from each other to maximize the possibility of sampling genets, not ramets. Latitude and longitude were recorded for each specimen.

4.2. DNA Sequencing

Leaf material was sent to data2bio (www.data2bio.com, accessed on 1 May 2023) for DNA isolation and tunable Genotyping-by-Sequencing (tGBS) to recognize single nucleotide polymorphisms (SNPs) across the populations. Using the restriction enzyme Bsp1286I, paired-end tGBS libraries were created [15] and subsequently sequenced with an Illumina HiSeq X (Illumina Inc., San Diego, CA, USA). Based on all sequence data, consensus reference sequences were generated with CD-HIT-454 [32] after sequencing depth was normalized to 50×, and sequencing errors were corrected using Fiona [33]. Low-quality reads were discarded (PHRED quality < 15 and error rates $\geq$ 3%) and trimmed, and GSNAP [34] was employed to map reads to the reference sequences based on the following parameters: $\leq$2 mismatches per 36 bp and less than five total per 75 bp for tails. SNPs were identified based on the following criteria: two most common alleles supported by at least 30% of the aligned bases, at least five unique reads, the sum of the one or two most common alleles covering at least 80% of the aligned reads, and no polymorphisms in the first or last three base pairs of each read. From the SNPs, two datasets were created: MCR90 with up to 10% missing data and MCR50 with up to 50% missing data.

4.3. Polyploidy Filtering

Because I. lacustris is a putative polyploid and many population genetic methods assume that species are (at most) diploid, polyRAD [24] was used to identify and filter loci that are diploid and tetraploid. The MCR90 and MCR50 datasets were filtered using the IteratePopStruct command to identify genotypes, and then the H_{ind}/H_E statistic [24,35] was employed to recognize diploid loci with $H_{ind}/H_E < 0.5$ and tetraploid loci with $H_{ind}/H_E > 0.75$. Datasets were created for each set of loci (Table 2). The number of SNPs in the diploid and tetraploid datasets does not equal the value in the initial datasets because of filtering with polyRAD.

4.4. Population Genomics

Observed and expected heterozygosity measurements and *F*-statistics were calculated with hierfstat [36,37], and AMOVA was conducted with poppr [38]. All 24 populations were examined, as were the populations divided into three and four geographic clusters, which are based on the optimal K values from preliminary analyses in fastSTRUCTURE (Table 2) and patterns of population structure from STRUCTURE and MavericK. fastSTRUCTURE [39] was employed to identify population structure, including the optimal number of clusters (K), and for these analyses, K = 1–24 were analyzed for the six SNP datasets, using Structure_threader [40], on the Kettering University High-Performance Computing Cluster (KUHPC). Ten replicates were run for each K, with a convergence criterion of 0.000001, a simple prior, and 100 test sets for cross-validation. The CLUMPAK main pipeline, which includes CLUMPP [41] and DISTRUCT [42], was employed to organize, cluster, and visualize the results of independent fastSTRUCTURE analyses, via 10,000 permutations of the LargeKGreedy algorithm [43]. To identify the optimal K value(s), the marginal likelihood that maximizes model complexity from fastSTRUCTURE and the MedMedK, MedMeanK, MaxMedK, and MaxMeanK values determined by StructureSelector [44,45] were examined. These latter four metrics are useful for uneven sampling and are based on recognizing the number of clusters that include, at minimum, one subpopulation. Differences among these metrics are the result of the arithmetic mean or median used and the median or maximum number of clusters identified [45].

For comparison, and given potential variation in ploidy at loci [27], STRUCTURE [25] and MavericK [26] were also used, with Structure_threader, for analyses with the three MCR90 datasets. With STRUCTURE, the following parameters were used with K = 1–24: 1,000,000 steps and 500,000 burnin, with alpha and lambda of 1, and with or without admixture. Ten replicates were run for each K. CLUMPAK and StructureSelector were also used for STRUCTURE analyses, with the best K also determined via the method of Evanno et al. [46] and Ln Pr (X | K). MavericK analyses were run for K = 1–12 with five replicates per K, without admixture, using the following parameters for each replicate: 50,000 steps and 5000 burnin for Markov Chain Monte Carlo (MCMC) sampling and an alpha of 1500 steps and 5000 burnin, with 50 rungs, for thermodynamic integration (TI) sampling, and 100 expectation-maximization repeats. With MavericK, graphs were visualized with R [47], and the optimal K value was determined using TI.

To explicitly include geographical data along with SNPs to investigate patterns of population genetics, tess3r [48] and conStruct [49] were used, and all datasets were analyzed with the former, but only the three MCR90 datasets with the latter. For tess3r, the alternating projected least squares method was undertaken for K = 1–24 for MCR90 and K = 1–12 for MCR50 datasets. Results for each K were visualized with bar graphs and maps in R [47], and the optimal K value was identified using the cross-validation plot for each dataset. For conStruct cross-validation, analyses were conducted with five replicates, for K = 1–8, using 10,000 MCMC iterations sampled every 1000 iterations and a training proportion of 0.5–0.8, depending on the dataset. Subsequently, analyses with K = 3–5 were conducted, with five replicates, using one chain run for 100,000 MCMC iterations sampled every 1000 iterations and with the spatial model.

In addition to analyses for explicit population structure, all datasets were analyzed with principal component analyses (PCA), correspondence analyses (CA), and discriminant analyses of principal components (DAPC) in adegenet [50], principal coordinate analyses (PCoA) in hierfstat [36,37], and isolation-by-distance (IBD) analyses in adegenet using separate Mantel tests for population and individuals, with 999 simulations for the Mantel test. For DAPC for each dataset, the Bayesian Information Criterion (BIC) was used to identify the optimal number of clusters, and cross-validation was employed to explore the most appropriate number of PCs to retain for analysis.

Loci under selection were determined with BayeScan [51] using 100,000 iterations, a burnin of 50,000 iterations, a thinning interval of 10, and a sample size of 5000, and for each analysis, 20 pilot runs were conducted, each with 5000 steps. Loci under selection were visualized in R using F_{ST} values and a false discovery rate of 0.05.

Demographic history and patterns of migration were explored using BA3-SNPs [52,53], DIYABC Random Forest (DIYABC-RF) [54], and abcranger [55], and only the three MCR90 datasets were used for these analyses, with the three and four aforementioned population clusters used (apart from all 24 populations investigated with MCR90 with BA3-SNPs). For BA3-SNPs, the datasets were each run for 50 million Markov Chain Monte Carlo (MCMC) iterations, with 20 million MCMC burnin iterations, and a sampling interval of 2500 iterations, and the initial parameters for allele frequencies, inbreeding coefficient, and migration rates were tuned to vary between 0.2–0.6. For DIYABC-RF, the optimal scenario for patterns of diversification were examined among all 15 arrangements of four bifurcating populations. For each scenario, population size was modelled to vary after populations split and one and two other times for when the second and first populations diverge (Supplemental Figure S17). For analyses, all genetic diversity, F_{ST} distances, Nei's distances, and admixture estimates were selected, and the analyses were run for 15 million simulations with a batch size of 1000. Using the results of the training, a random forest analysis was conducted with abcranger [55] using 1000 trees to identify the number of trees supporting each model and to estimate the parameters of the model, with and without linear discriminant analysis, for partial least squares (PLS) estimation on the optimal model for each dataset.

4.5. Conservation Units

Conservation and management units were identified following the three-step method of Funk et al. [16], in which (1) evolutionarily significant units (ESUs) are recognized using all loci, (2) management units (MUs) are delimited with non-outlier loci, and (3) adaptive groups are determined using outlier loci. For the three steps, fastSTRUCTURE [39], PCA, and DAPC were used [50]. The first step was described above for datasets with all loci, and the other two steps were conducted using the same parameters for the three analyses and were based on two datasets (loci under and not under selection as determined via BayeScan [51]) for each MCR50 dataset and the all loci dataset of MCR90 (Table 2). The optimal K value was identified using STRUCTURESELECTOR [44], the marginal likelihood that maximizes model complexity from fastSTRUCTURE [39], and the BIC for DAPC with adegenet [50]. Based on the results of these analyses, ESUs, MUs, and adaptive groups were identified (Supplemental Figure S18).

5. Conclusions

The present study provides evidence of genomic variation and local adaptation across the geographic range of the species, which is novel given the negligible genetic diversity previously recovered for *I. lacustris* [8–10]. However, as Van Kley and Wujek [6] stated thirty years ago, "Despite a preference for a somewhat disturbed habitat, *Iris lacustris* will not grow where the habitat has been destroyed by residential, resort, or industrial development".

Therefore, the conservation genetic results are of limited value if management steps are not taken to ensure that individuals of *I. lacustris* have the opportunity to be successful in situ. This includes not only ensuring intermediate light conditions and limited litter [5,6], but also that as much genetic diversity across the entire geographic range of the species is conserved and managed appropriately. Indeed, given the local genetic diversity recognized among the nine adaptive units, it would be prudent to strive to conserve representatives from these areas. This is particularly important because the populations that are best able to adapt to the changing climate in the Great Lakes region is presently unknown [56]. Therefore, to ensure the longevity of this charismatic species, appropriate long-term management is necessary. Future work that includes the populations of *I. lacustris* from Ontario can extend the presented results to investigate the ways in which these populations relate to those in the United States. Given the international geographic range of the species, conservation efforts that are binational would be particularly useful.

Supplementary Materials: The following supporting information can be downloaded at: https://www.mdpi.com/article/10.3390/plants12132557/s1, Figure S1. Results for Isolation-by-Distance (IBD) for the six datasets. The x-axis is geographic distance, and the y-axis is genetic distance. Figure S2. Structure bar graphs from STRUCTURE for the six datasets analyzed in the present study for K (clusters) = 3–5. Individual ancestry denoted by color. Populations are denoted below each graph. Figure S3. Structure bar graphs from MaverickK, without admixture, for the three MCR90 datasets analyzed in the present study for K (clusters) = 3–5. Individual ancestry denoted by color. Populations are denoted below each graph. Figure S4. tess3r maps of population assignation for the six datasets analyzed in the present study for K (clusters) = 3–5. Individual ancestry denoted by color. Figure S5. Maps and bar graphs of population assignation for the three MCR90 datasets analyzed in the present study for K (clusters) = 3–5. Individual ancestry denoted by color. Figure S6. Results for best K from StructureSelector for analyses with fastStructure for (A) MCR90 all loci, (B) MCR90 diploid loci, and (C) MCR90 tetraploid loci. Figure S7. Results for best K from StructureSelector for analyses with fastStructure for (A) MCR50 all loci, (B) MCR50 diploid loci, and (C) MCR50 tetraploid loci. Figure S8. Results for best K from StructureSelector for analyses with Structure without admixture for (A) MCR90 all loci, (B) MCR90 diploid loci, and (C) MCR90 tetraploid loci. Figure S9. Results for best K from StructureSelector for analyses with Structure with admixture for (A) MCR90 all loci, (B) MCR90 diploid loci, and (C) MCR90 tetraploid loci. Figure S10. Results for best K from MaverickK, based on thermodynamic integration (TI), for analyses without admixture (A) MCR90 all loci, (B) MCR90 diploid loci, and (C) MCR90 tetraploid loci. Figure S11. Results for cross-validation scores for tess3r analyses for (A) MCR90 all loci, (B) MCR90 diploid loci, (C) MCR90 tetraploid loci, (D), MCR50 all loci, (E) MCR50 diploid loci, and (F) MCR50 tetraploid loci. Figure S12. Results for cross-validation scores for conStruct validation analyses for (A) MCR90 all loci, (B) MCR90 diploid loci, and (C) MCR90 tetraploid loci to identify best K (clusters). Graphs with blue and green dots are for spatial and non-spatial models, respectively, and graph with only blue dots displays predictive accuracy for spatial model with confidence intervals. Figure S13. Results for Bayesian Information Criterion (BIC), to identify best K (clusters), from discriminant analysis of principal components (DAPC) for (A) MCR90 all loci, (B) MCR90 diploid loci, (C) MCR90 tetraploid loci, (D), MCR50 all loci, (E) MCR50 diploid loci, and (F) MCR50 tetraploid loci. Figure S14. Results from StructureSelector for best K (clusters) fastStructure analyses for loci under selection for (A) MCR90 all loci, (B) MCR50 all loci, (C) MCR50 diploid loci, and (D) MCR50 tetraploid loci. Figure S15. Results from StructureSelector for best K (clusters) fastStructure analyses for loci not under selection for (A) MCR90 all loci, (B) MCR50 all loci, (C) MCR50 diploid loci, and (D) MCR50 tetraploid loci. Figure S16. Results for Bayesian Information Criterion (BIC), to identify best K (clusters), from discriminant analysis of principal components (DAPC) analyses for (A) MCR90 all loci under selection, (B) MCR50 all loci under selection, (C) MCR50 diploid loci under selection, (D), MCR50 tetraploid loci under selection, (E) MCR90 all loci not under selection, (F) MCR50 all loci not under selection, (G) MCR50 diploid loci not under selection, (H), MCR50 tetraploid loci not under selection. Figure S17. 15 branching scenarios evaluated in DIYABC. Pop 1 is East, Pop 2 is Mid 1, Pop 3 is Mid 2, Pop 4 is West. See Table 1 for population assignation to each population. Change in color represents potential change in population size. Scenario 3 is optimal for all and tetraploid loci, and scenario 7 is optimal for diploid loci. Figure S18. Nine Adaptive Units recognized from population genetic analyses using loci under selection. Map of locations sampled in present study. Dark gray entire lines denote division between East, Mid1, Mid2, and West clusters (also recognized as Management Units). The dashed gray line separates Mid1 and Mid2 populations, and Mid includes both groups of populations together. Light gray lines separate Wisconsin (USA), Michigan (USA), and Ontario (Canada). Scale bar, in red, represents 50 kilometers. Table S1. AMOVA results for all datasets. Table S2. K values for the MCR90 datasets for STRUCTURE and Maverick.

Author Contributions: Conceptualization, J.I.C. and S.T.-C.; methodology, J.I.C. and S.T.-C.; formal analysis, J.I.C.; data curation, J.I.C.; writing—original draft preparation, J.I.C.; writing—review and editing, J.I.C. and S.T.-C.; visualization, J.I.C. and S.T.-C.; funding acquisition, J.I.C. and S.T.-C. All authors have read and agreed to the published version of the manuscript.

Funding: This research was funded by a Kettering University Faculty Research Fellowship, the Michigan Natural Features Inventory (MNFI), and Weber State University. The APC was funded by MDPI.

Data Availability Statement: Data files in VCF format are available at the Dryad repository (https://doi.org/10.5061/dryad.xwdbrv1jh).

Acknowledgments: The authors thank R. Hackett who collected the vast majority of the leaf material, and R. Bowman collected leaf material from Wisconsin. R. Hackett, P. Higman, C. Tansy, J. Dingledine, and S. Hicks provided invaluable conversation about the Dwarf Lake Iris. R. Hackett's comments on a draft of the manuscript were very helpful. Three reviewers provided helpful comments to improve the manuscript. We appreciate L. Coffey and data2bio for tGBS sequencing. L. Clark provided helpful comments on using polyRAD.

Conflicts of Interest: The authors declare no conflict of interest.

References

1. Nuttall, T. *The Genera of North American Plants: And a Catalogue of the Species, to the Year 1817*; D. Heartt: Philadelphia, PA, USA, 1817; Volume 1.
2. U.S. Fish and Wildlife Service. *Status Review—Dwarf Lake Iris (Iris lacustris)*; East Lansing Field Office: East Lansing, MI, USA, 2022; p. 10.
3. Voss, E.G. *Michigan Flora*, 3rd ed.; Cranbrook Institute of Science: Bloomfield Hills, MI, USA; University of Michigan Herbarium: Ann Arbor, MI, USA, 1972; Volume 1, p. 488.
4. Brotske, V. Pollination, Seed Dispersal, Germination, and Seedling Survival in the Federally Threatened Dwarf Lake Iris (Iris Lacustris). Master's Thesis, University of Wisconsin-Green Bay, Green Bay, WI, USA, 2018.
5. U.S. Fish and Wildlife Service. *5-Year Review Dwarf Lake Iris (Iris lacustris)*; U.S. Fish and Wildlife Service: East Lansing, MI, USA, 2011; p. 21.
6. Van Kley, J.E.; Wujek, D.E. Habitat and ecology of Iris lacustris (the dwarf lake iris). *Mich. Bot.* **1993**, *32*, 209–222.
7. State of Michigan. State Facts and Symbols. Available online: https://www.michigan.gov/som/about-michigan/state-facts-and-symbols (accessed on 15 January 2023).
8. Simonich, M.T.; Morgan, M.D. Allozymic uniformity in Iris lacustris (dwarf lake iris) in Wisconsin. *Can. J. Bot.* **1994**, *72*, 1720–1722. [CrossRef]
9. Orick, M.W. Enzyme Polymorphism and Genetic Diversity in the Great Lakes Endemic Iris lacustris Nutt. (Dwarf Lake Iris). Master's Thesis, Eastern Michigan University, Ypsilanti, MI, USA, 1992.
10. Hannan, G.L.; Orick, M.W. Isozyme diversity in Iris cristata and the threatened glacial endemic I. lacustris (Iridaceae). *Am. J. Bot.* **2000**, *87*, 293–301. [CrossRef]
11. Guo, J.; Wilson, C.A. Molecular phylogeny of crested Iris based on five plastid markers (Iridaceae). *Syst. Bot.* **2013**, *38*, 987–995. [CrossRef]
12. Soltis, P.S.; Soltis, D.E. The role of genetic and genomic attributes in the success of polyploids. *Proc. Natl. Acad. Sci. USA* **2000**, *97*, 7051–7057. [CrossRef] [PubMed]
13. Luttikhuizen, P.C.; Stift, M.; Kuperus, P.; Van Tienderen, P.H. Genetic diversity in diploid vs. tetraploid Rorippa amphibia (Brassicaceae). *Mol. Ecol.* **2007**, *16*, 3544–3553. [CrossRef] [PubMed]
14. Van de Peer, Y.; Ashman, T.-L.; Soltis, P.S.; Soltis, D.E. Polyploidy: An evolutionary and ecological force in stressful times. *Plant Cell* **2021**, *33*, 11–26. [CrossRef] [PubMed]
15. Ott, A.; Liu, S.; Schnable, J.C.; Yeh, C.-T.E.; Wang, K.-S.; Schnable, P.S. tGBS® genotyping-by-sequencing enables reliable genotyping of heterozygous loci. *Nucleic Acids Res.* **2017**, *45*, e178. [CrossRef]
16. Funk, W.C.; McKay, J.K.; Hohenlohe, P.A.; Allendorf, F.W. Harnessing genomics for delineating conservation units. *Trends Ecol. Evol.* **2012**, *27*, 489–496. [CrossRef]

17. Millar, M.A.; Byrne, M. Variable clonality and genetic structure among disjunct populations of Banksia mimica. *Conserv. Genet.* **2020**, *21*, 803–818. [CrossRef]
18. Edgeloe, J.M.; Severn-Ellis, A.A.; Bayer, P.E.; Mehravi, S.; Breed, M.F.; Krauss, S.L.; Batley, J.; Kendrick, G.A.; Sinclair, E.A. Extensive polyploid clonality was a successful strategy for seagrass to expand into a newly submerged environment. *Proc. R. Soc. B* **2022**, *289*, 20220538. [CrossRef] [PubMed]
19. Sessa, E.B. Polyploidy as a mechanism for surviving global change. *New Phytol.* **2019**, *221*, 5–6. [CrossRef] [PubMed]
20. Fant, J.B.; Havens, K.; Keller, J.M.; Radosavljevic, A.; Yates, E.D. The influence of contemporary and historic landscape features on the genetic structure of the sand dune endemic, Cirsium pitcheri (Asteraceae). *Heredity* **2014**, *112*, 519–530. [CrossRef]
21. Kincare, K.; Larson, G.J. Evolution of the Great Lakes. In *Michigan Geography and Geology*; Schaetzl, R.J., Darden, J.T., Brandt, D., Eds.; Pearson Custom Publishing: Boston, MA, USA, 2009; pp. 174–190.
22. Larson, G.; Schaetzl, R. Origin and evolution of the Great Lakes. *J. Great Lakes Res.* **2001**, *27*, 518–546. [CrossRef]
23. Chung, M.Y.; López-Pujol, J.; Lee, Y.M.; Oh, S.H.; Chung, M.G. Clonal and genetic structure of Iris odaesanensis and Iris rossii (Iridaceae): Insights of the Baekdudaegan Mountains as a glacial refugium for boreal and temperate plants. *Plant Syst. Evol.* **2015**, *301*, 1397–1409. [CrossRef]
24. Clark, L.V.; Lipka, A.E.; Sacks, E.J. polyRAD: Genotype calling with uncertainty from sequencing data in polyploids and diploids. *G3 Genes Genomes Genet.* **2019**, *9*, 663–673. [CrossRef]
25. Pritchard, J.K.; Stephens, M.; Donnelly, P. Inference of population structure using multilocus genotype data. *Genetics* **2000**, *155*, 945–959. [CrossRef] [PubMed]
26. Verity, R.; Nichols, R.A. Estimating the number of subpopulations (K) in structured populations. *Genetics* **2016**, *203*, 1827–1839. [CrossRef] [PubMed]
27. Stift, M.; Kolář, F.; Meirmans, P.G. STRUCTURE is more robust than other clustering methods in simulated mixed-ploidy populations. *Heredity* **2019**, *123*, 429–441. [CrossRef]
28. Chafin, T.K.; Regmi, B.; Douglas, M.R.; Edds, D.R.; Wangchuk, K.; Dorji, S.; Norbu, P.; Norbu, S.; Changlu, C.; Khanal, G.P. Parallel introgression, not recurrent emergence, explains apparent elevational ecotypes of polyploid Himalayan snowtrout. *R. Soc. Open Sci.* **2021**, *8*, 210727. [CrossRef]
29. Salvado, P.; Aymerich Boixader, P.; Parera, J.; Vila Bonfill, A.; Martin, M.; Quélennec, C.; Lewin, J.M.; Delorme-Hinoux, V.; Bertrand, J.A.M. Little hope for the polyploid endemic Pyrenean Larkspur (Delphinium montanum): Evidences from population genomics and Ecological Niche Modeling. *Ecol. Evol.* **2022**, *12*, e8711. [CrossRef] [PubMed]
30. Barnes, B.V.; Wagner, W.H., Jr. *Michigan Trees. A Guide to the Trees of Michigan and the Great Lakes Region*; University of Michigan Press: Ann Arbor, MI, USA, 1981.
31. Walker, W.S.; Barnes, B.V.; Kashian, D.M. Landscape ecosystems of the Mack Lake burn, northern Lower Michigan, and the occurrence of the Kirtland's warbler. *For. Sci.* **2003**, *49*, 119–139. [CrossRef]
32. Fu, L.; Niu, B.; Zhu, Z.; Wu, S.; Li, W. CD-HIT: Accelerated for clustering the next-generation sequencing data. *Bioinformatics* **2012**, *28*, 3150–3152. [CrossRef]
33. Schulz, M.H.; Weese, D.; Holtgrewe, M.; Dimitrova, V.; Niu, S.; Reinert, K.; Richard, H. Fiona: A parallel and automatic strategy for read error correction. *Bioinformatics* **2014**, *30*, i356–i363. [CrossRef] [PubMed]
34. Wu, T.D.; Nacu, S. Fast and SNP-tolerant detection of complex variants and splicing in short reads. *Bioinformatics* **2010**, *26*, 873–881. [CrossRef] [PubMed]
35. Clark, L.V.; Mays, W.; Lipka, A.E.; Sacks, E.J. A population-level statistic for assessing Mendelian behavior of genotyping-by-sequencing data from highly duplicated genomes. *BMC Bioinform.* **2022**, *23*, 101. [CrossRef]
36. De Meeûs, T.; Goudet, J. A step-by-step tutorial to use HierFstat to analyse populations hierarchically structured at multiple levels. *Infect. Genet. Evol.* **2007**, *7*, 731–735. [CrossRef]
37. Goudet, J. HIERFSTAT, a package for R to compute and test hierarchical F-statistics. *Mol. Ecol. Notes* **2005**, *5*, 184–186. [CrossRef]
38. Kamvar, Z.N.; Tabima, J.F.; Grünwald, N.J. Poppr: An R package for genetic analysis of populations with clonal, partially clonal, and/or sexual reproduction. *PeerJ* **2014**, *2*, e281. [CrossRef]
39. Raj, A.; Stephens, M.; Pritchard, J.K. fastSTRUCTURE: Variational inference of population structure in large SNP data sets. *Genetics* **2014**, *197*, 573–589. [CrossRef]
40. Pina-Martins, F.; Silva, D.N.; Fino, J.; Paulo, O.S. Structure_threader: An improved method for automation and parallelization of programs structure, fastStructure and MavericK on multicore CPU systems. *Mol. Ecol. Res.* **2017**, *17*, e268–e274. [CrossRef]
41. Jakobsson, M.; Rosenberg, N.A. CLUMPP: A cluster matching and permutation program for dealing with label switching and multimodality in analysis of population structure. *Bioinformatics* **2007**, *23*, 1801–1806. [CrossRef]
42. Rosenberg, N.A. DISTRUCT: A program for the graphical display of population structure. *Mol. Ecol. Notes* **2004**, *4*, 137–138. [CrossRef]
43. Kopelman, N.M.; Mayzel, J.; Jakobsson, M.; Rosenberg, N.A.; Mayrose, I. CLUMPAK: A program for identifying clustering modes and packaging population structure inferences across K. *Mol. Ecol. Resour.* **2015**, *15*, 1179–1191. [CrossRef]
44. Li, Y.L.; Liu, J.X. STRUCTURESELECTOR: A web-based software to select and visualize the optimal number of clusters using multiple methods. *Mol. Ecol. Resour.* **2018**, *18*, 176–177. [CrossRef]
45. Puechmaille, S.J. The program STRUCTURE does not reliably recover the correct population structure when sampling is uneven: Subsampling and new estimators alleviate the problem. *Mol. Ecol. Resour.* **2016**, *16*, 608–627. [CrossRef] [PubMed]

46. Evanno, G.; Regnaut, S.; Goudet, J. Detecting the number of clusters of individuals using the software STRUCTURE: A simulation study. *Mol. Ecol.* **2005**, *14*, 2611–2620. [CrossRef] [PubMed]
47. R Developement Core Team. A Language and Environment for Statistical Computing. 2009. Available online: http://www.R-project.org (accessed on 5 January 2023).
48. Caye, K.; Jay, F.; Michel, O.; François, O. Fast inference of individual admixture coefficients using geographic data. *Ann. Appl. Stat.* **2018**, *12*, 586–608. [CrossRef]
49. Bradburd, G.S.; Coop, G.M.; Ralph, P.L. Inferring continuous and discrete population genetic structure across space. *Genetics* **2018**, *210*, 33–52. [CrossRef] [PubMed]
50. Jombart, T.; Ahmed, I. adegenet 1.3-1: New tools for the analysis of genome-wide SNP data. *Bioinformatics* **2011**, *27*, 3070–3071. [CrossRef]
51. Foll, M.; Gaggiotti, O. A genome-scan method to identify selected loci appropriate for both dominant and codominant markers: A Bayesian perspective. *Genetics* **2008**, *180*, 977–993. [CrossRef] [PubMed]
52. Mussmann, S.M.; Douglas, M.R.; Chafin, T.K.; Douglas, M.E. BA3-SNPs: Contemporary migration reconfigured in BayesAss for next-generation sequence data. *Methods Ecol. Evol.* **2019**, *10*, 1808–1813. [CrossRef]
53. Wilson, G.A.; Rannala, B. Bayesian inference of recent migration rates using multilocus genotypes. *Genetics* **2003**, *163*, 1177–1191. [CrossRef] [PubMed]
54. Collin, F.D.; Durif, G.; Raynal, L.; Lombaert, E.; Gautier, M.; Vitalis, R.; Marin, J.M.; Estoup, A. Extending approximate Bayesian computation with supervised machine learning to infer demographic history from genetic polymorphisms using DIYABC Random Forest. *Mol. Ecol. Resour.* **2021**, *21*, 2598–2613. [CrossRef] [PubMed]
55. Collin, F.-D.; Estoup, A.; Marin, J.-M.; Raynal, L. Bringing ABC inference to the machine learning realm: AbcRanger, an optimized random forests library for ABC. In Proceedings of the JOBIM 2020, Montpellier, France, 30 June 2020.
56. Byun, K.; Chiu, C.-M.; Hamlet, A.F. Effects of 21st century climate change on seasonal flow regimes and hydrologic extremes over the Midwest and Great Lakes region of the US. *Sci. Total Environ.* **2019**, *650*, 1261–1277. [CrossRef]

plants

Article

Conservation Genetics of the Endangered Lompoc Yerba Santa (*Eriodictyon capitatum* Eastw., Namaceae), including Phylogenomic Insights into the Evolution of *Eriodictyon*

C. Matt Guilliams * and Kristen E. Hasenstab-Lehman

Santa Barbara Botanic Garden, Santa Barbara, CA 93105, USA; klehman@sbbotanicgarden.org
* Correspondence: mguilliams@sbbotanicgarden.org; Tel.: +1-805-682-4726

Abstract: *Eriodictyon capitatum* (Namaceae) is a narrowly distributed shrub endemic to western Santa Barbara County, where it is known from only 10 extant California Natural Diversity Database element occurrences (EOs). Owing to low numbers of plants in nature, a limited overall extent, and multiple current threats, *E. capitatum* is listed as Endangered under the Federal Endangered Species Act and as Rare under the California Native Plant Protection Act. In the present study, high-throughput DNA sequence data were analyzed to investigate genetic diversity within and among all accessible EOs; to determine the extent of genetic isolation among EOs; to examine clonality within EOs; and to examine the taxonomic circumscriptions of *E. capitatum*, *E. altissimum*, *E. angustifolium*, and *E. californicum* through phylogenomic analysis. Population genetic analyses of *E. capitatum* reveal a pattern of strong genetic differentiation by location/EO. The clonality assessment shows that certain small EOs may support relatively few multilocus genotypes. The phylogenomic analyses strongly support the present-day taxonomic circumscriptions of both *E. altissimum* and *E. capitatum*, showing them to be reciprocally monophyletic and sister with strong support. Taken together, these results paint a picture of an evolutionarily and morphologically distinct species known from relatively few, genetically isolated stations.

Keywords: *Eriodictyon*; Namaceae; ddRADseq; endangered; endemic; clonality

Citation: Guilliams, C.M.; Hasenstab-Lehman, K.E. Conservation Genetics of the Endangered Lompoc Yerba Santa (*Eriodictyon capitatum* Eastw., Namaceae), including Phylogenomic Insights into the Evolution of *Eriodictyon*. *Plants* **2024**, *13*, 90. https://doi.org/10.3390/plants13010090

Academic Editors: Brenda Molano-Flores and James Cohen

Received: 2 November 2023
Revised: 19 December 2023
Accepted: 20 December 2023
Published: 27 December 2023

1. Introduction

Eriodictyon Benth. is a small genus of perennial herbs and shrubs endemic to western North America. It is typically delimited to include 11 species and 14 minimum-rank taxa, inclusive of species, subspecies, and varieties [1,2]. Based on phylogenetic evidence, *Nama rothrockii* A. Gray may be closely related to *Eriodictyon* as well, although no combination for this plant in *Eriodictyon* exists at the present time [1]. Members of the genus largely occur in the California Floristic Province [3], which extends from southwestern Oregon, United States, southward to northwestern Baja California, Mexico. The genus also has a second center of distribution in Arizona, Nevada, and Utah. Figure 1A shows the distribution of *Eriodictyon* based on specimen records available from the Global Biodiversity Information Facility (GBIF).

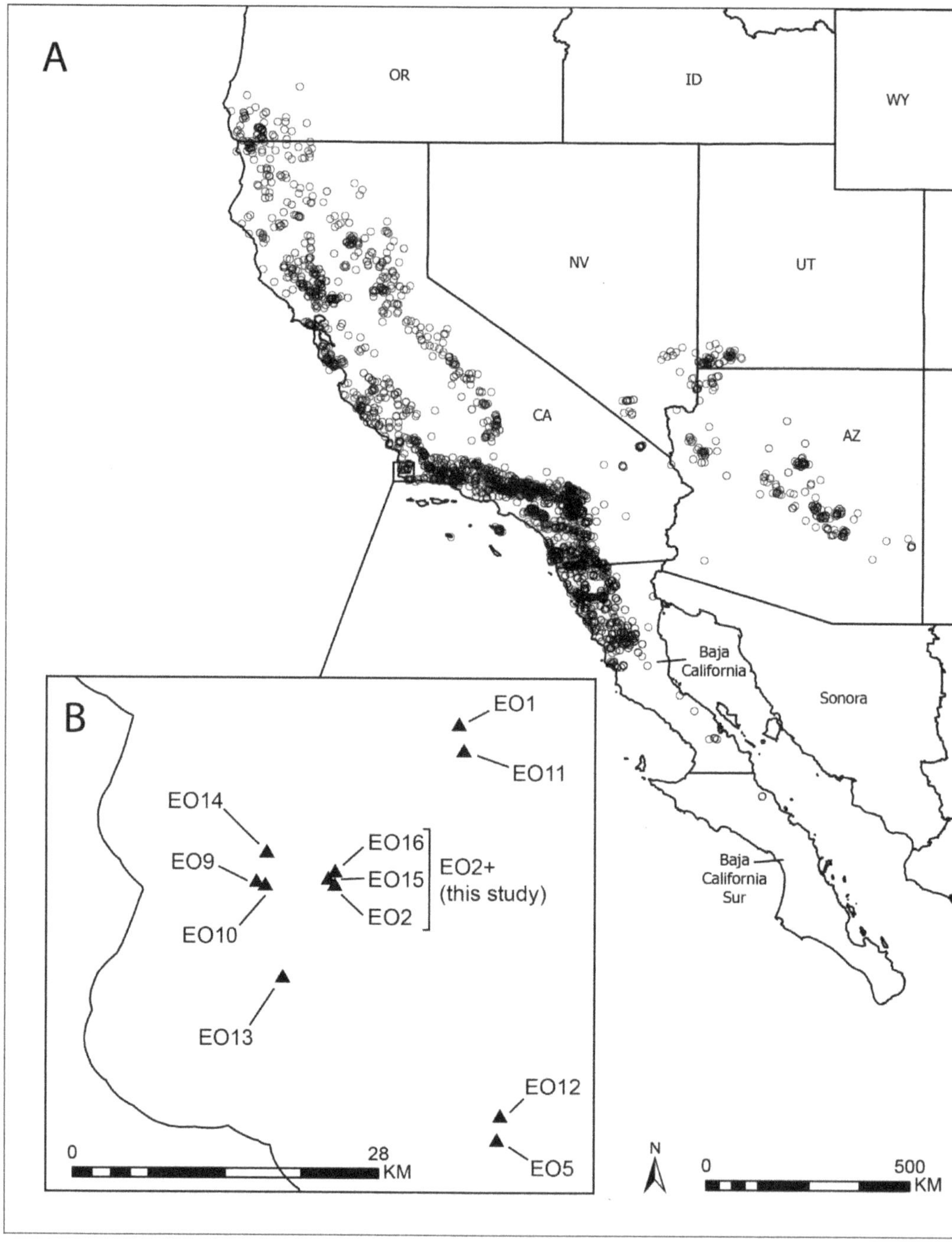

Figure 1. (**A**) Map of western North America showing distribution of *Eriodicyton* based on specimen records from the Global Biodiversity Information Facility (GBIF); (**B**) Inset map of Central California showing *E. capitatum* element occurrences (EOs) based on Kofron et al. (2022) [4].

Eriodictyon is one of three genera in the Namaceae, a recently recognized family in the order Boraginales [5,6]. Recognition of this new family was motivated by previous molecular phylogenetic studies that revealed the Hydrophyllaceae s.l. to be non-monophyletic [1,7,8]. A family-level classification of the Boraginales and the rationale for recognizing Namaceae as distinct from Hydrophyllaceae can be found in Luebert et al. [5]. Currently recognized members of *Eriodictyon* are distinctive in the Namaceae as being rhizomatous shrubs, or in one case, a rhizomatous perennial herb [2]. Stems are usually erect and between one and four meters tall. Leaves are cauline, alternate, and linear, oblong, lanceolate, elliptical, or oblanceolate in shape. Inflorescences are usually borne at branch or twig apices and can be open or dense. Flowers have white to purple corollas that are usually either funnel- or urn-shaped. Plants are often glandular, usually with at least one plant organ producing an aromatic, sticky exudate. Fruits in *Eriodictyon* are typically small capsules with valvate dehiscence.

Eriodictyon species are considered to be short-lived, pioneer, or early successional taxa based on the observation that they often thrive in ecological settings that experience regular disturbance, such as fire-prone vegetation communities and roadsides [9–11]. In such settings, a single genetic individual may spread by rhizomes and produce a colony (genet) of one or more genetically identical stems (ramets). Following a disturbance that destroys some or all of the above-ground stems in a genet (e.g., fire, roadside clearing), new stems rapidly develop from existing rhizomes. Sexual reproduction by seed has been observed in some *Eriodictyon* taxa, with germination rates greatly enhanced by fire cues such as heat and charate [11–13]. Plants in the genus are reported to be obligately outcrossing; however, this is a life history feature that may result in minimal seed production in uniclonal stands [14].

Eriodictyon capitatum Eastw. is a narrowly distributed shrub endemic to western Santa Barbara County (Figure 1B), where it grows in coastal settings in central coast maritime chaparral, bishop pine forest, and coastal scrub. Plants are usually less than three meters tall [2], but they have been observed to be up to five meters tall in some cases [4]. Leaves in *E. capitatum* are linear (Figure 2), an uncommon trait in the genus that it shares with only *E. altissimum* P.V. Wells of coastal San Luis Obispo County and *E. angustifolium* Nutt. of arid southeastern California, Nevada, Utah, and Arizona. *Eriodictyon angustifolium* also occurs disjunctly in northern Baja California, Mexico. As implied by the name, the inflorescences of *E. capitatum* are dense, head-like clusters of several flowers. The flowers have densely long-hairy calyces and lavender, funnel-form corollas. The combination of linear leaves, head-like inflorescences, and lavender, funnel-form corollas, is diagnostic for *E. capitatum* in the genus.

An evolutionary hypothesis involving *E. capitatum* was posited by Wells [9], who speculated that the geographically adjacent, Central California endemic *E. altissimum* may have arisen through historical hybridization between *E. capitatum* and the widespread *E. californicum* (Hook. & Arn.) Torr. He noted that *E. altissimum* has linear leaves morphologically similar to those of *E. capitatum* but open, glabrous inflorescences similar to those of *E. californicum*. He acknowledged that neither of these taxa occur in the vicinity of *E. altissimum* in the present day, an observation that would seem to weaken support for the hybrid origin hypothesis. To date, Wells' idea has not been tested phylogenetically.

Owing to low numbers of plants in nature, a limited overall areal extent, and multiple current threats, *E. capitatum* is listed as Endangered under the Federal Endangered Species Act, as Rare under the California Native Plant Protection Act, and has been given the California Rare Plant Rank of 1B.2 by the California Native Plant Society. As a rare plant, the taxon is tracked by the State of California in the California Natural Diversity Database (CNDDB). *Eriodictyon capitatum* is known from only 10 extant CNDDB element occurrences (EOs). An EO is defined as a specific location where a taxon of conservation concern has been documented as occurring. By convention, observations of individuals and/or populations of focal taxa are grouped together into one EO when the distance between them is less than $\frac{1}{4}$ mile. As a result, EOs may be composed of several or even dozens of biological populations or be limited to one or only a few individuals. Six of the extant *E.*

capitatum EOs are on Vandenberg Space Force Base, and four are on private property. See Kofron et al. [4] for detailed information about each *E. capitatum* EO.

Figure 2. (**A**) Photograph of *Eriodictyon capitatum* in typical shrubland vegetation; (**B**) Head-like inflorescence of *E. capitatum*.

Elam [14] studied several aspects of *E. capitatum* in six populations as part of her doctoral research. Note that these six populations are now treated as belonging to three present-day EOs. She used starch gel electrophoresis of isozymes to examine clonality, inferring a wide range of clonality levels among the populations based upon gel banding patterns. In two populations, all sampled ramets had identical banding patterns and were assumed to belong to the same genet. In other populations, the numbers of inferred genets were much greater than one. A total of 17 unique isozyme banding patterns were detected among 26 sampled ramets in one population on Hollister Ranch (65 percent unique patterns), each assumed to represent a distinct genet. Seed production per sampled ramet was assessed by direct counts over two years. Seed production varied significantly between populations in both years, with population 3 on Vandenberg Space Force Base producing considerably more seed (40.3 in 1992 and 72.6 in 1993) than the other sampled populations (0.4–2.3 in 1992 and 0.5–9.1 in 1993). Self-incompatibility was assessed within multi-clonal populations by hand self- and cross-pollinations between ramets, as determined by earlier isozyme banding patterns. Inflorescences were bagged after hand pollination. Mean fruit and mean seed production for each ramet were quantified. The percent of flowers setting seed was significantly higher ($t = 5.18$, $p < 0.001$) in hand cross pollinated inflorescences (mean = 53.1) than hand self-pollinated inflorescences (mean = 1.9). Seeds per fruit were also significantly different between treatments ($t = 4.47$, $p < 0.002$), with a mean of 1.77 seeds per fruit for hand cross-pollinations and only 0.03 seeds per fruit for hand self-pollinations. Finally, the relationship between mean seed production and clonal diversity per population was assessed, but statistical tests were either not significant or only marginally significant in the two study years.

Although the plant's rarity and listing status have resulted in considerable conservation focus, much remains to be learned about *E. capitatum*. While Elam's work provided critical insight into several aspects of the species' biology, it focused on only three present-day EOs and used an older approach to assess genetic diversity. Therefore, it would be useful to examine the magnitude of genetic diversity in present-day EOs using an updated approach. It would also be useful to better understand the extent to which EOs are genetically distinct from one another. Although Elam examined clonality in certain *E. capitatum* EOs, clonality has not been assessed using DNA sequence data. Finally, the phylogenetic and taxonomic distinctiveness of *E. capitatum* with respect to putative close relatives *E. altissimum*, *E. angustifolium*, and *E. californicum* has never been evaluated using molecular tools (but see Vasile et al. [15] for a recent phylogenetic analysis that included some members of *Eriodictyon*). Doing so would permit the evaluation of Wells' *E. altissimum* hybrid origin hypothesis. Leveraging the utility of high-throughput sequencing to resolve some or all of these data gaps would be useful in the case of *E. capitatum*, allowing resource agencies and land managers to use this information for conservation planning.

Here we generated a SNP dataset using high-throughput sequencing and used it to (1) investigate the genetic diversity within and among all accessible *E. capitatum* EOs; (2) determine the extent of genetic isolation among EOs of *E. capitatum*; (3) examine clonality within EOs of *E. capitatum* (i.e., how many unique genets are there within sampled ramets of an element occurrence); and (4) use phylogenomic analysis to evaluate the evidence for the current taxonomic circumscriptions of *E. capitatum*, *E. altissimum*, *E. angustifolium*, and *E. californicum*, thereby assessing Wells' *E. altissimum* hybrid origin hypothesis.

2. Results

2.1. Population Genomic Analyses

Dataset 1 included 85 samples with 200,186 SNPs and an aligned matrix of 2,778,920 bps. The percent of missing data was 31.07% in the SNP matrix. Summary statistics by sample are provided in Supplementary Table S1.

Table 1 provides population genomic summary statistics for each EO averaged across all loci, including: number of individuals from each population/EO (*N*), mean individuals genotyped at each locus (*n*), number of private alleles (Private), mean frequency of the major allele (P), observed heterozygosity (*Ho*), expected heterozygosity (*He*), nucleotide diversity (Pi), and the mean Wright's inbreeding coefficient (F_{IS}). The number of individuals sequenced per EO (*N*) ranged from 12 to 20. The average number of individuals genotyped at each locus (*n*) ranged from about 7 to 15.7. Private alleles per EO ranged from 8572 to 32,246. The mean frequency of the major allele (P) was quite high, ranging from 0.917 to 0.977. Observed heterozygosity (*Ho*) was lower than expected heterozygosity (*He*) in three EOs (EO1, EO2+, and EO5) and greater or about equal to expected heterozygosity in EO9, EO13, and EO14. Nucleotide diversity (Pi) ranged from 0.045 in the small La Salle population (EO13) to 0.127 in the largest EO, Orcutt Hill (EO1). Mean Wright's inbreeding coefficient (F_{IS}) ranged from −0.009 to 0.205. Calculated F_{ST} values are provided in Table 2. Values range from 0.114 between EO1 and EO2 to 0.417 between EO13 and EO14.

Table 1. Summary population genomic statistics for each element occurrence (EO) averaged across polymorphic loci. Statistics include the number of individuals from each population (*N*), the mean individuals genotyped at each locus (*n*), the number of private alleles (Private), the mean frequency of the major allele (P), observed heterozygosity (*Ho*), expected heterozygosity (*He*), nucleotide diversity (Pi), and the mean Wright's inbreeding coefficient (F_{IS}).

EO	*N*	*n*	Private	P	*Ho*	*He*	Pi	F_{IS}
Orcutt Hill—EO1	12	7.819	31,142	0.917	0.059	0.118	0.127	0.186
Pine Canyon—EO2+	16	10.325	32,246	0.919	0.056	0.116	0.123	0.205
Hollister Ranch—EO5	13	6.961	21,888	0.936	0.057	0.087	0.095	0.093
35th St.—EO9	12	9.641	15,293	0.942	0.089	0.071	0.078	−0.009
La Salle—EO13	12	8.875	8572	0.967	0.046	0.041	0.045	0.001
Air Field—EO14	20	15.675	14,296	0.977	0.034	0.029	0.031	−0.004

Table 2. F_{ST} values for each pair of element occurrences (EOs).

	EO2+	EO5	EO9	EO13	EO14
EO1	0.114	0.170	0.187	0.214	0.226
EO2+		0.163	0.178	0.168	0.215
EO5			0.266	0.319	0.326
EO9				0.354	0.357
EO13					0.417

Principal components (PCs) 1, 2, and 3 explained 18.6, 11.2, and 8.2 percent of the variability in the PCoA analysis, respectively. Scatterplots of PC2 versus PC1 and PC3 versus PC1 are given in Figure 3A,B. In nearly all cases, samples cluster tightly by EO, and clusters of samples by EO are usually not overlapping with other samples. There are exceptions in both cases. In Figure 3A,B, there are two samples from La Salle EO13 that do not cluster with the other samples from this EO. In Figure 3B, there is one sample from Orcutt Hill EO1 that appears near Pine Canyon EO2. In Figure 3A, the samples from Orcutt Hill EO1 and Hollister Ranch EO5 overlap strongly but form non-overlapping clusters in Figure 3B.

Table 3 shows the summary statistics associated with the STRUCTURE analysis. The highest deltaK value was associated with genetic subdivisions (K) = 7 (557.149). Figure 4 shows the STRUCTURE barplot for K = 7, averaged across replicates. Inferred genetic subdivisions are each represented by a color. Vertical bars represent individual samples, which are labeled along the *x*-axis. Samples are grouped by EO. Five out of six EOs contain samples that are assigned entirely to the same genetic subdivision. Only Pine Canyon EO2 has samples assigned to multiple genetic subdivisions. Two of these genetic subdivisions are unique to Pine Canyon EO2, suggesting a likely genetic substructure within this EO.

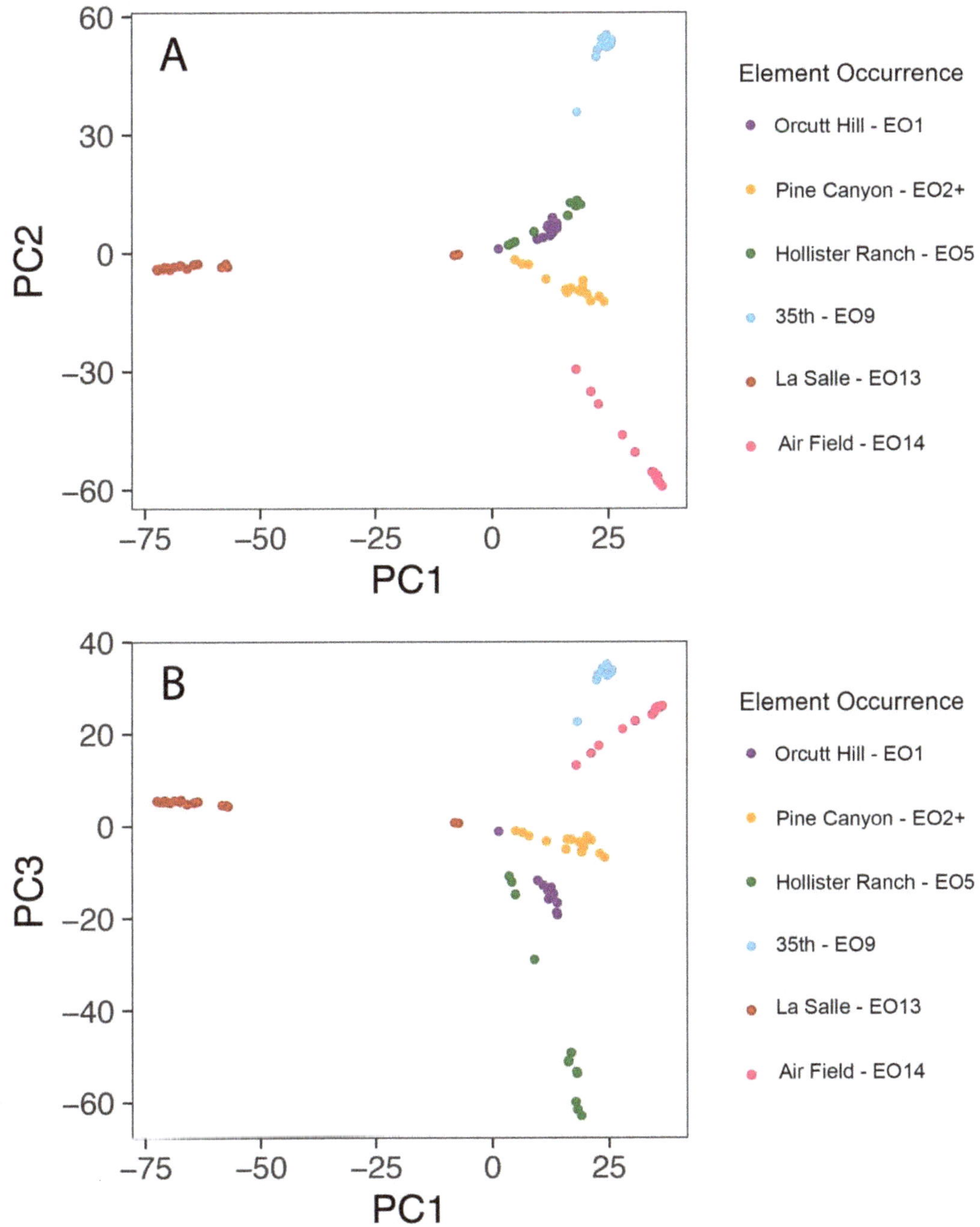

Figure 3. Scatter plots of (**A**) PC2 vs. PC1 and (**B**) PC3 vs. PC1.

Table 3. Summary statistics associated with the STRUCTURE analysis.

K	Nreps	lnPK	lnPPK	deltaK	estLnProbMean	estLnProbStdev
1	10	0.000	0.000	0.000	-5.277×10^5	2.517×10^3
2	10	8.039×10^4	2.439×10^4	5.419	-4.473×10^5	4.502×10^3
3	10	5.600×10^4	1.607×10^4	4.925	-3.913×10^5	3.264×10^3
4	10	3.993×10^4	8.172×10^2	0.193	-3.514×10^5	4.224×10^3
5	10	3.911×10^4	1.418×10^4	1.601	-3.123×10^5	8.859×10^3
6	10	2.493×10^4	1.243×10^4	8.368	-2.873×10^5	1.485×10^3
7	10	1.250×10^4	1.154×10^6	557.149	-2.748×10^5	2.072×10^3
8	10	-1.142×10^6	1.993×10^6	1.248	-1.417×10^6	1.597×10^6
9	10	-3.135×10^6	3.679×10^6	5.829	-4.552×10^6	6.311×10^5
10	10	5.434×10^5	0.000	0.000	-4.009×10^6	2.164×10^6

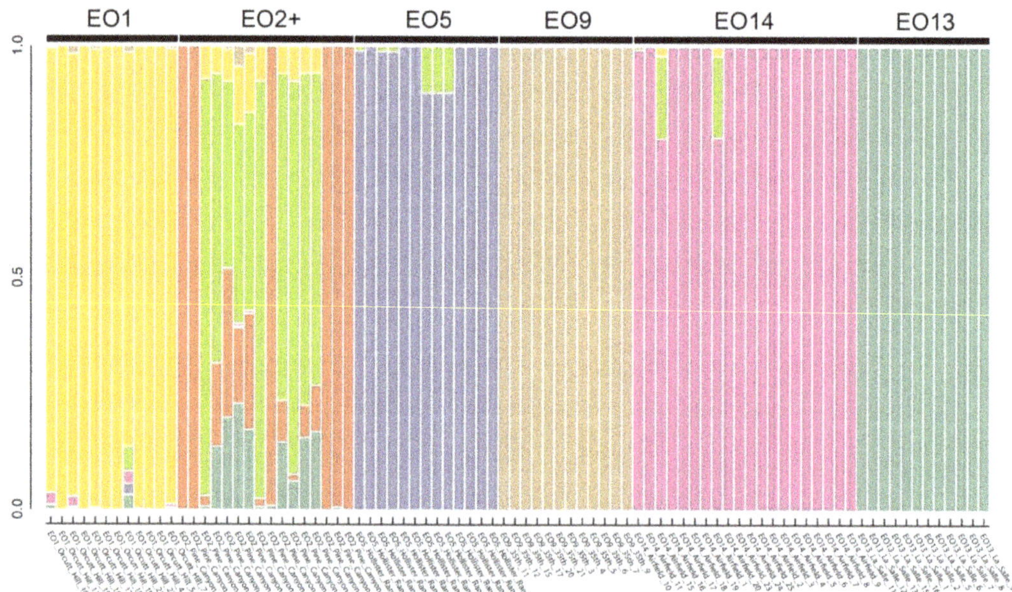

Figure 4. STRUCTURE barplot for K = 7, averaged across replicates. Inferred genetic subdivisions are each represented by a color.

The tree diagrams resulting from the phylogenetic analysis of Dataset 1 using ML in RAxML are shown in Figures 5 and 6. Figure 5 shows an unrooted tree without sample names or the majority of bootstrap support values, so that overall patterns by EO are more apparent. Groupings of samples by EO are indicated with colored ellipses. In nearly all cases, deep relationships in the tree are strongly supported (e.g., ML bootstrap = 100). Samples form well-supported clades by EO in all cases but Pine Canyon. For Pine Canyon samples, most form a single large grouping with poor support (ML bootstrap = 60), sister to a grouping of EO13 samples + two additional Pine Canyon samples. This latter grouping of EO13 + two Pine Canyon samples is also poorly supported (ML bootstrap = 57). For this reason, the placement of these two Pine Canyon samples is somewhat equivocal.

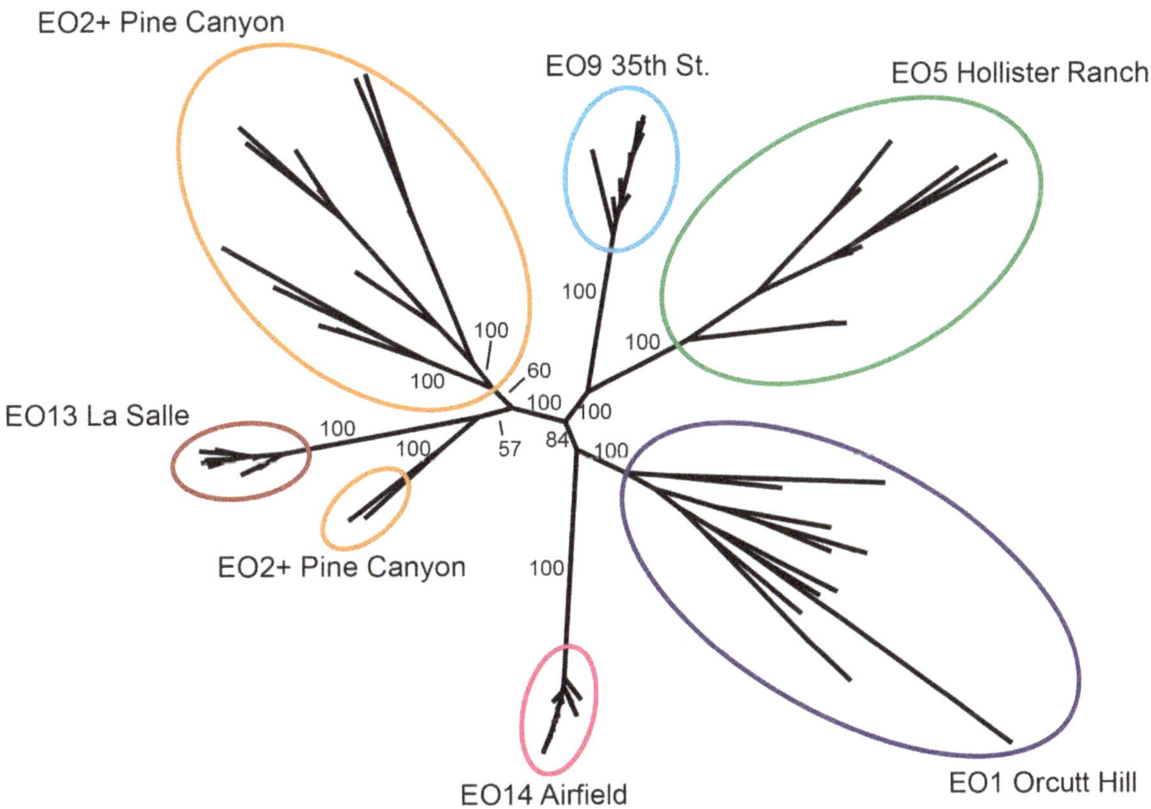

Figure 5. Unrooted phylogenetic tree inferred using maximum likelihood in RAxML. Internal nodes are annotated with maximum likelihood bootstrap support values. Colored ellipses denote sampled element occurrences.

Figure 6 shows the same phylogenetic tree as Figure 5, but rooted arbitrarily on the EO1 clade. In this figure, all sample/tip labels are shown, as are the ML bootstrap values for each branch. Samples by EO are denoted by labeled vertical bars. Note that grouping patterns are identical between Figures 5 and 6, having been based on the same ML tree. Relationships and statistical support within each EO are more readily apparent, however, as are branch lengths. For example, note relatively long branch lengths within EO1 and EO2, and relatively short branch lengths within EO14 and EO9.

2.2. *Clonality*

The filter_stats function of the R package poppr v2 [16,17] was used to produce the plot in Figure 7. This plot shows a histogram of all pairwise genetic distances between samples (gray bars) along with the number of multilocus genotypes (called "multilocus lineages" on the y-axis) that would result for all genetic distance cutoff values (x-axis) under the three clustering methods. A large gap in the inferred multilocus genotypes under the farthest neighbor method (red circles) is evident between genetic distance cutoff values of approximately 0.0225 and 0.025.

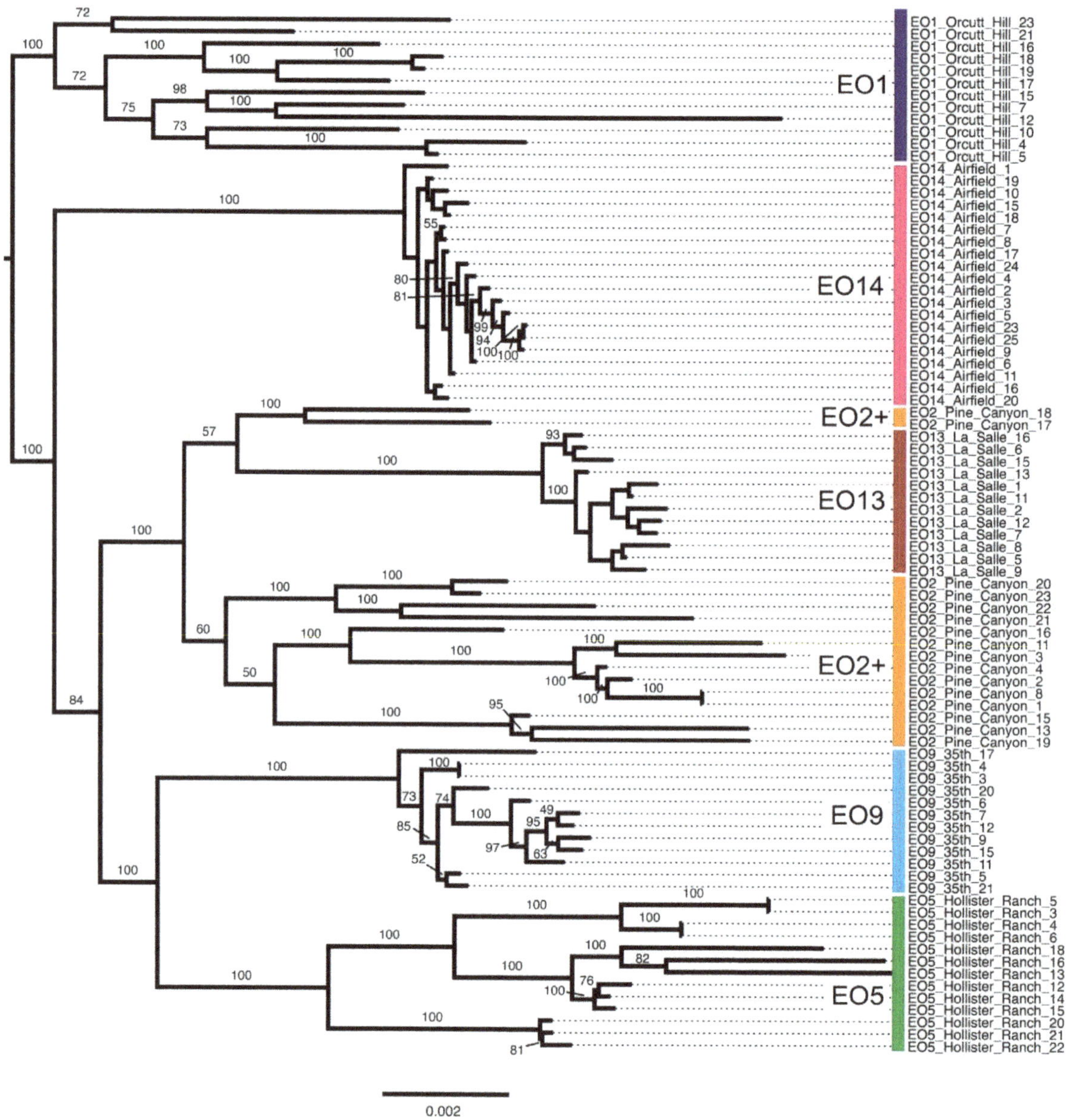

Figure 6. Phylogenetic tree inferred using maximum likelihood in RAxML, arbitrarily rooted on the EO1 clade of samples. Nodes are annotated with maximum likelihood bootstrap support values above 50.

The cutoff.predictor function in the R package poppr was used as one approach for identifying a threshold genetic distance at which individual samples would be assigned to multilocus genotypes under the conservative farthest neighbor clustering method. This function returned a threshold of 0.001747, which was much lower than the evident gap in inferred multilocus genotypes under the farthest neighbor method (red circles).

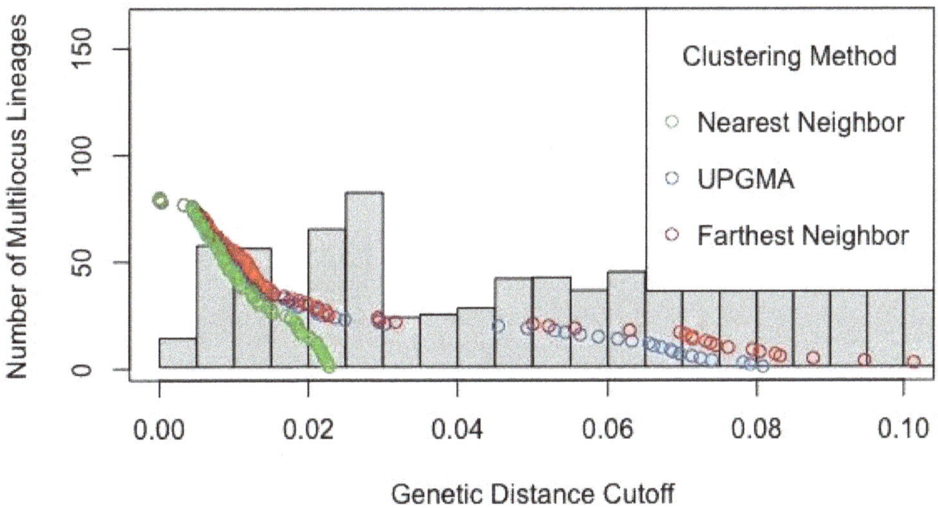

Figure 7. Filter_stats plot from R package poppr showing the number of multilocus lineages that result from different genetic distance cutoff values under three different clustering methods: nearest neighbor (green circles), UPGMA (blue circles), and farthest neighbor (red circles). A histogram of all pairwise genetic distances is shown as gray bars.

Multilocus genotype assignments were made using the mlg.filter function in the R program poppr v2 using two thresholds. First, the threshold resulting from the cutoff.predictor function of 0.001747 was used, despite appearing illogically low. Using this threshold, the 81 samples that passed filtering steps were assigned to 78 total multilocus genotypes. Because this result is essentially uninformative and likely based upon a spurious and arbitrarily low threshold, the results following the use of this threshold are not discussed further. Second, a threshold corresponding to the gap inferred in the multilocus genotypes of 0.025 was used (see Figure 7). Using this threshold, the 81 samples that passed filtering steps were assigned to 25 total multilocus genotypes.

The number of multilocus genotypes inferred under the 0.025 genetic similarity threshold differed among EOs. Table 4 lists the number of multilocus genotypes inferred by EO. Supplementary Table S2 provides the multilocus genotype assignment for each sample. In no case were multilocus genotypes shared between or among EOs.

Table 4. Multilocus genotypes inferred by element occurrence (EO).

EO	# Samples	# Multilocus Genotypes
Orcutt Hill—EO1	11	9
Pine Canyon—EO2+	16	9
Hollister Ranch—EO5	13	3
VSFB 35th St.—EO9	11	2
VSFB La Salle—EO13	12	1
VSFB Airfield—EO14	18	1

2.3. Phylogenomics

The full Phylip dataset constructed in ipyrad for the purpose of inferring phylogenetic relationships contained 41 samples and was 2,248,102 base pairs long. The ML phylogenetic tree resulting from the RAxML analysis is shown in Figure 8. In general, the tree topology is well supported, with most clades supported with bootstrap values of 100. Bootstrap values less than 100 are indicated on the tree. Only five branches have bootstrap values below 70. Two of these low values occur at shallow phylogenetic depths of divergence among samples of the same taxon (e.g., 68 within a clade of *Nama rothrockii* samples). The other three low ML bootstrap values occur at deeper depths of divergence. Of note is a low bootstrap value (ML bootstrap = 16) near the base of *Eriodictyon*, which, given the surrounding strong pattern of statistical support, likely indicates multiple +- equally likely placements for samples of *E.* [*Turricula*] *parryi* (A. Gray) Green.

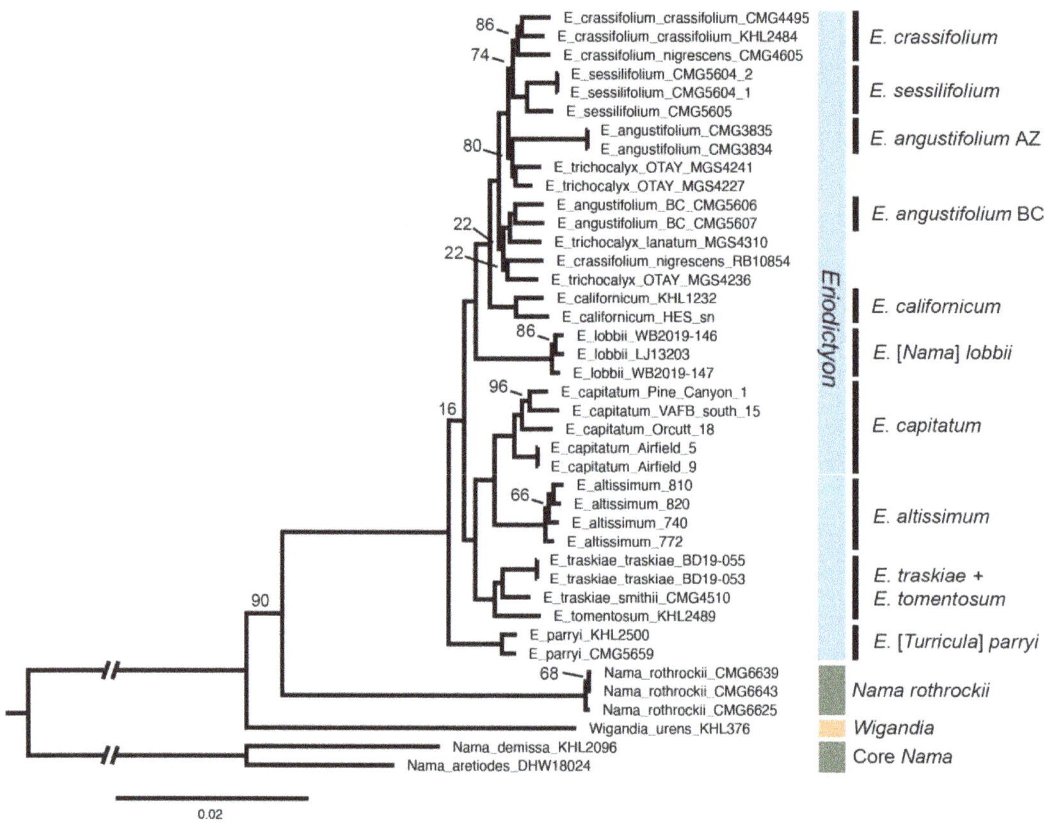

Figure 8. Phylogeny of *Eriodictyon* inferred using maximum likelihood in RAxML. Maximum likelihood support values are 100, except as noted.

The *Eriodictyon* clade has the highest possible statistical support (ML bootstrap = 100). Most *Eriodictyon* samples in the analysis form strongly supported clades by taxon. This includes *E. altissimum*, *E. californicum*, *E. capitatum*, *E. crassifolium* Benth. (all but one sample), *E.* [*Nama*] *lobbii*, *E.* [*Turricula*] *parryi*, *E. sessilifolium* Greene, and *E. traskiae* Eastw. Of these clades of samples by taxon, only the *E. crassifolium* clade is supported by a bootstrap value of less than 100 (ML bootstrap = 86).

Of the *Eriodictyon* taxa in this analysis, only the *E. angustifolium* and *E. trichocalyx* samples do not form clades by taxon. *Eriodictyon angustifolium* samples form two strongly supported clades by region: Arizona and Baja California. These *E. angustifolium* clades by region are not closely related in this analysis. Similarly, samples of *E. trichocalyx* appear in two different parts of the tree, although support is weak in one of these regions.

Phylogenetic relationships among *E. altissimum*, *E. angustifolium*, *E. californicum*, and *E. capitatum* are relatively well resolved in this analysis, and support for current taxonomic circumscriptions is robust for all but *E. angustifolium*. Both narrow-leaved Central Coast taxa, *E. altissimum* and *E. capitatum*, have samples that form well-supported clades (each with ML bootstrap = 100). The *E. altissimum* clade is sister to the *E. capitatum* clade with strong support (ML bootstrap = 100), and this is sister to a well-supported clade of central coast and southern California taxa. The two included samples of *E. californicum* form a well-supported clade (ML bootstrap = 100) that is relatively distantly related to *E. altissimum* and *E. capitatum*. Although narrow-leaved *E. angustifolium* samples do not form a single clade, both groupings of samples of *E. angustifolium* are relatively distantly related to *E. altissimum* and *E. capitatum*.

Outgroup sampling for this project permitted the evaluation of the placement of *E. lobbii*, previously included in the genus *Nama*, *E. parryi*, previously included in *Turricula*, and *Nama rothrockii*. Samples of *E.* [*Nama*] *lobbii* form a clade with strong support (ML bootstrap = 100) that is resolved within *Eriodictyon*. The two included samples of *E.* [*Turricula*] *parryi* form a clade with strong support (ML bootstrap = 100) that is placed sister to the remainder of *Eriodictyon* in the best ML tree, but with poor statistical support; the *E. parryi* clade is sometimes placed within *Eriodictyon* in bootstrap replicates. Samples of *Nama rothrockii* also form a clade with strong support (ML bootstrap = 100); this was recovered on a relatively long branch sister to a strongly supported (ML bootstrap = 100) clade of *Eriodictyon* samples.

3. Discussion

3.1. Population Genomic Analyses

Genetic diversity among the sampled EOs is variable, but the overall patterns are consistent with EO census sizes and varying areal extents. The EO with the highest nucleotide diversity (0.127) and second largest number of private alleles (31,142) is Orcutt Hill (EO1), which has the second highest census count (>7000 ramets in 2018) and a large areal extent. Similarly, with the second highest nucleotide diversity (0.123) and largest number of private alleles (32,246), the Pine Canyon EO (EO2+) also supports the largest number of ramets (9794 stems in part in 2015) over a relatively large area. In contrast, the Air Field location (EO14) has the lowest nucleotide diversity (0.031) and the second-smallest number of private alleles (14,296), which corresponds with a low census count (78 ramets in 2018) and small areal extent. Similarly, the newly discovered La Salle location (EO13) has the second lowest nucleotide diversity (0.045) and the lowest number of private alleles (8572), a finding consistent with the low ramet count (258 ramets in 2018) and small areal extent.

Despite proportionally higher genetic diversity in the larger EOs, these locations also had higher inferred levels of inbreeding, as suggested by lower observed heterozygosity estimates relative to expected heterozygosity (e.g., 0.059 observed and 0.118 expected in EO1) and higher mean Wright's inbreeding coefficients (e.g., 0.186 and 0.205 for EO1 and EO2, respectively). This suggests that despite having a large number of above-ground *E. capitatum* stems at these locations, some factor (e.g., clonality) might be affecting estimates of heterozygosity and inbreeding.

Interpretation of F_{ST} values is straight-forward in some cases but not in others. The lowest F_{ST} value (0.114) was between Orcutt Hill (EO1) and Pine Canyon (EO2+). Although not as close together as some pairs of EOs, Pine Canyon is the closest sampled EO to Orcutt Hill. In contrast, Hollister Ranch (EO5) has a relatively low F_{ST} with respect to both Orcutt Hill (EO1) and Pine Canyon (EO2+), despite relatively large distances between EOs in both cases (F_{ST} 0.170 and 0.163, respectively). The largest F_{ST} values, suggesting the largest pairwise degree of genetic differentiation, involve EOs 9, 13, and 14, in most cases, which are the western-most EOs of *E. capitatum*. The F_{ST} value for the EO9–EO13 comparison was 0.345; for the EO9–EO14 comparison, the value was 0.357; and for the EO13–EO14 comparison, the value was 0.417. This finding is unexpected given the relative geographic proximity of EO9, EO13, and EO14.

Visualizing genetic distances using a PCoA ordination approach revealed strong genetic similarities among samples within EOs. It also showed intriguing patterns among

EOs. In the scatterplot of PC2 versus PC1 (Figure 3A), the clusters of samples from 35th EO9, La Salle EO13, and Air Field EO14 were mostly well-separated from all other EOs, suggesting minimal gene flow between each of these and all other EOs. The other three EOs—Orcutt Hill EO1, Pine Canyon EO2+, and Hollister Ranch EO5—were relatively tightly clustered together in ordination space. This finding is surprising for Hollister Ranch EO5, given its relative geographic distance from Orcutt Hill EO1 and Pine Canyon EO2+. This finding is consistent with the calculated F_{ST} values, however. In the scatterplot of PC3 versus PC1 (Figure 3B), the La Salle EO13 was again strongly separated from all other EOs, but 35th EO9 and Air Field EO14 were close together in the plot. As in the first scatterplot, Orcutt Hill EO1, Pine Canyon EO2+, and Hollister Ranch EO5 were relatively close together in the plot.

The STRUCTURE results found genetic subdivisions in the data that corresponded almost perfectly with location (EO). Only Pine Canyon EO2+ had samples assigned primarily to more than one genetic subdivision. In this EO, some samples were assigned to one EO-specific subdivision colored orange, while the other samples were assigned primarily to an EO-specific subdivision colored bright green and, to a lesser extent, orange. These latter samples were also assigned with less likelihood to genetic subdivisions found in other EOs, including Orcutt Hill EO1 and La Salle EO13. These results strongly support the genetic distinctiveness of each *E. capitatum* EO and suggest a potential genetic substructure in Pine Canyon EO2+.

The patterns in the RAxML phylogenetic trees applied to Dataset 1 reinforce the findings of earlier analyses. Basic population genetic summary statistics revealed that each EO contained moderate and sometimes unique genetic diversity (based on nucleotide diversity and private alleles), and pairwise F_{ST} values showed moderate genetic divergence in all combinations. These results are supported by the inference of robustly supported clades of *E. capitatum* samples by EO and long branch lengths in some tips of the tree (samples). Beyond corroborating other findings, the tree diagram presented in Figure 7 may be useful in the future for selecting genetically dissimilar stems within an EO, e.g., for use in hand-crossing experiments where an emphasis would be on crossing stems/ramets that belong to different genets, as was conducted by Elam [14].

3.2. Clonality

Examination of clonality using high-throughput sequence data, which by its nature may include non-trivial amounts of missing data, requires careful assessment. Assigning samples to the same genetic individual based on genetic identity between or among samples is not practicable given certain properties of the data that result from certain high-throughput sequencing approaches, such as ddRADseq. Kamvar et al. [16,17] recommend a thresholding approach in which samples are collapsed to multilocus genotypes when pairwise genetic distances fall below a certain value. This was the approach followed for this study.

Identification of the appropriate methods for determining this threshold for ddRADseq data in particular appears to have been little studied to date. Kamvar et al. [16,17] provide some guidance in their papers describing the use of their R package poppr. In the present study, one of the approaches advocated by Kamvar et al. (the cutoff.predictor function) did not yield what appears to be a biologically meaningful threshold (0.001747). Using this threshold, the 81 included samples were collapsed to only 78 multilocus genotypes (here interpreted as genets). This outcome is consistent with the close physical proximity of some of the samples included here. It is also well below what Guilliams and Hasenstab-Lehman [18] recovered using the same function for the close relative, *Eriodictyon altissimum* (0.034700). Instead of relying on the threshold produced by the cutoff.predictor function, we estimated the position of the calculated gap in the number of inferred multilocus genotypes under the farthest neighbor approach using Figure 7 (0.025). These approaches should result in similar thresholds, as occurred in Guilliams and Hasenstab-Lehman [18], so it is unclear why the cutoff.predictor function failed here.

Applying the threshold based on Figure 7, samples within EOs were collapsed to 25 multilocus genotypes. Spatially extensive EOs that support a large number of ramets, such as Orcutt Hill (EO1) and Pine Canyon (EO2+), have a large number of multilocus genotypes. Conversely, spatially restricted EOs that support relatively few ramets have few multilocus genotypes. For example, the VSFB La Salle EO (13) collapsed to one multilocus genotype, which is consistent with the number of ramets and areal extent of this small EO. Exhaustive sampling of all ramets would be required to estimate the total number of multilocus genotypes within a given EO, but the results described here provide important preliminary information that may be useful for conservation planning.

3.3. Phylogenomics

In general, the ddRADseq approach was successful in inferring evolutionary relationships in the *Eriodictyon*. Most branches in the tree had the highest possible statistical support (i.e., ML bootstrap value of 100), and only five branches had ML bootstrap values below 70. The *Eriodictyon* clade had the highest possible statistical support, and in general, samples were resolved in clades by taxon.

Phylogenetic relationships were confirmed for taxa historically of uncertain placement. Both *E.* [*Turricula*] *parryi* and *E.* [*Nama*] *lobbii* have been treated recently in *Eriodictyon* [2] due to the findings of Ferguson [1]. In that study, *Eriodictyon* formed a strongly supported clade (parsimony BS = 100), with *E.* [*Turricula*] *parryi* being the sister to all other samples of *Eriodictyon*. Ferguson inferred *E.* [*Nama*] *lobbii* to be sister to *E. californicum*, but with low to moderate statistical support (parsimony BS = 62) and incomplete taxonomic sampling of *Eriodictyon*. While Ferguson's findings were well-supported in general, reliance upon a single chloroplast locus (ndhF)—which was common at the time—and incomplete sampling allowed some doubt to persist. Here, *E. lobbii* is a recovered sister to a subclade of *Eriodictyon*. *Eriodictyon parryi* is recovered as a sister to the rest of *Eriodictyon*, as determined by Ferguson, but low support indicates other alternative potential phylogenetic placements in *Eriodictyon*. For this reason, treatment of *E. parryi* in *Eriodictyon* rather than *Turricula* seems warranted.

Nama rothrockii was sister to *Turricula* + *Eriodictyon* in Ferguson's analysis, but with somewhat low statistical support (parsimony BS = 58). *Nama rothrockii* is a perennial herb, an uncommon trait in the Namaceae of California, which it shares with *E. lobbii*. For this reason, and given the placement of *E. lobbii* in *Eriodictyon* in Ferguson's analysis, it was possible that *N. rothrockii* would be recovered in *Eriodictyon* as well. In the analysis presented here, *N. rothrockii* samples form a clade with the highest possible support (ML bootstrap value = 100) that is sister to *Eriodictyon* (inclusive of *E.* [*Turricula*] *parryi*) with strong support (ML bootstrap value = 90). Given the strong phylogenetic placement of *N. rothrockii* as more closely related to *Eriodictyon* than *Nama*, a new name for *N. rothrockii* will be required so that only monophyletic groups are recognized taxonomically (Guilliams and Hasenstab-Lehman, in prep).

The results presented here shed light on the taxonomic circumscriptions and evolutionary history of *E. altissimum*, *E. angustifolium*, *E. californicum*, and *E. capitatum*. Explicitly [9] or implicitly connected by a mosaic of morphological similarity in leaf and flower features, these taxa were of special interest in this study due to the rarity and listing status of *E. altissimum* and *E. capitatum*. Wells [9] speculated that *E. altissimum* may have arisen through hybridization between *E. californicum*, with which it shares inflorescence (open panicle, glabrous axes) and flower (glabrous calyx) features, and *E. capitatum*, with which it shares leaf features (e.g., linear leaves). Here we find maximum statistical support for the present taxonomic circumscriptions of *E. altissimum* and *E. capitatum*, which are reciprocally monophyletic and sister in our analysis. *Eriodictyon californicum* is not closely related to either taxon in our analysis, nor is the linear-leaved *E. angustifolium*.

4. Materials and Methods

4.1. Sampling

Sampling for Objectives 1 to 3 of this study focused on obtaining high-quality, silica-dried tissues from throughout as much of the range of *E. capitatum* as possible. For *E. capitatum*, 12 samples were included from Orcutt Hills (EO1), 16 samples were included from Pine Canyon (EO2+), 13 samples were included from Hollister Ranch (EO5), 12 samples were included from 35th St. (EO9), 12 samples were included from La Salle (EO13), and 20 samples were included from the Air Field (EO14). Note that EO15 and EO16, which are geographically proximal to EO2, were recently designated. For simplicity, we included them in EO2 (as EO2+) in this study. It was not possible to obtain samples from the Dangermond location (EO12) for this study. For Objective 4, silica-dried tissues were collected from throughout the range of the genus. Outgroup sampling for Objective 4 included *Wigandia* Kunth (one sample), core *Nama* L. (two samples), and *Nama rothrockii* (three samples). A typical tissue collection included 1–2 fresh green leaves placed in a clean, labeled coin envelope. Coin envelopes were aggregated in small batches into small ziplock bags containing silica gel. All tissues have been deposited in the Tissue Bank at the Santa Barbara Botanic Garden. Vouchers for the study have been deposited at the Clifton Smith Herbarium (SBBG) at the Santa Barbara Botanic Garden. Data for each sample included in the study are given in Supplementary Table S1.

4.2. DNA Extraction

Dried silica material was ground with a multi-sample Bead Beater tissue homogenizer into a fine powder and extracted using a modified CTAB protocol [19] with the following change: incubation in the CTAB extraction buffer with proteinase K at 65 degrees for 3–4 h, with overnight precipitations. Half the samples went through a final cleaning step with a Zymo DNA Clean and Concentrator-25 kit (Zymo Research, Irvine, CA, USA). Extractions were quantified on a Qubit fluorometer using the Qubit Double Stranded High Sensitivity Assay Kit (Invitrogen, Carlsbad, CA, USA) to check for a suitable genomic DNA quantity. DNA quality was assessed by visualization on an agarose gel following gel electrophoresis.

4.3. Library Preparation

Libraries were prepared for high-throughput sequencing using a restriction site-associated DNA sequencing (RADseq) protocol. The RADseq approach is a genomic DNA reduction technique that isolates sequencing regions of genomic DNA near a set of restriction enzyme cut sites. The approach is cost-effective and can be repeated in large numbers of samples to produce a reduced subset of the genome in each individual. After sequencing, the data are re-assembled into loci, anchored by the presence of the restriction enzyme cut site [20,21], and subsequently, single nucleotide polymorphisms (SNPs) are identified across those loci. Double digestion RADseq (ddRADseq) was selected for its ease of use and cost-effective implementation for generating a large SNP dataset from non-model organisms [22,23]. In ddRADseq, two restriction enzymes are used to fragment genomic DNA, followed by size selection of the fragments. This results in sequencing libraries with loci randomly distributed throughout the genome of the study system. This method has been employed in numerous studies and has typically resulted in hundreds to thousands of loci sufficient to address typical population genetics studies in model and non-model organisms.

ddRADseq libraries were prepared in the genetics laboratory at the Santa Barbara Botanic Garden. Library preparation and barcode design follow Tripp et al. [24], with the following modifications: Total genomic DNA was fragmented using the MseI and EcoRI restriction enzymes. A total of 150–500 ng of genomic DNA was added to a reaction solution consisting of: 8.2 μL molecular grade water, 1.15 μL Tango Buffer (Fisher Scientific, Carlsbad, CA, USA), 0.6 μL of 1.0 M NaCl, 0.3 μL (1.0 mg/mL) Bovine Serum Albumin (BSA), 0.28 μL High Fidelity EcoRI (Fisher Scientific), and 0.12 μL MseI (Fisher Scientific).

Digestion reactions were incubated at 37 °C for 15 min, followed by an incubation step at 65 °C for 45 min.

Barcodes and adaptors containing an Illumina PCR priming site and the EcoRI cut site were prepared by integrated DNA technologies (Coralville, IA, USA) and follow the design of Tripp et al. [24]. Each ligation reaction consisted of the entire double restriction digestion reaction containing the fragmented genomic DNA to which we added 1.0 μL of 1.0 μM EcoRI adaptor + barcodes, 0.072 μL water, 0.1 μL 10× T4 buffer, 0.05 μL of 1.0 M NaCl, 1.0 mg/mL BSA, 1.0 10 nM MseI adaptor, and 0.165 μL T4 DNA ligase. Reactions were mixed, centrifuged, and incubated for 16 h at 16 °C, then heat-inactivated at 65 °C for 10 min. These restriction-ligation reactions were diluted 1:10 using 0.1× TE buffer.

We ran two separate 20 μL PCR reactions per restriction-ligation product [22]. PCR reactions contained: 8.6 μL molecular grade water, 4.0 μL Phusion High Fidelity Buffer (New England Biolabs, Ipswich, MA, USA), 0.5 μL of 10 μM Illumina primer 1 (IDT; (A*A*TGATACGGCGACCACCGAGATCTACACTCTTTCCCTACACGACGCT CTTCC-GATCT), 0.5 μL of 10 μM Illumina primer 2 (IDT; C*A*AGCAGAAGACGGCATACGA GCTCTTCCGATCTGTAAG), 1.6 μL of 2.5 mM dNTPs, 0.1 μL Phusion High Fidelity DNA polymerase (New England Biolabs, Ipswich, MA, USA), and 5 μL diluted restriction-ligation reaction. Each PCR reaction used the following cycling parameters: 98 °C for 60 s; 25 cycles of 98 °C for 20 s; 60 °C for 30 s; 72 °C for 40 s; 72 °C for 10 m; 4 °C hold. Gel electrophoresis and imaging were used as a qualitative assay to ensure PCR amplification of fragments at the desired 300–400 bp range for each sample. Successful PCR amplifications were cleaned with Zymo DNA Clean and Concentrator kits (Zymo Research, Irvine, CA, USA), then pooled.

4.4. Size Selection, Library Quantification, and Sequencing

The genomic library was sent to the University of California Riverside (UCR) Institute for Integrative Genome Biology Core Instrumentation Facility for size selection. Libraries were size-selected on a 1.5% agarose gel cassette for fragments between 350 and 550 bp in length. The libraries were quality checked with a Bioanalyzer 2100 (Agilent, Santa Clara, CA, USA) at UCR to ensure library quality and concentration prior to sequencing each pooled library on a NextSeq 500 (Illumina, La Jolla, CA, USA), each as a single lane of 1 × 75 single end base pair (bp) reads under the rapid run setting at UCR.

4.5. Data Processing, SNP Calling

Raw sequence reads were demultiplexed by UCR using custom scripts. Read pools were cleaned and quality checked using FastQC [25]. To assemble loci and generate phylip files for downstream phylogenomic analyses, cleaned sequence data were further processed with ipyrad v. 0.9.69 [26,27] on an iMac Pro with 10 cores.

Sequence assembly was performed using the de novo assembly in ipyrad, using the following parameters: ddrad datatype, phred quality score minimum of 33, with the parameters clustering threshold at 0.85, mindepth of 6 and maximum barcode mismatch of 0, 35 bps minimum length of sequences after the adaptor trim, a maximum of 2 alleles per site in consensus sequences, 0.02 max_Ns_consens, 0.05 heterozygotes, 0.2 SNPs per locus, and 8 max_indels.

We generated two different datasets in ipyrad to address different components of the study. Dataset 1 included only samples of *E. capitatum* and was filtered to include loci with a single SNP per locus. This dataset was used in analyses associated with Objectives 1 to 3. In Dataset 2, sampling was expanded to include nearly all other *Eriodictyon* taxa, along with outgroups in the Namaceae. This dataset was used to accomplish Objective 4 of the study.

4.6. Population Genomic Analyses

Population genomic summary statistics were calculated by element occurrence in the program Stacks v. 2.60 [28,29] from the VCF output from ipyrad for Dataset 1. Each EO was treated as a population for the purposes of running these analyses. Settings used to calculate the population genetics statistics included: --min-populations = 2, indicating the minimum number of populations a locus must be present in to process a locus; -R, --min-samples-overall = 50 for the minimum percentage of individuals across the dataset required to process a locus; -H was applied to prune unshared SNPS to reduce haplotype-wise missing data; --write-random-snps restricts data analysis to one random SNP per locus. Statistics include the number of individuals from each population/EO (N), the mean individuals genotyped at each locus (n), the number of private alleles (Private), the mean frequency of the major allele (P), observed heterozygosity (Ho), expected heterozygosity (He), nucleotide diversity (Pi), and the mean Wright's inbreeding coefficient (F_{IS}).

Genetic differentiation between pairwise combinations of *E. capitatum* EOs was examined using F_{ST}. F_{ST} is a common measure of genetic differentiation, with higher values indicating a greater degree of genetic differentiation between populations and lower values indicating a greater number of shared alleles. These values were calculated under the Stacks pipeline using the "populations" program.

Multi-variate statistical methods were used to examine patterns in the genetic dataset. These methods are largely exploratory in nature and do not have strong assumptions about an underlying genetic model, such as the presence of Hardy-Weinberg equilibrium or the absence of linkage disequilibrium [30]. A principal coordinates analysis (PCoA) was performed on the genlight matrix in the R package dartR [31]. PCoA is a statistical procedure that transforms a large number of variables into fewer composite variables, or PCs. These composite variables can be used to identify possible structures or clusters of genotypes within and among populations of individuals in the dataset.

To further assess the population structure of *E. capitatum*, two different analyses were performed. First, a phylogenetic analysis was performed on Dataset 1 using maximum likelihood (ML) in the program RAxML [32]. The analysis was performed on the CIPRES Science Gateway v3.3 [33]. Statistical confidence was assessed using ML bootstrapping. Second, Bayesian clustering was implemented in the program STRUCTURE [34] with ipyrad v. 0.9.77 analysis tools [35] in a Jupyter notebook [36] on the VCF output from populations and converted to hdf5 file format. STRUCTURE identifies genetic subdivisions in the data and then assigns samples to these subdivisions using an admixture model, assuming correlation of allele frequencies without prior knowledge of sample locality, for subdivisions (K) = 1–10, with $n = 10$ for each K value. STRUCTURE was run using an imap dictionary to color individuals; minmap = 0.5, which filters SNPs to only include those that have data for a 50% proportion of samples in every group; and mincov = 50 for the entire dataset. We set the MCMC chain to a burn-in of 20,000, followed by 100,000 MCMC iterations. To obtain the most likely value of K, the LnP(K) and deltaK were evaluated under the Evanno method [37]. Results from the separate 10 MCMC analyses were summarized in a barplot, with each genetic cluster assigned a different color. Each sample is colored by the estimated proportion of genotypes shared with each cluster.

4.7. Clonality

Clonality in *E. capitatum* was assessed using Dataset 1. For this, the multilocus genotypes within the samples were inferred using the R package poppr v2 [16,17]. Multilocus genotypes are unique combinations of alleles across at least two loci [16]. There are several ways to construct multilocus genotypes from a dataset. Naïve string matching is one approach that collapses samples together only when they are identical. This approach is not appropriate for calling multilocus genotypes using high-throughput sequencing, however, as samples may vary slightly owing to hypervariable loci and common artifacts of high-throughput sequencing such as missing data [17].

The resulting VCF file was read into R and converted to a genlight object. Individuals with low read counts (<5000) were removed from the dataset prior to converting the genlight object to a snpclone object for use in the R package poppr. The threshold of genetic similarity below which samples were collapsed into multilocus genotypes was determined in two ways. First, a threshold was calculated using the cutoff.predictor function in poppr. The cutoff.predictor function identifies the largest gap between inferred numbers of multilocus genotypes for all thresholds and can be run under nearest neighbor, UPGMA, and farthest neighbor clustering methods. The farthest neighbor method was used as it is the most conservative [17]. In addition, the initial largest gap between the inferred numbers of multilocus genotypes for all thresholds was also identified using a plot-based approach. The mlg.filter function was used to assign multilocus genotypes.

4.8. Phylogenomics

Phylogenetic analyses were performed using maximum likelihood (ML) in the program RAxML [32] on the CIPRES Science Gateway v3.3 [33]. Statistical confidence was assessed using ML bootstrapping, with bootstrapping halted automatically by the program. Analyses were performed with the RAxML HPC2 on the XSEDE tool using default parameters. The resulting trees were visualized using the program FigTree v1.4.3.

5. Conclusions and Recommendations

Eriodictyon capitatum is a narrowly distributed shrub endemic to western Santa Barbara County (Santa Barbara, CA, USA), where it is known from only 10 EOs. Here, high-throughput DNA sequence data were analyzed to investigate genetic diversity within and among all accessible EOs; to determine the extent of genetic isolation among EOs; to examine clonality within EOs; and to examine the taxonomic circumscriptions of *E. capitatum*, *E. altissimum*, *E. angustifolium*, and *E. californicum* through phylogenomic analysis. Population genetic analyses of *E. capitatum* revealed a pattern of strong genetic differentiation by location/EO. The clonality assessment showed that certain small EOs may support relatively few multilocus genotypes. The phylogenomic analyses strongly supported the present-day taxonomic circumscriptions of both *E. altissimum* and *E. capitatum*, showing them to be reciprocally monophyletic and sister with strong support. Taken together, these results paint a picture of an evolutionarily and morphologically distinct species known from relatively few, genetically isolated stations.

The results of this study were used to develop a list of conservation recommendations. Most broadly, both *E. capitatum* and *E. altissimum* were strongly supported as monophyletic in the phylogenomic analysis and should continue to be managed as evolutionarily distinct rare plant taxa under applicable federal, state, and local laws. Similarly, *E. capitatum* EOs were found to be genetically differentiated using population genetic and phylogenetic approaches. The preservation of each EO should be prioritized to conserve the overall genetic diversity of *E. capitatum*. Care should be taken to avoid the unintended movement of genetic material (e.g., pollen, seeds) between EOs. Conversely, because plants of *E. capitatum* are largely self-incompatible [14], it may be desirable to develop an ex situ research program to explore the feasibility of using hand-pollination crosses between genets, sourced from the same and/or different EOs, to bolster seed production. If such a study in the greenhouse resulted in increased seed production, then a potential in situ program might be designed in collaboration with government agency personnel and land managers.

The results presented here are consistent with general patterns of plant biodiversity in California as they pertain to high levels of endemism in coastal Central California. The California Floristic Province (CA-FP) has long been recognized as a global biodiversity hotspot, owing to the region's large proportion of endemic taxa and high degree of habitat loss [38,39]. In a state-wide analysis of endemism, Stebbins and Major [40] suggested CA-FP subdivisions of roughly similar size and tabulated their properties. The range of *E. capitatum* lies near the border of their Central Coast and Southern California subdivisions, both of which were reported to harbor the highest levels of endemism in the state under multiple of their metrics. A recent, spatially explicit study based on georeferenced herbarium specimens [41] estimated that species-rank endemism in the CA-FP is approximately 36.9% (1846 endemic species of 5006 total native species), and endemism of minimum-rank taxa (inclusive of subspecies and varieties) was even higher, at 42.5% (2612 endemic minimum-rank taxa of 6143 native minimum-rank taxa). The Central Western California Region (CW), where *E. capitatum* occurs, contains a large number of CA-FP endemic species (740) and the greatest number of endemic species when scaled to unit area (20 endemic species per 1000 km^2) [41]. Using a different spatially explicit approach, Baldwin and colleagues [42] showed high species richness for portions of this same region, along with concentrations of grid cells with high values of weighted endemism. Although the results of the present study pertain to only two of the many endemic plant species of California's Central Coast, insights into the history of these two plants support regional findings and may be more broadly representative of overall patterns of biodiversity in the state.

Supplementary Materials: The following supporting information can be downloaded at: https://www.mdpi.com/article/10.3390/plants13010090/s1, Supplementary Table S1: Sample information; Supplementary Table S2: Inferred multilocus genotypes.

Author Contributions: Conceptualization, C.M.G. and K.E.H.-L.; methodology, C.M.G. and K.E.H.-L.; formal analysis, C.M.G.; data curation, K.E.H.-L.; writing—original draft preparation, C.M.G.; writing—review and editing, C.M.G. and K.E.H.-L.; visualization, C.M.G.; supervision, C.M.G.; project administration, C.M.G.; funding acquisition, C.M.G. All authors have read and agreed to the published version of the manuscript.

Funding: This research was partially funded by the United States Fish and Wildlife Service (USFWS), Solicitation Number 14F0118Q0229.

Data Availability Statement: All sequence data generated for this project have been uploaded to the NCBI Sequence Read Archive (https://www.ncbi.nlm.nih.gov/sra (accessed on 31 October 2023)). Sample accession numbers can be found in Supplementary Table S1.

Acknowledgments: We thank several USFWS biologists who were supportive of this project, including Julie Vanderwier, Mark Elvin, Chris Kofron, and Kristie Scarrazo. We also thank the USFWS for supporting the Santa Barbara Botanic Garden's (SBBG) 10(a)(1)(A) recovery permit to perform work on the federally listed *Eriodictyon altissimum* and *E. capitatum*. For assistance in the field, we thank Heather Schneider (SBBG), Lisa Andreano (California State Parks), Chris Kofron (USFWS), Connie Rutherford (USFWS), and Mike Walgren (California State Parks). We thank the Guidetti Family for access to Indian Knob to collect *E. altissimum*. We thank Luanne Lum for her assistance at Vandenberg Space Force Base. We thank Anne Parsons for facilitating access to and field assistance on Hollister Ranch. We thank the Pacific Coast Energy Company for granting access to the Orcutt Hill EO1 location. Lawrence Janeway and Wendy Boes provided tissues of *E. lobbii*. Scott Eliason provided access to the San Bernardino National Forest for the collection of *Nama rothrockii* and *E. trichocalyx*. Kyle Gunther and Michael Simpson provided samples of *Eriodictyon* from Otay Mountain in San Diego County. We thank José Delgadillo Rodriguez and Exequiel Ezcurra for allowing the collection of *Eriodictyon* in Mexico under their plant collection permits. We thank José Flores and Annie Ayers for their help in preparing Figure 1. Finally, we thank two anonymous reviewers for their comments, which improved the manuscript, and James Cohen and Brenda Molano-Flores for their invitation to contribute to this special issue.

Conflicts of Interest: The authors declare no conflicts of interest.

References

1. Ferguson, D.M. Phylogenetic analysis and relationships in Hydrophyllaceae based on *NdhF* sequence data. *Syst. Bot.* **1998**, *23*, 253–268. [CrossRef]
2. Hannon, G.L. Eriodictyon. In Jepson Flora Project (Eds.) Jepson eFlora, Revision 4. 2021. Available online: http://ucjeps.berkeley.edu/eflora/eflora_display.php?tid=9648 (accessed on 31 August 2023).
3. Howell, J.T. The California flora and its province. *Leafl. West. Bot.* **1957**, *8*, 133–138.
4. Kofron, C.P.; Termondt, S.E.; Lum, L.H.; Schneider, H.E.; Hasenstab-Lehman, K.E.; Guilliams, C.M. Status of Lompoc yerba *Eriodictyon capitatum* (Namaceae): An endangered plant in Santa Barbara County, southern California. *Bull. South. Calif. Acad. Sci.* **2022**, *121*, 139–159. [CrossRef]
5. Luebert, F.; Cecchi, L.; Frohlich, M.W.; Gottschling, M.; Guilliams, C.M.; Hasenstab-Lehman, K.E.; Hilger, H.H.; Miller, J.S.; Mittelbach, M.; Nazaire, M.; et al. Familial classification of the Boraginales. *Taxon* **2016**, *65*, 502–522. [CrossRef]
6. Molinari-Novoa, E.A. Two new Lamiid families for the Americas. *Weberbauerella* **2016**, *1*, 1–4.

7. Weigend, M.; Luebert, F.; Gottschling, M.; Couvreur, T.L.P.; Hilger, H.H.; Miller, J.S. From capsules to nutlets-phylogenetic relationships in the Boraginales. *Cladistics* **2014**, *30*, 508–518. [CrossRef] [PubMed]
8. Refulio-Rodriguez, N.F.; Olmstead, R.G. Phylogeny of Lamiidae. *Am. J. Bot.* **2014**, *101*, 287–299. [CrossRef] [PubMed]
9. Wells, P.V. A subarborescent *Eriodictyon* (Hydrophyllaceae) from San Luis Obispo County, California. *Madroño* **1962**, *16*, 184–186.
10. Carlquist, S.; Eckhart, V.; Michener, D. Wood anatomy of Hydrophyllaceae. I. *Eriodictyon*. *Aliso* **1983**, *10*, 397–412. [CrossRef]
11. United States Fish and Wildlife Service. *Eriodictyon altissimum (Indian Knob Mountainbalm)*; United States Fish and Wildlife Service Final Species Report; United States Fish and Wildlife Service: Washington, DC, USA, 2013.
12. Schneider, H.E.; Carson, S.A.; Termondt, S.E. Smoke-induced germination in the endangered *Eriodictyon capitatum* (Namaceae). *Madroño* **2021**, *68*, 87–98. [CrossRef]
13. Keeley, J.E. Role of fire in seed germination of woody taxa in California chaparral. *Ecology* **1987**, *68*, 434–443. [CrossRef]
14. Elam, D.R. Genetic Variation and Reproductive Output in Plant Populations: Effects of Population Size and Incompatibility. Ph.D. Thesis, University of California, Riverside, Riverside, CA, USA, 1994.
15. Vasile, M.A.; Jeiter, J.; Weigend, M.; Luebert, F. Phylogeny and historical biogeography of Hydrophyllaceae and Namaceae, with a special reference to *Phacelia* and *Wigandia*. *Syst. Biodivers.* **2020**, *18*, 757–770. [CrossRef]
16. Kamvar, Z.N.; Tabima, J.F.; Grünwald, N.J. Poppr: An R package for genetic analysis of populations with clonal, partially clonal, and/or sexual reproduction. *PeerJ* **2014**, *2*, e281. [CrossRef] [PubMed]
17. Kamvar, Z.N.; Brooks, J.C.; Grünwald, N.J. Novel R tools for analysis of genome-wide population genetic data with emphasis on clonality. *Front. Genet.* **2015**, *6*, 208. [CrossRef] [PubMed]
18. Guilliams, C.M.; Hasenstab-Lehman, K. *Conservation Genomics of the Endangered Indian Knob Mountainbalm (Eriodictyon altissimum, Namaceae)*; Santa Barbara Botanic Garden: Santa Barbara, CA, USA, 2021.
19. Doyle, J.J.; Doyle, J.L. A rapid DNA isolation procedure for small quantities of fresh leaf tissue. *Phytochem. Bull.* **1987**, *19*, 11–15. [CrossRef]
20. Baird, N.A.; Etter, P.D.; Atwood, T.S.; Currey, M.C.; Shiver, A.L.; Lewis, Z.A.; Selker, E.U.; Cresko, W.A.; Johnson, E.A. Rapid SNP discovery and genetic mapping using sequenced RAD markers. *PLoS ONE* **2008**, *3*, e3376. [CrossRef]
21. Etter, P.D.; Preston, J.L.; Bassham, S.; Cresko, W.A.; Johnson, E.A. Local de novo assembly of RAD paired-end contigs using short sequencing reads. *PLoS ONE* **2011**, *6*, e18561. [CrossRef]
22. Parchman, T.L.; Gompert, Z.; Mudge, J.; Schilkey, F.D.; Benkman, C.W.; Buerkle, C.A. Genome-wide association genetics of an adaptive trait in lodgepole pine. *Mol. Ecol.* **2012**, *21*, 2991–3005. [CrossRef]
23. Peterson, B.K.; Weber, J.N.; Kay, E.H.; Fisher, H.S.; Hoekstra, H.E. Double digest RADseq: An inexpensive method for de novo SNP discovery and genotyping in model and non-model species. *PLoS ONE* **2012**, *7*, e37135. [CrossRef]
24. Tripp, E.A.; Tsai, Y.H.E.; Zhuang, Y.; Dexter, K.G. RADseq Dataset with 90% missing data fully resolves recent radiation of *Petalidium* (Acanthaceae) in the ultra-arid deserts of Namibia. *Ecol. Evol.* **2017**, *7*, 7920–7936. [CrossRef]
25. Andrews, S. FastQC-A Quality Control Tool for High Throughput Sequence Data. Available online: http://www.Bioinformatics.Babraham.Ac.Uk/Projects/Fastqc/ (accessed on 26 October 2019).
26. Eaton, D.A.R. PyRAD: Assembly of de novo RADseq loci for phylogenetic analyses. *Bioinformatics* **2014**, *30*, 1844–1849. [CrossRef] [PubMed]
27. Eaton, D.A.R.; Ree, R.H. Inferring phylogeny and introgression using RADseq data: An example from flowering plants (*Pedicularis*: Orobanchaceae). *Syst. Biol.* **2013**, *62*, 689–706. [CrossRef] [PubMed]
28. Rochette, N.C.; Rivera-Colón, A.G.; Catchen, J.M. Stacks 2: Analytical methods for paired-end sequencing improve RADseq-based population genomics. *Mol. Ecol.* **2019**, *28*, 4737–4754. [CrossRef] [PubMed]
29. Catchen, J.; Hohenlohe, P.A.; Bassham, S.; Amores, A.; Cresko, W.A. Stacks: An analysis tool set for population genomics. *Mol. Ecol.* **2013**, *22*, 3124–3140. [CrossRef] [PubMed]
30. Jombart, T. *Adegenet*: A R package for the multivariate analysis of genetic markers. *Bioinformatics* **2008**, *24*, 1403–1405. [CrossRef] [PubMed]
31. Gruber, B.; Unmack, P.J.; Berry, O.F.; Georges, A. Dartr: An R package to facilitate analysis of SNP data generated from reduced representation genome sequencing. *Mol. Ecol. Resour.* **2018**, *18*, 691–699. [CrossRef] [PubMed]
32. Stamatakis, A. RAxML version 8: A tool for phylogenetic analysis and post-analysis of large phylogenies. *Bioinformatics* **2014**, *30*, 1312–1313. [CrossRef]
33. Miller, M.A.; Pfeiffer, W.; Schwartz, T. Creating the CIPRES Science Gateway for inference of large phylogenetic trees. In Proceedings of the 2010 Gateway Computing Environments Workshop, GCE 2010, New Orleans, LA, USA, 14 November 2010.
34. Pritchard, J.K.; Stephens, M.; Donnelly, P. Inference of population structure using multilocus genotype data. *Genetics* **2000**, *155*, 945–959. [CrossRef]
35. Eaton, D.A.R.; Overcast, I. Ipyrad: Interactive assembly and analysis of RADseq datasets. *Bioinformatics* **2020**, *36*, 2592–2594. [CrossRef]
36. Kluyver, T.; Ragan-Kelley, B.; Pérez, F.; Granger, B.; Bussonnier, M.; Frederic, J.; Kelley, K.; Hamrick, J.; Grout, J.; Corlay, S.; et al. Jupyter Notebooks—A publishing format for reproducible computational workflows. In Proceedings of the Positioning and Power in Academic Publishing: Players, Agents and Agendas-Proceedings of the 20th International Conference on Electronic Publishing, ELPUB 2016, Göttingen, Germany, 7–9 June 2016.

37. Evanno, G.; Regnaut, S.; Goudet, J. Detecting the number of clusters of individuals using the software STRUCTURE: A simulation study. *Mol. Ecol.* **2005**, *14*, 2611–2620. [CrossRef]
38. Myers, N.; Mittermeler, R.A.; Mittermeler, C.G.; Da Fonseca, G.A.B.; Kent, J. Biodiversity hotspots for conservation priorities. *Nature* **2000**, *403*, 853–858. [CrossRef] [PubMed]
39. Raven, P.H.; Axelrod, D.I. *Origin and Relationships of the California Flora*; University of California Press: Berkeley, CA, USA, 1978; Volume 72.
40. Stebbins, G.L.; Major, J. Endemism and speciation in the California flora. *Ecol. Monogr.* **1965**, *35*, 1–35. [CrossRef]
41. Burge, D.O.; Thorne, J.H.; Harrison, S.P.; O'Brien, B.C.; Rebman, J.P.; Shevock, J.R.; Alverson, E.R.; Hardison, L.K.; Rodríguez, J.D.; Junak, S.A.; et al. Plant diversity and endemism in the California Floristic Province. *Madroño* **2016**, *63*, 3–206. [CrossRef]
42. Baldwin, B.G.; Thornhill, A.H.; Freyman, W.A.; Ackerly, D.D.; Kling, M.M.; Morueta-Holme, N.; Mishler, B.D. Species richness and endemism in the native flora of California. *Am. J. Bot.* **2017**, *104*, 487–501. [CrossRef]

plants

MDPI

Article

Population Genetics, Genetic Structure, and Inbreeding of *Commiphora gileadensis* (L.) C. Chr Inferred from SSR Markers in Some Mountainous Sites of Makkah Province

Hassan Mansour [1,2,*], Khalid H. Alamer [1] and Zaki M. Al-Hasawi [1,3]

1 Biological Sciences Department, Faculty of Science and Arts, King Abdulaziz University, Rabigh 21911, Saudi Arabia; kalamer@kau.edu.sa (K.H.A.); zalhasawy@kau.edu.sa (Z.M.A.-H.)
2 Department of Botany and Microbiology, Faculty of Science, Suez Canal University, Ismailia 41522, Egypt
3 Department of Biological Sciences, Faculty of Science, King Abdulaziz University, Jeddah 21589, Saudi Arabia
* Correspondence: hmansor@kau.edu.sa; Tel.: +966-506741048

Abstract: *Commiphora gileadensis* (L.) C. Chr is a perennial plant existing mainly in the southern and western mountains of the Arabian Peninsula. In the Makkah province, the remaining populations are threatened by many factors such as overcutting, overgrazing, and urban developments. These dangers are expected to be aggravated by the progression of aridification factors arising from climate change. To overcome the decline in remaining populations of this valuable species, a timely evaluation of the population's genetic variables and genetic structure is vital for the conservation of existing *C. gileadensis* populations. In this study, we used 61 SSR primers to achieve this objective. Only 50 loci showed polymorphisms, which led to further analysis of the population genetics for 600 genotypes that were collected from 50 populations of *C. gileadensis* found in 10 different sites in the Makkah region: Gebel Al Muliesaa, Wadi Albathna, Wadi Houra, Wadi Albaidaa, Wadi Elebiedia, Gebel Kniethl, Wadi Sayaa, Wadi Elbarasa, Wadi Alfawara, and Wadi Alkharar. The results showed an obvious decrease in genetic diversity variables in all studied populations. The range of *PPL* was between 8 and 40; additionally, the low H_T value of 0.804 and the high value of inbreeding, F_{is} = 0.238, reflected a severe lack of heterozygotes. High levels of F_{ST} and G_{ST} and low gene flow indicate considerable segregation among the *C. gileadensis* populations, which creates a barrier to gene migration. Our data suggest the need for conservation planning for *C. gileadensis* in order to avoid the species' forthcoming extinction. Efforts should be largely oriented around managing water consumption, prohibiting overcutting and overgrazing, and establishing appropriate seed banks.

Keywords: *Commiphora gileadensis*; conservation; populations; genetic diversity; Makkah

Citation: Mansour, H.; Alamer, K.H.; Al-Hasawi, Z.M. Population Genetics, Genetic Structure, and Inbreeding of *Commiphora gileadensis* (L.) C. Chr Inferred from SSR Markers in Some Mountainous Sites of Makkah Province. *Plants* **2023**, *12*, 2506. https://doi.org/10.3390/plants12132506

Academic Editors: Brenda Molano-Flores and James Cohen

Received: 25 March 2023
Revised: 26 June 2023
Accepted: 27 June 2023
Published: 30 June 2023

1. Introduction

Many members of the genus *Commiphora* are listed as endangered due to the overcollection of their populations for utilisation in medicinal and economical purposes (Burseraceae [Myrrh family]) [1]. *Commiphora gileadensis* (L.) C. Chr is considered one of the most economically and medicinally valuable trees grown in the Northern Hemisphere. Its distribution is mainly centred in the Red Sea region in the southern and western mountains of Saudi Arabia and other mountainous habitats in neighbouring countries within the Southern Arabian Peninsula and East Africa [2,3]. The well-known name of this tree in Saudi Arabia is "Besham" or Balsam, and its economic value is largely due to its use in the perfume industry [4,5] as well as its many medicinal applications, e.g., as a remedy for respiratory system diseases [6].

Commiphora gileadensis is found in few remaining populations, and it is mainly associated with the bottom of rocky mountains in the south and western regions of Saudi Arabia where its distribution is centred in Makkah province (Figure 3). The species was originally estimated to have between 1800 and 3000 populations that declined to the current estimate of only 60 populations (personal observation). The apparent reduction in the population numbers and population size of *C. gileadensis* could be attributed to progressive climate change conditions in the region [7] which, exacerbated by anthropogenic impacts [8,9], are anticipated to lead to a greater decline in the sizes of existing *C. gileadensis* populations and other associated plant taxa in arid habitats such as the Makkah region.

Commiphora gileadensis plants could be affected by a considerable decline in genetic diversity as a consequence of genetic drift, which is a key reason for the apparent low fitness and thus severe inability of many populations to adapt to ambient environmental confrontations [10–12]. Therefore, elucidating the population genetic variables and genetic structures of the remaining plant populations of *C. gileadensis* is crucial to conserve and restore this valuable plant species [13] and may support conservation plans for other plant taxa [14].

One of the main limiting factors for plant species with low population sizes is their potential for outcrossing and performing successful seed setting, especially under the stress of arid conditions. For *C. gileadensis*, self-incompatibility represents an extra stress that endangers its existence. This species has a floral structure like other members of Burseraceae and is recognizable by small, actinomorphic, and slightly odoriferous flowers; these features are considered to promote obligate outcrossing [15]. This reproductive system can severely impact population genetics corresponding to the survival of plant populations in arid habitats and can cause the loss of polymorphic genes, genetic drift, as well as progressive inbreeding [16,17].

As a result of the ambient climate change conditions connected with human over-utilisation in the Makkah province, the existing plant taxa—including remaining populations of *C. gileadensis*—are prone to the risk of extinction because of the ongoing decline in population size and potential loss of genetic diversity. Genetic analysis is pivotal for detecting genetic variation [18] using SSR (simple sequence repeats). Loci are considered to be of great value for measuring the gene diversity and genetic structures in collapsing plant populations of *C. gileadensis* due to their high potential to detect repeat regions with variable sequences in the target genome. These loci are well characterized as co-dominant molecular markers [19–23]. SSR markers were successfully applied to assess population genetics and genetic structure in other rare plant species [24–26].

Our research aims to elucidate the genetic diversity and genetic structure patterns of *C. gileadensis* populations under the xeric conditions of the Makkah province; by applying microsatellite loci, we can extract the required data for proposing mandatory conservation plans that are crucial to prevent the imminent threat of extinction for *C. gileadensis* and other associated plant species grown in this region.

2. Results

A total of 50 loci showed polymorphisms. The percentage of polymorphic loci (Table 1) was at its maximum value (40) in the Walb 5 and Walbd5 populations in Wadi Albathna and Wadi Albaidaa, respectively, whereas the minimum percentage of polymorphic loci (8) was detected in the Welbi1 population in Wadi Elebiedia. High selfing was indicated by our results for *C. gileadensis*, as the average inbreeding coefficient (F_{is}) was 0.238, verifying an obvious deficit of heterozygotes (Table 1).

Table 1. The measurements of population genetic variables of *C. gileadensis* populations across studied sites in Makkah province.

Population	N_a	N_e	I	No. of Private Alleles	Ho	He	P	F_{is}
Gmul 1	1.200	1.103	0.092	0.066	0.073	0.058	18.00	−0.221
Gmul 2	1.320	1.144	0.139	0.057	0.083	0.085	26.00	0.136
Gmul 3	1.440	1.281	0.209	0.087	0.143	0.131	32.00	−0.073
Gmul 4	1.380	1.166	0.145	0.066	0.078	0.084	26.00	0.163
Gmul 5	1.340	1.172	0.155	0.052	0.083	0.097	30.00	0.064

Table 1. *Cont.*

Population	N_a	N_e	I	No. of Private Alleles	Ho	He	P	F_{is}
Walb 1	1.140	1.062	0.066	0.000	0.040	0.042	14.00	−0.031
Walb 2	1.300	1.163	0.151	0.028	0.042	0.098	28.00	0.583
Walb 3	1.240	1.144	0.122	0.020	0.047	0.079	20.00	0.317
Walb 4	1.220	1.080	0.085	0.000	0.013	0.052	22.00	0.691
Walb 5	1.480	1.229	0.217	0.000	0.112	0.138	40.00	0.152
Whr 1	1.340	1.180	0.162	0.057	0.133	0.104	30.00	−0.123
Whr 2	1.340	1.161	0.140	0.064	0.057	0.084	26.00	0.335
Whr 3	1.180	1.061	0.067	0.059	0.050	0.041	18.00	0.009
Whr 4	1.280	1.124	0.117	0.039	0.065	0.071	22.00	0.162
Whr 5	1.160	1.059	0.068	0.052	0.035	0.042	16.00	0.234
Walbd1	1.120	1.056	0.054	0.028	0.043	0.034	10.00	−0.147
Walbd2	1.100	1.037	0.040	0.000	0.018	0.025	10.00	0.092
Walbd3	1.240	1.133	0.120	0.000	0.037	0.079	22.00	0.438
Walbd4	1.120	1.037	0.047	0.000	0.007	0.028	12.00	0.796
Walbd5	1.400	1.129	0.148	0.028	0.031	0.089	40.00	0.767
Welbi1	1.080	1.039	0.039	0.028	0.010	0.025	8.00	0.429
Welbi2	1.240	1.079	0.091	0.000	0.032	0.053	20.00	0.305
Welbi3	1.260	1.152	0.139	0.000	0.037	0.092	26.00	0.528
Welbi4	1.260	1.112	0.107	0.020	0.038	0.066	20.00	0.393
Welbi5	1.240	1.118	0.115	0.000	0.087	0.072	20.00	−0.204
Gknt 1	1.420	1.256	0.189	0.070	0.142	0.113	30.00	−0.245
Gknt 2	1.420	1.187	0.162	0.044	0.072	0.093	30.00	0.330
Gknt 3	1.340	1.227	0.160	0.062	0.055	0.098	20.00	0.415
Gknt 4	1.420	1.241	0.195	0.020	0.077	0.121	30.00	0.347
Gknt 5	1.360	1.208	0.155	0.070	0.080	0.091	24.00	0.160
Welbr1	1.240	1.141	0.108	0.034	0.057	0.065	16.00	0.032
Welbr2	1.180	1.079	0.080	0.000	0.030	0.049	14.00	0.428
Welbr3	1.240	1.149	0.129	0.000	0.000	0.087	24.00	1.000
Welbr4	1.280	1.166	0.147	0.044	0.052	0.097	26.00	0.417
Welbr5	1.200	1.110	0.105	0.000	0.025	0.069	20.00	0.619
Wsay1	1.160	1.077	0.073	0.048	0.050	0.047	16.00	0.007
Wsay2	1.120	1.071	0.062	0.000	0.020	0.041	12.00	0.667
Wsay3	1.260	1.121	0.118	0.028	0.055	0.075	24.00	0.440
Wsay4	1.160	1.090	0.083	0.020	0.033	0.055	16.00	0.357
Wsay5	1.180	1.119	0.102	0.020	0.042	0.069	18.00	0.434
Walf 1	1.240	1.135	0.113	0.028	0.088	0.070	18.00	−0.233
Walf 2	1.100	1.051	0.048	0.000	0.037	0.031	10.00	0.089
Walf 3	1.240	1.124	0.113	0.044	0.077	0.070	20.00	−0.099
Walf 4	1.140	1.084	0.065	0.020	0.027	0.042	10.00	0.409
Walf 5	1.220	1.130	0.117	0.039	0.095	0.078	22.00	−0.139
Walk 1	1.240	1.120	0.106	0.020	0.073	0.065	18.00	−0.139
Walk 2	1.140	1.062	0.056	0.000	0.022	0.034	10.00	0.412
Walk 3	1.180	1.101	0.095	0.000	0.072	0.062	18.00	−0.158
Walk 4	1.140	1.052	0.050	0.028	0.015	0.031	10.00	0.416
Walk 5	1.220	1.136	0.114	0.069	0.070	0.073	18.00	−0.024
Overall mean	1.245	1.125	0.112	0.029	0.055	0.070	20.60	0.238

The mean number of alleles per locus (N_a) varied between 1.48 (Walb 5 population) and 1.08 (Welbi1 population), resulting in the mean number of effective alleles per locus (N_e) and the Shannon index (I). The highest number of private alleles was 0.087 and was calculated in the Gmul 3 population, while no private alleles were detected in Walb 1, Walb 4, Walbd5, Welbi2, Welbi3, Welbi5, Welbr2, Welbr3, Welbr5, Walk 2, and Walk 3 populations. Expected heterozygosity (H_e) ranged from 1.281, 0.217, and 0.138, respectively, in the Gmul 3 and Walb 5 populations to 1.037, 0.039, and 0.025, respectively, in the Welbi1 and Walbd2 populations (Table 1). The average total heterozygosity (H_T) for all loci and populations was equal to 0.804.

The PCoA results (Figure 1) indicated that five out of seven principal components were significant (eigenvalue > 1) and considered as 99.9 of the sum variation. The main five significant components were N_a, N_e, I, the number of private alleles, and H_o.

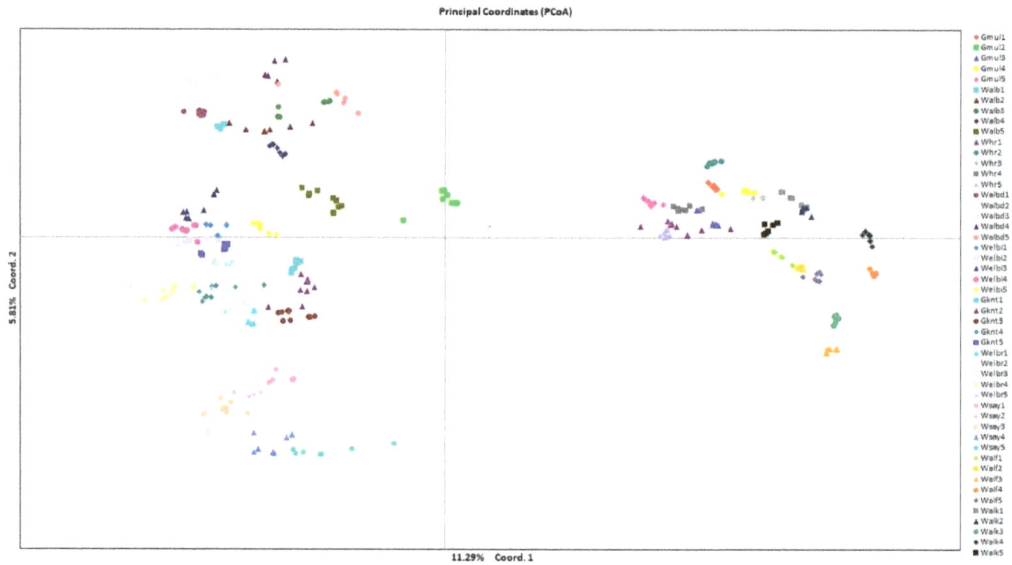

Figure 1. Principal coordinate analysis to categorize *C. gileadensis* populations based on pairwise genetic distance among 600 individual genotypes belonging to 50 studied populations with coloured and polygon codes for each population acronym.

The analysis categorized the studied populations of *C. gileadensis* into four groups. The upper-right group comprised individuals who belonged to populations of Gebel Al Muliesaa, Wadi Alkharar, and Wadi Houra; the upper-left group contained individuals in populations from Wadi Albathna, Wadi Elebiedia, and Wadi Albaidaa; the lower-left group contained populations from Gebel Kniethl, Wadi Elbarasa, and Wadi Sayaa; and the lower-right group contained Wadi Alfawara populations.

Evanno's method [27] indicated that K = 2 was optimal among the 50 populations of *C. gileadensis* (Figure 2).

The AMOVA showed substantial genetic differentiation among the studied *C. gileadensis* populations where $F_{ST} = 0.896$ and was higher with $R_{ST} = 0.980$. The maximum genetic differentiation was observed between different populations (98, $p = 0.001$), while the minimum value (1, $p = 0.010$) was measured among individuals in the same population. The gene flow of Nm = 0.024 is a low value for the gene migration among population per generation. G-statistic measurements showed a higher value of $F_{ST} = 0.913$ and G_{st} value = 0.908 compared to the F_{ST} resulting from AMOVA, which confirmed high genetic differentiation among different populations.

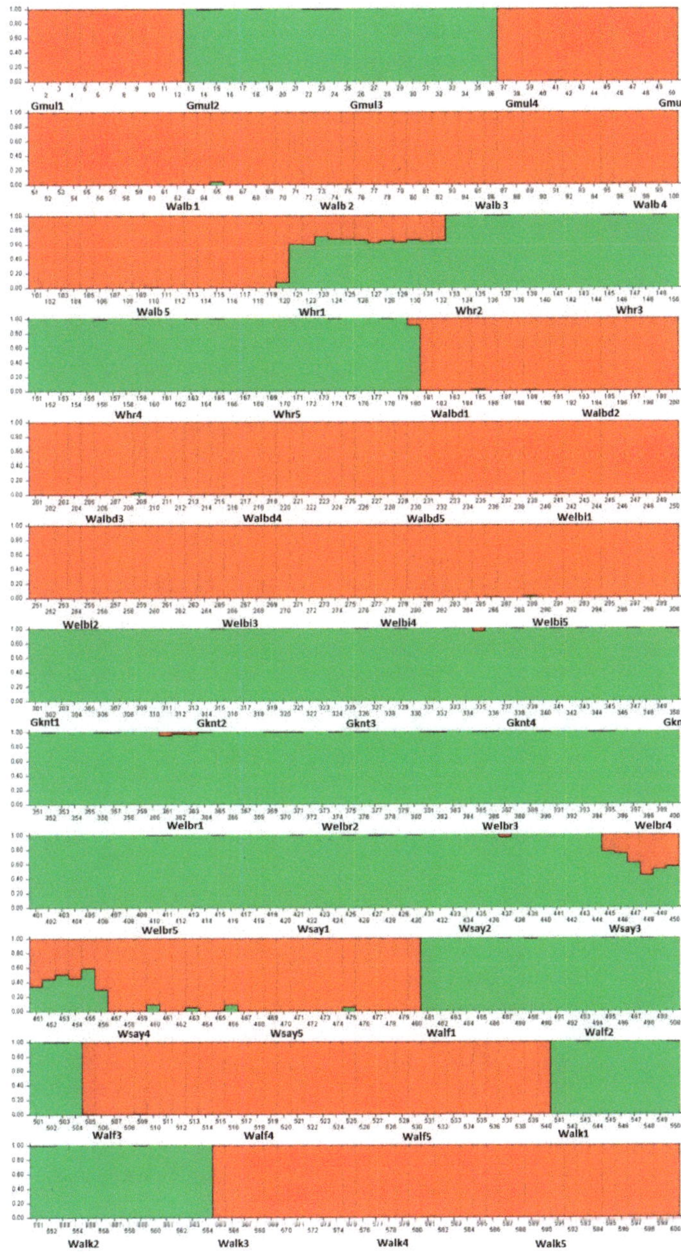

Figure 2. Population structure of 600 genotypes of *C. gileadensis* belonging to 50 populations that were grouped into two subgroups based on 50 SSR markers (K = 2); the acronyms of each population were written next to the genotypes where they belonged. The green colour referred for the first subgroup and the red colour referred to the second subgroup.

3. Discussion

In this study, all the genetic diversity variables of *C. gileadensis* showed a modest to extreme decline in measurements among all the enduring populations analysed and was in agreement with other studies on closely related plant species of Burseraceae. This indicates an extensive decline in gene diversity and meets the corresponding severe environmental circumstances [28–34] confirmed by the comparison of the high genetic differentiation F_{ST} values, which measured for *C. gileadensis* (F_{ST} = 0.896) with other rare plant species in similar plant habitats in South Sinai, and they revealed considerable values of genetic differentiation, e.g., *Primula boveana* (F_{ST} = 0.737) [35] and *Cotoneaster orbicularis* (F_{ST} = 0.634) [32].

The main reason behind the lack of genetic diversity in the existing populations may be their small population size. A leading factor underlying this decline could be that *C. gileadensis* populations are subjected to high genetic drift and inbreeding, which exacerbate the problem of decline in polymorphic alleles as an imminent result of encountering environmental conditions [36–38].

The distribution outline of genetic diversity locations among the studied populations revealed a considerable range of variability, with relatively modest polymorphisms measured in the populations of Gebel Al Muliesaa, Wadi Alkharar, and Wadi Houra, which could be attributed to the relative abundance of water reserves in these regions. Wadi Albathna and Wadi Albaidaa are located at the foot of the Bany Ayoub Mountain region, where water reserves are more abundant due to frequent floods caused by rain in this location. Wadi Elbarasa has low-altitude valleys characterized by water aggregations that allow for the growth of a few plant populations. The slight gorges in Gebel Kniethl (750–1000 m a.s.l) facilitate the growth of plant populations [39], revealing the same connection between genetic diversity maintenance in rare plant populations and water availability in desert habitats.

The PCoA determined the association between the genetic diversity in *C. gileadensis* and the abundance of water reserves. The PCoA subdivided populations with relatively high polymorphisms, showing sites with water abundance on the right and left upper parts of the PCoA axis: Gebel Al Muliesaa, Wadi Alkharar, and Wadi Houra were in the upper-right group and populations from Wadi Albathna, Wadi Elebiedia, and Wadi Albaidaa were in the upper-left group. The STRUCTURE analysis results for the existence of only two subpopulations for the studied 500 genotypes of *C. gileadensis* and the notable values of genetic differentiation among populations, calculated using AMOVA and confirmed with G-statistics, indicated considerable isolation among the population sites. The main reason for this high isolation could be the increasing activities of human inhabitants, which include overcutting for cosmetic and medicinal purposes. This can also include excessive overgrazing by camels and sheep herds owned by local tribes in the area or other tribes inhabiting the southern region of Saudi Arabia with drier climatic conditions [40] that become extremely hazardous during spring. Moreover, water reserves are at risk of high depletion due to the recent increasing human populations associated with growing industries and petroleum refinery companies in the region [7,41].

Larger decreases in the population size of *C. gileadensis* and further isolation are anticipated with increasing temperatures and water deficiency conditions, as indicated by the fluctuations of rain frequency in these sites [8,9]. Moreover, the constant influence of increasing temperatures could pose a greater risk to the reproductive capabilities of *C. gileadensis* flowers—as well as have a negative impact on pollination potential—and is thus expected to increase selfing [33,42], as revealed by the excessive low values of the measured inbreeding coefficient (F_{is}).

C. gileadensis is characterized by drupe-type fruit with one seed that is considered relatively heavy for wind dispersal (sizes range from 3.5 to 4.8 cm), which supports our computed low level of gene flow among *C. gileadensis* populations. For this reason, the extensive anthropogenic and climatic causes of isolation have promoted the elevation of the genetic differentiation value among the remaining populations of *C. gileadensis*. This phenomenon was clearly outlined in the PCoA. The values of the calculated gene flow decreased considerably from the values required for preventing an increase in genetic drift [43]. The concurrent influences of genetic drift and gene flow could aggravate the future drop in gene diversity among the remaining populations of *C. gileadensis*.

4. Materials and Methods

4.1. Plant Materials

The sampling of *C. gileadensis* in the mountains and plains of the Makkah province included fifty populations from ten different sites (Table 2, Figure 3). From every site, five populations were selected for sampling. All studied populations were located at the bottom of the rocky mountains between the Alabwaa village and the Makkah metropolitan area. The highest number of individuals was found in a population located in Wadi Albaidaa at the bottom of the Ayoub mountains east of Abwaa (Figure 2). The lowest number of individuals (16) was the Whr1 population in the Wadi Houra site, which was the result of extensive human activities in the area—such as overcutting—and overgrazing by sheep and cattle herds over the whole site.

Table 2. Information of the studied sites and populations of *Commiphora gileadensis* in Makkah province (acronym of 3–4 letters refers to population name, and the number refers to different populations within same site).

Population Acronym	Population Site	Longitude (E)	Latitude (N)	Altitude (m.)	Total No. of Individuals
Gmul 1	Gebel Al Muliesaa	39°16′43″	23°26′03″	216	22
Gmul 2					30
Gmul 3					19
Gmul 4					20
Gmul 5					15
Walb 1	Wadi Albathna	38°57′32″	23°29′22″	371	19
Walb 2					29
Walb 3					38
Walb 4					33
Walb 5					30
Whr 1	Wadi Houra	40°04′18″	21°48′16″	612	16
Whr 2					26
Whr 3					30
Whr 4					35
Whr 5					22
Walbd1	Wadi Albaidaa	39°48′54″	23°21′07″	600	48
Walbd2					36
Walbd3					30
Walbd4					42
Walbd5					33
Welbi1	Wadi Elebiedia	39°51′43″	22°21′31″	551	44
Welbi2					40
Welbi3					33
Welbi4					49
Welbi5					31
Gknt 1	Gebel Kniethl	40°06′19″	21°34′45″	762	38
Gknt 2					31
Gknt 3					22
Gknt 4					37
Gknt 5					35
Welbr	Wadi Elbarasa	39°37′07″	22°00′09″	320	44
Welbr					22
Welbr					29
Welbr					26
Welbr					35
Wsay1	Wadi Sayaa	39°53′32″	22°28′42″	629	39
Wsay2					36
Wsay3					30
Wsay4					22
Wsay5					28

Table 2. *Cont.*

Population Acronym	Population Site	Longitude (E)	Latitude (N)	Altitude (m.)	Total No. of Individuals
Walf 1	Wadi Alfawara	40°09′02″	21°50′42″	721	32
Walf 2					40
Walf 3					33
Walf 4					23
Walf 5					18
Walk 1	Wadi Alkharar	40°06′26″	21°16′48″	747	34
Walk 2					33
Walk 3					21
Walk 4					32
Walk 5					29

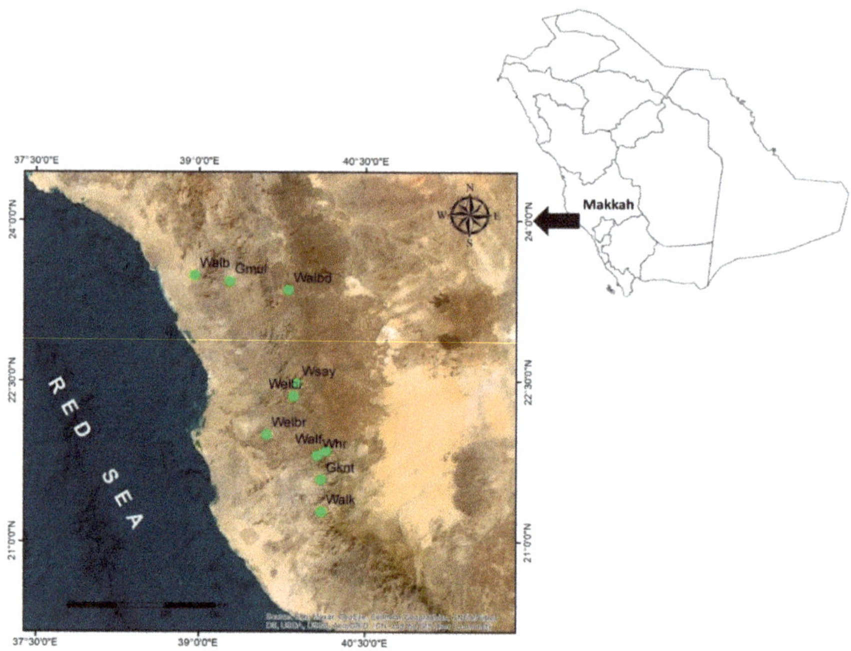

Figure 3. The studied sites of Commiphora gileadensis in the Makkah province, Kingdom of Saudi Arabia (Gmul: Gebel Al Muliesaa; Walb: Wadi Albathna; Whr: Wadi Houra; Walbd: Wadi Albaidaa; Welbi: Wadi Elebiedia; Gknt: Gebel Kniethl; Welbr: Wadi Elbarasa; Wsay: Wadi Sayaa; Walf: Wadi Alfawara; Walk: Wadi Alkharar).

Twelve individuals were sampled from each of the studied populations. Two to three leaflets were preserved directly in liquid nitrogen and then placed at −20 °C in a freezer for further DNA isolation.

4.2. Genomic DNA Extraction and PCR Tests

Isolation of DNA from the preserved leaflet samples for 600 plant individuals was performed using a DNeasy Plant Mini Kit (Qiagen, Germantown, MD, USA). Fifty loci revealing polymorphisms were recognized using sixty-one formerly published primers for other species belonging to the Burseraceae family [28–31]. The polymorphic primer tests were performed for all sampled individuals (Table S1). A master mix for PCR trials was prepared according to procedures outlined in [34]. PCR tests were carried out using a C1000 Thermal Cycler (BioRad, Hercules, CA, USA). The PCR reaction conditions were as follows: initial denaturation at 95 °C for 5 min; followed by 45 cycles at 94 °C for 35 s for denaturation; 55 °C for 40 s for annealing; and 72 °C for 5 min for final extension.

The products of the PCR reactions were sequenced using a 3130xl Genetic Analyzer (Applied Biosystems, Foster City, CA, USA) with LIZ500 as a size standard. The sequences of amplified fragments were determined using GeneMapper 4.0 (Applied Biosystems, Foster City, CA, USA), and the lengths of the amplified fragments ranged from 112 to 300 bp in accordance with [44].

4.3. Population Genetic Analysis

The variables of genetic diversity, genetic structure, inbreeding, and G-statistics were calculated using GenAlEx 6.5 [45], measuring the genetic differentiation among the populations with RST for microsatellite loci [46].

The genetic structure for 50 populations of *C. gileadensis* was conducted using the Bayesian clustering method in STRUCTURE version 2.3.4 [47], with the admixture model implemented and K values (number of potential clusters) ranging from one to ten. The burn-in period and Markov Chain Monte Carlo (MCMC) conducted 100,000 iterations [27]. The optimal K value was determined using the method of Evanno et al. [27] as implemented in STRUCTURE Harvester [48].

AMOVA was applied with 999 permutations to assess genetic differentiation among populations [49,50]. Gene flow (*Nm*) was calculated via the private allele method [51]. The analysis of heterozygosity (H_o), the expected heterozygosity (H_e) under Hardy–Weinberg equilibrium, and Wright's fixation index ($F = 1 - H_o/H_e$) were tested for each locus in each population to test deviations from the Hardy–Weinberg equilibrium and thereby determine inbreeding in existing populations of *C. gileadensis*. Principal coordinate analysis (PCoA) was performed using GenAlEx 6.5 [45] based on pairwise genetic distance data among 600 individual genotypes belonging to 50 studied populations.

5. Conclusions and Recommendations

Our current research represents the first assessment of distribution patterns of population genetic variables and genetic structures among and within the remaining populations of *C. gileadensis* in habitats of the Makkah region; it was carried out in order to contribute to the conservation management and protection for *C. gileadensis* as an economically and medicinally valuable plant species. This species is confronted by the danger of forthcoming extinction due to an extreme loss of gene polymorphism associated with excessive interpopulation genetic differentiation and severe inbreeding.

The conservation plan of *C. gileadensis* should be based on long-term and progressive actions and should be oriented mainly to prevent the continuous degradation of its populations and its habitats. Many efforts could be considered in this area. Firstly, the results of the present study suggest that we should establish enclosures of a wire-fence type to protect populations that are subjected to severe low genetic polymorphisms [52], as indicated in our study for Wadi Alkharar, Wadi Albaidaa, Wadi Alfawara, and Gebel Al Muliesaa. Such enclosures are crucial to prevent the dangers of overgrazing from camels and sheep in these locations; they should be monitored regularly to observe vegetation parameters and detect any further alterations in the protected populations. Secondly, management plans for water resources should be proposed to promote water utilisation, enhancing the reuse of wastewater and the effective usage and storage of rainwater and underground water. These plans should be incorporated into undergraduate learning programs in educational institutes, in addition to increasing public awareness for managing water consumption among inhabitants of the Makkah region.

Thirdly, the apparent decline in genetic diversity and the considerable genetic differentiation in these populations could be reclaimed via performing a test of interpopulation crosses among the populations with the highest genetic differentiation values (F_{ST}) and higher selfing, potentially increasing the population fitness [53]. Many factors should be considered during the design of the interpopulation crosses to alleviate the consequences of outbreeding depression, such as the genetic distance between the concerned populations, their latitude, and the genetic diversity magnitude, e.g., crosses between extremely low genetic diversity populations resulted in the outbreeding value in F1 [54]. Moreover, a test bank for the collected seeds from the studied remaining populations can be established [53]. The collected seeds should be planted in greenhouses as nurseries, and seedlings with strong vegetative characteristics could be planted into populations with the lowest genetic diversity resembling that of the populations from which the parent seeds were collected; this would reduce potential consequences, including further inbreeding and an extreme decline of gene flow. Parts of the collected viable seeds should be preserved using suitable procedures for seed maintenance in well-equipped test banks; these test banks would be of high value for future *C. gileadensis* conservation efforts in its habitats. The recommended measures for the conservation of *C. gileadensis* could be applicable to the other rare species in the genus *Commiphora* which grow in similar arid habitats.

Supplementary Materials: The following supporting information can be downloaded at: https://www.mdpi.com/article/10.3390/plants12132506/s1, Table S1: The tested primers status for *Commiphora gileadensis*.

Author Contributions: Conceptualization, H.M.; methodology, H.M. and Z.M.A.-H.; software, H.M.; formal analysis, K.H.A.; investigation, H.M. and Z.M.A.-H.; resources, H.M., K.H.A. and Z.M.A.-H.; data curation, K.H.A.; writing—original draft, H.M., K.H.A. and Z.M.A.-H.; writing—review and editing, H.M.; visualization, H.M. and K.H.A.; project administration, H.M., K.H.A. and Z.M.A.-H.; funding acquisition, H.M., K.H.A. and Z.M.A.-H. All authors have read and agreed to the published version of the manuscript.

Funding: This research was funded by the Ministry of Education and Deanship of Scientific Research (DSR), at King Abdulaziz University, Jeddah, Saudi Arabia: under grant no. IFPIP: 728-662-1443.

Data Availability Statement: Not applicable.

Acknowledgments: This present research work was funded by Institutional Fund Projects under grant no. (IFPIP: 728-662-1443). The authors gratefully acknowledge technical and financial support provided by the Ministry of Education and King Abdulaziz University, DSR, Jeddah, Saudi Arabia.

Conflicts of Interest: The authors declare no conflict of interest.

References

1. Mathur, M.; Mathur, P.; Purohit, H. Ecological niche modelling of a critically endangered species *Commiphora wightii* (Arn.) Bhandari using bioclimatic and non-bioclimatic variables. *Ecol. Process* **2023**, *12*, 8. [CrossRef]
2. Miller, A.G.; Morris, M. *Plants of Dhofar, the Southern Region of Oman: Traditional, Economic and Medicinal Uses*; The Office of the Advisor for Conservation of the Environment, Diwan of Royal Court Sultanate of Oman: Muscat, Oman, 1988.
3. Wood, J.R.I. *A Handbook of the Yemen Flora. With Color Illustrations by Hugo Haig-Thomas*; Royal Botanic Gardens: Kew, UK, 1997.
4. Shen, T.; Li, G.H.; Wang, X.N.; Lou, H.X. The genus Commiphora: A review of its traditional uses, phytochemistry and pharmacology. *J. Ethnopharmacol.* **2012**, *142*, 319–330. [CrossRef]
5. Mahr, D. *Commiphora*: An Introduction to the Genus. *Cactus Succul. J.* **2012**, *84*, 140–154. [CrossRef]
6. González-Tejero, M.R.; Casares-Porcel, M.; Sánchez-Rojas, C.P.; Ramiro-Gutiérrez, J.M.; Molero-Mesa, J.; Pieroni, A.; Giusti, M.E.; Censorii, E.; De Pasquale, C.; Della, A.; et al. Medicinal plants in the Mediterranean area: Synthesis of the results of the project Rubia. *J. Ethnopharmacol.* **2008**, *116*, 341–357. [CrossRef]
7. Tarawneh, Q.Y.; Chowdhury, S. Trends of Climate Change in Saudi Arabia: Implications on Water Resources. *Climate* **2018**, *6*, 8. [CrossRef]
8. Issar, A.S. The impact of global warming on the water resources of the Middle East: Past, present and future. In *Climate Changes and Water Resources in the Middle East and North Africa*; Zereini, F., Hötzl, H., Eds.; Springer: Heidelberg, Germany, 2008.
9. Soultan, A.; Wikelski, M.; Safi, K. Risk of biodiversity collapse under climate change in the Afro-Arabian region. *Sci. Rep.* **2019**, *9*, 955. [CrossRef]
10. Luijten, S.H.; Dierick, A.; Gerard, J.; Oostermeijer, B.; Raijmann, L.E.J.; Den Nijs, H.C.M. Population size, genetic variation, and reproductive success in a rapidly declining, self-incompatible perennial (*Arnica montana*) in the Netherlands. *Conserv. Biol.* **2000**, *14*, 1776–1787.
11. Hansson, B.; Westerberg, L. On the correlation between heterozygosity and fitness in natural populations. *Mol. Ecol.* **2002**, *11*, 2467–2474. [CrossRef]

12. Bastiaan, S.; Hamish, G.S. Effects of genetic drift and gene flow on the selective maintenance of genetic variation. *Genetics* **2003**, *194*, 235–244.
13. Shalabi, L.F.; Otaif, F.S. *Commiphora* Jacq (Burseraceae) in Saudi Arabia, Botanical, Phytochemical and Ethnobotanical Notes. *Ecologies* **2022**, *3*, 38–57. [CrossRef]
14. Hatmaker, E.A.; Staton, M.E.; Dattilo, A.J.; Hadziabdic, D.; Rinehart, T.A.; Schilling, E.E.; Trigiano, R.N.; Wadl, P.A. Population Structure and Genetic Diversity within the endangered species *Pityopsis ruthii* (Asteraceae). *Front. Plant Sci.* **2018**, *9*, 943. [CrossRef]
15. Raju, A.J.S.; Lakshmi, P.V.; Ramana, K.V.; Chandra, P.H. Entomophily, ornithophily and anemochory in the self-incompatible *Boswellia ovalifoliolata* Bal. & Henry (Burseraceae), an endemic and endangered medicinally important tree species. *J. Threat. Taxa* **2012**, *4*, 2673–2684.
16. Keller, L.F.; Waller, D.M. Inbreeding effects in wild populations. *Trends Ecol. Evol.* **2002**, *17*, 230–241. [CrossRef]
17. Vilas, C.; San Miguel, E.; Amaro, R.; García, C. Relative contribution of inbreeding depression and eroded adaptive diversity to extinction risk in small populations of shore campion. *Conserv. Biol.* **2005**, *20*, 229–238. [CrossRef]
18. Kadam, U.S.; Lossie, A.C.; Schulz, B.; Irudayaraj, J. Gene expression analysis using conventional and imaging methods. In *DNA and RNA Nanobiotechnologies in Medicine: Diagnosis and Treatment of Diseases*; Springer: Berlin/Heidelberg, Germany, 2013; pp. 141–162.
19. Kim, K.S.; Sappington, T.W. Microsatellite Data Analysis for Population Genetics. In *Microsatellites. Methods in Molecular Biology (Methods and Protocols)*; Kantartzi, S., Ed.; Humana Press: Totowa, NJ, USA, 2013; Volume 1006, pp. 271–295.
20. Upadhyay, A.; Kadam, U.S.; Chacko, P.; Karibasappa, G.S. Microsatellite and RAPD analysis of grape (*Vitis* spp.) accessions and identification of duplicates/misnomers in germplasm collection. *Indian J. Hortic.* **2010**, *67*, 8–15.
21. Upadhyay, A.; Kadam, U.S.; Chacko, P.M.; Aher, L.; Karibasappa, G.S. Microsatellite analysis to differentiate clones of Thompson seedless grapevine. *Indian J. Hortic.* **2010**, *67*, 260–263.
22. Hinge, V.R.; Shaikh, I.M.; Chavhan, R.L.; Deshmukh, A.S.; Shelake, R.M.; Ghuge, S.A.; Dethe, A.M.; Suprasanna, P.; Kadam, U.S. Assessment of genetic diversity and volatile content of commercially grown banana (*Musa* spp.) cultivars. *Sci. Rep.* **2022**, *12*, 1–6. [CrossRef]
23. Chavhan, R.L.; Sable, S.; Narwade, A.V.; Hinge, V.R.; Kalbande, B.B.; Mukherjee, A.K.; Chakrabarty, P.K.; Kadam, U.S. Multiplex molecular marker-assisted analysis of significant pathogens of cotton (*Gossypium* sp.). *Biocatal. Agric. Biotechnol.* **2023**, *47*, 102557. [CrossRef]
24. Szczecińska, M.; Sramko, G.; Wołosz, K.; Sawicki, J. Genetic Diversity and Population Structure of the Rare and Endangered Plant Species *Pulsatilla patens* (L.) Mill in East Central Europe. *PLoS ONE* **2016**, *11*, e0151730. [CrossRef]
25. Yu, Y.L.; Wang, H.C.; Yu, Z.X.; Schinnerl, J.; Tang, R.; Geng, Y.P.; Chen, G. Genetic diversity and structure of the endemic and endangered species *Aristolochia delavayi* growing along the Jinsha River. *Plant Divers* **2021**, *43*, 225–233. [CrossRef]
26. Chen, L.; Pan, T.; Qian, H.; Zhang, M.; Yang, G.; Wang, X. Genetic Diversity and Population Structure Revealed by SSR Markers on Endemic Species *Osmanthus serrulatus* Rehder from Southwestern Sichuan Basin, China. *Forests* **2021**, *12*, 1365. [CrossRef]
27. Evanno, G.; Regnaut, S.; Goudet, J. Detecting the number of clusters of individuals using the software structure: A simulation study. *Mol. Ecol.* **2005**, *14*, 2611–2620. [CrossRef]
28. Misiewicz, T.M.; Barbosa, C.E.; Fine, P.V. Microsatellite primers for an Amazonian lowland tropical tree, *Protium subserratum* (Burseraceae). *Am. J. Bot.* **2012**, *99*, e465-7. [CrossRef]
29. Maradani, B.S.; Gudasalamani, R.; Setty, S.; Chandrasekaran, R. Development of microsatellite markers for the resin-yielding, non-timber forest product species *Boswellia serrata* (Burseraceae). *Appl. Plant Sci.* **2018**, *6*, e01180. [CrossRef]
30. Koffi, K.G.; Heuertz, M.; Jans, R.; Hardy, O.J.; Vendramin, G.G.; Duminil, J. Characterization of new microsatellite loci isolated from *Santiria trimera* (Burseraceae). *Am. J. Bot.* **2012**, *99*, e334-6. [CrossRef]
31. Rimlinger, A.; Marie, L.; Avana, M.L.; Bouka, G.U.; Zekraoui, L.; Mariac, C.; Carrière, S.M.; Duminil, J. New microsatellite markers for *Dacryodes edulis* (Burseraceae), an indigenous fruit tree species from Central Africa. *Mol. Biol. Rep.* **2020**, *47*, 2391–2396. [CrossRef]
32. Mansour, H.; Sliwinska, E. Genetic Diversity and Inbreeding Level of Cotoneaster orbicularis Schltdl. in The Sinai Mountains Revealed by Microsatellite Markers and Flow Cytometry. *Egypt. J. Bot.* **2017**, *2*, 351–361. [CrossRef]
33. Mansour, H.; Alsamadany, H.; Al-Hasawi, Z.M. Genetic diversity and genetic structure of *Salvadora persica* L., rare plant species in Rabigh province, Saudi Arabia: Implications for conservation. *J. Taibah Univ. Sci.* **2020**, *14*, 881–888. [CrossRef]
34. Mansour, H.; Alsamadany, H.; Al-Hasawi, Z.M. Molecular Assessment of Genetic Diversity and Genetic Structure of *Rhanterium epapposum* Oliv. in Scarce Populations in Some Regions of Western Saudi Arabia. *Plants* **2022**, *11*, 1560. [CrossRef]
35. Jimenez, A.; Mansour, H.; Keller, B.; Conti, E. Low genetic diversity and a high level of inbreeding in the Sinai primrose (*Primula boveana*), a species on the brink of extinction. *Plant Syst. Evol.* **2014**, *300*, 1199–1208. [CrossRef]
36. Blomqvist, D.; Pauliny, A.; Larsson, M.; Flodin, L.A. Trapped in the extinction vortex? Strong genetic effects in a declining vertebrate population. *BMC Evol. Biol.* **2010**, *10*, 33. [CrossRef]
37. Jacquemyn, H.; Roldán-Ruiz, I.; Honnay, O. Evidence for demographic bottlenecks and limited gene flow leading to low genetic diversity in a rare thistle. *Conserv. Genet.* **2010**, *11*, 1979–1987. [CrossRef]
38. Smyser, T.J.; Duchamp, J.E.; Johnson, S.A.; Larkin, J.L.; Rhodes, O.E., Jr. Consequences of metapopulation collapse: Comparison of genetic attributes between two Allegheny woodrat metapopulations. *Conserv. Genet.* **2012**, *13*, 849–858. [CrossRef]

39. Al-Gharaibeh, M.M.; Hamasha, H.R.; Rosche, C.; Lachmuth, S.; Wesche, K.; Hensen, I. Environmental gradients shape the genetic structure of two medicinal *Salvia* species in Jordan. *Plant Biol.* **2017**, *19*, 227–238.
40. Al-Rowaily, S.L.; El-Bana, M.I.; Al-Bakre, D.A.; Assaeed, A.M.; Hegazy, A.K.; Ali, M.B. Effects of open grazing and livestock exclusion on floristic composition and diversity in natural ecosystem of Western Saudi Arabia. *Saudi. J. Biol. Sci.* **2015**, *22*, 430–437. [CrossRef]
41. Harter, T.; Davis, H.; Mathews, M.; Meyer, R. Shallow ground water quality on dairy farms with irrigated forage crops. *J. Contam. Hydrol.* **2022**, *55*, 287–315. [CrossRef] [PubMed]
42. Root, T.L.; Price, J.T.; Hall, K.R.; Schneider, S.H.; Rosenzweig, C.; Pounds, J.A. Fingerprints of global warming on wild animals and plants. *Nature* **2003**, *421*, 57–60. [CrossRef]
43. Spieth, P.T. Gene flow and genetic differentiation. *Genetics* **1974**, *78*, 961–965. [CrossRef]
44. Arif, I.A.; Khan, H.A.; Shobrak, M.; Al Homaidan, A.A.; Al Sadoon, M.; Al Farhan, A.H.; Bahkali, A.H. Interpretation of electrophoretograms of seven microsatellite loci to determine the genetic diversity of the Arabian Oryx. *Genet. Mol. Res.* **2010**, *9*, 259–265. [CrossRef]
45. Peakall, R.; Smouse, P.E. GenAlEx v.6.5: Genetic analysis in Excel. Population genetic software for teaching and research. *Bioinformatics* **2012**, *28*, 2537–2539. [CrossRef]
46. Slatkin, M. A measure of population subdivision based on microsatellite allele frequencies. *Genetics* **1995**, *139*, 457–462. [CrossRef] [PubMed]
47. Pritchard, J.; Stephens, M.; Rosenberg, N.; Donnelly, P. Association mapping in structured populations. *Am. J. Hum. Genet.* **2000**, *67*, 170–180. [CrossRef]
48. Earl, E.A.; Von Holdt, B.M. Structure Harvester: A website and program for visualizing STRUCTURE output and implementing the Evanno method. *Conserv. Genet. Res.* **2012**, *4*, 359–361. [CrossRef]
49. Excoffier, L.; Smouse, P.E.; Quattro, J.M. Analysis of molecular variance inferred from metric distances among DNA haplotypes: Application to human mitochondrial DNA restriction data. *Genetics* **1992**, *131*, 479–491. [CrossRef] [PubMed]
50. Michalakis, Y.; Excoffier, L. A generic estimation of population subdivision using distances between alleles with special reference for microsatellite loci. *Genetics* **1996**, *142*, 1061–1064. [CrossRef]
51. Barton, N.H.; Slatkin, M. A Quasi-equilibrium theory of the distribution of rare alleles in a subdivided population. *Heredity* **1986**, *56*, 409–415. [CrossRef] [PubMed]
52. Koyama, A.; Uchida, K.; Ozeki, M.; Iwasaki, T.; Nakahama, N.; Suka, T. Conservation of endangered and rare plants requires strategies additional to deer-proof fencing for conservation of sub-alpine plant diversity. *Appl. Veg. Sci.* **2021**, *24*, e12553. [CrossRef]
53. Oakley, C.G.; Lundemo, S.; Ågren, J.; Schemske, D.W. Heterosis is common and inbreeding depression absent in natural populations of Arabidopsis thaliana. *J. Evol. Biol.* **2019**, *32*, 592–603. [CrossRef]
54. Holsinger, K.E.; Gottlieb, L.D. Conservation of rare and endangered plants: Principles and prospects. In *Genetics and Conservation of Rare Plants*; Falk, D.A., Holsinger, K.E., Eds.; Oxford University Press: New York, NY, USA, 1991; pp. 195–208.

plants

MDPI

Article

Taxonomic Identification and Molecular DNA Barcoding of Collected Wild-Growing Orchids Used Traditionally for Salep Production

Aphrodite Tsaballa [1], George Kelesidis [1], Nikos Krigas [1], Virginia Sarropoulou [1], Panagiotis Bagatzounis [2] and Katerina Grigoriadou [1,*]

[1] Hellenic Agricultural Organization Demeter (ELGO-DIMITRA), Institute of Plant Breeding and Genetic Resources, Thermi, 57001 Thessaloniki, Greece; atsampalla@elgo.gr (A.T.); gkthgk@gmail.com (G.K.); nkrigas@elgo.gr (N.K.); vsarrop@gmail.com (V.S.)

[2] 'Spices Bagatzounis' Company: El Greco, Natural Herbs & Teas, Vatero, 50100 Kozani, Greece; panos@bagatzounis.com

* Correspondence: katgrigoriadou@elgo.gr; Tel.: +30-2310-471110

Abstract: Molecular DNA barcoding combined with botanical taxonomy can be used for the identification and conservation of collected Greek orchids used for salep production as well as in the regulation of fair salep trade. A modified CTAB protocol was used for DNA extraction, amplification of barcoding regions (*ITS*, *matK*, *rbcL*, *trnH-psbA*), and sequencing. Sequencing data were assembled using Bioedit software, and the BLAST algorithm was used on the NCBI database for species identification at the genus level. Molecular barcoding data based on genetic similarity identification was in full coherence with taxonomic classification based on morphological data. The combination of *ITS* and *matK* exhibited a greater capacity to identify a species among the Greek salep samples. Out of the 53 samples examined, 52.9% were classified as *Dactylorhiza* spp. and 33.3% as *Anacamptis* spp., whereas only 6 samples were identified as *Orchis* spp. (11.8%). Given that a superior-quality salep beverage comes from tubers of the latter, the number of samples classified as such in northwestern Greece is unexpectedly low. A database of 53 original reference sequences from wild-growing samples of Greek origin was generated, providing a valuable resource for the identification of other salep samples from different regions. The DNA barcoding results unveiled that salep samples from northwestern Greece are related to nine members of four different genera of Orchidaceae. All species are nationally protected and covered by the CITES convention, while many of these orchids are included in the EU Directive 92/43/EEC appendix as "Other Important Species". Thus, expedited coordinated management actions are needed to ensure their survival in the future.

Keywords: *Dactylorhiza*; *Orchis*; *Anacamptis*; *Himantoglossum*; Orchidaceae; non-timber forest products (NTFP); wild products; salep; illegal trade; conservation value

Citation: Tsaballa, A.; Kelesidis, G.; Krigas, N.; Sarropoulou, V.; Bagatzounis, P.; Grigoriadou, K. Taxonomic Identification and Molecular DNA Barcoding of Collected Wild-Growing Orchids Used Traditionally for Salep Production. *Plants* **2023**, *12*, 3038. https://doi.org/10.3390/plants12173038

Academic Editors: Brenda Molano-Flores and James Cohen

Received: 16 June 2023
Revised: 21 August 2023
Accepted: 22 August 2023
Published: 24 August 2023

1. Introduction

Apart from including attractive plants, the family Orchidaceae, comprising about 736 genera and 28,000 species worldwide [1], is rather extraordinary since its members have been included in Appendices of the Convention on International Trade in Endangered Species, namely, CITES (https://cites.org/eng, accessed on 1 May 2023) [2]. To date, many orchid species are highly represented in either conventional or electronic commerce over the internet, thus being traded legally or not for their high ornamental value or as a source of components for cosmeceuticals and herbal medicines and as food [3,4]. Consequently, numerous orchid species are long-known in the folk tradition of many nations around the world [5–7].

The commercial product commonly called 'salep' ('salepi' in Greek) is the most famous folk preparation comprising orchids [8]. For thousands of years, terrestrial orchids in the Eastern Mediterranean and the Balkans have been harvested to ground their tubers to produce the famous salep powder [7,9–18]. This valuable substance is traditionally served as a hot drink, which is consumed especially during winter, or it is used as a basic ingredient for the derivate ice cream called 'salep dondurma' in Turkey or 'Kaimaki' in Greece [9–11,14,19]. Although this preparation originated from the Eastern Mediterranean and the Balkans, it became famous across Europe during the Renaissance period following the trend after the publication of *Gerard's Herbal* in 1633 [7]. To date, it has been reported that thousands of orchid bulbs are widely harvested every year for salep in different regions [19–21]. In the past, orchid tubers were commonly collected in northern Greece for the preparation of salep beverages and gelatin for porridge, representing a working-class staple [22]; however, overharvesting still occurs. This harvesting remains common in Turkey, the Balkans, and Iran [7,10,15,19]. In Greece, all orchid species are protected at the national level by the Greek Presidential Decree 67/1981 and are covered by the CITES convention. Their collection from the wild is forbidden and banned and is thus considered illegal [23].

Currently, the most popular form of salep is a hot beverage that is consumed to soothe the throat and ease stomach aches and intestinal cramps and as a remedy for colds and coughing [14]. Several orchid species are commonly collected from northern Greece for salep production such as *Anacamptis coriophora* (L.) R.M. Bateman, Pridgeon & M.W. Chase, *Anacamptis morio* (L.) R.M. Bateman, Pridgeon & M.W. Chase, *Anacamptis papilionacea* (L.) R.M. Bateman, Pridgeon & M.W. Chase, *Orchis anthropophora* (L.) All., and *Orchis italica* Poir. [22], or *Anacamptis pyramidalis* (L.) Rich., *Dactylorhiza sambucina* (L.) Soó, *Dactylorhiza saccifera* (Brongn.) Soó, *Orchis militaris* L., *Orchis provincialis* Balb. ex Lam. & DC., and *Orchis simia* Lam. [14]. One way or another, the starch contained in these salep orchids may interact with arabin and tragacanthin substances when diluted in water or milk, creating a thick and sticky liquid due to the contained substance vassarin; the latter seems to have a soothing effect against coughs and asthma [5]. Among different orchid species, the best salep is considered to be the one derived from *Orchis mascula* [14]. Apart from strengthening and stimulating the body and soothing stomach aches, *Orchis mascula* salep has presumed lung anticancer activity, strong antioxidant activity, and beneficial effects on hyperlipidemia and hypertension [24]. Undoubtedly, salep, or salepi in Greek, is an ideal high-energy drink due to the additional spices it contains (e.g., cinnamon) and its high content of carbohydrates and valuable elements such as phosphorus and calcium [24].

Given the ethnic importance of salep and the concomitant concerns for orchid conservation [7,10,15,19], several investigations have already focused on this nutritious and valuable yet controversial product with largely unproven health or aphrodisiac effects [8]. Previous review studies [8] report as many as 46 tuberous and 1 rhizomatous orchid species sourced from the wild for salep in the European context. However, other studies from Turkey have reported that up to 90 different taxa (species and subspecies) may be harvested for this aim [10], and 38 species from 7 genera were reported to be harvested in Iran [12,13]. In one way or another, several members of the genera *Orchis*, *Anacamptis*, and *Ophrys* and also *Dactylorhiza*, *Himantoglossum*, *Neotinea*, *Platanthera*, and *Serapias* are reported as common among these studies [8].

The above-mentioned studies coupled with the complex taxonomy of Orchidaceae [25–28] imply that advanced taxonomic identification is needed to elucidate salep composition from different regions. The use of DNA sequences as "barcodes" constitutes a fast, dependable, low-cost, and straightforward solution for the identification of species [29,30]. In land plants, short regions of nuclear DNA, such as the internal transcribed spacer (*ITS*), and chloroplast DNA, such as *rbcL*, *maturase K* (*matK*), *psbA-trnH*, and the transfer ribonucleic acid leucine (*trnL*), are broadly used as markers for the molecular identification of plant species [31]. However, in the complex and large family of Orchidaceae, more regions of chloroplastic DNA have been additionally used for the classification of species such as *psaB*, *trnL-F*, and nuclear Xdh [25–27]. In general, the simultaneous utilization of coding and non-coding regions is necessary for the successful identification of species in the Orchidaceae family [26,32–34]. Previous studies have tested several barcoding regions and suggested that the combination of *ITS* and *matK* markers can successfully identify members of the large genus *Dendrobium* in the family Orchidaceae [28].

Other studies have used *nrITS*, *trnL-F*, and *matK* for barcoding DNA extracted from 150 collected tubers of Iranian orchids used to produce salep [35], revealing that most Iranian tubers belonged to species in the genera *Orchis*, *Anacamptis*, and *Dactylorhiza*. Specifically, *nrITS* and the *trnL-F* spacer proved to be easier to amplify and sequence than the *matK*, providing a better-discriminating ability that eventually led to the recognition of the species that are most threatened by overharvesting in different regions of Iran [35]. Thus the latter has shown how DNA barcoding may aid the onset of conservation strategies.

Genomics technologies and approaches are of vital importance to maintain and conserve biodiversity as they can provide reliable results, even if DNA is too degraded and difficult to sequence using next-generation tools [36]. However, the analytical methods and specific sequences for DNA barcodes are still limited for large families such as Orchidaceae [37]. With the development of molecular biology and bioinformatics, a more improved integrated analytic method for DNA barcoding can be established to identify and distinguish different species [38].

The scope of this study was first to check whether widely used DNA barcoding markers already used in other Orchidaceae studies can be used for the identification of collected wild-growing orchids used for salep by the Sarakatsani ethnic Greek population subgroup in northwestern Greece [14]. The second goal was to assign the collected orchids to specific genera or species using the combination of molecular DNA barcoding and botanical taxonomic identification. The final objective was to set up a specific, easy, and straightforward DNA barcoding protocol that can aid the identification and conservation of Greek orchids, with the aim to incorporate molecular genomic techniques such as DNA barcoding and other taxonomic methods in the regulation of fair salep trade.

2. Results and Discussion

DNA was extracted from 19 fresh and 32 dried plant tissue samples of wild-growing Greek orchids depending on the availability. The barcoding regions *ITS* and *matK* were successfully amplified, sequenced, and used for the identification of almost all samples, while three barcoding regions (*ITS*, *matK* and *trnH-psbA*, or *rbcL*) were successfully used for only four samples. Although in most samples the *trnH-psbA* region was amplified and sequenced, the results were in some cases divergent from the results of the other two barcoding regions that were totally in agreement. Divergent results are accompanied by low BLAST similarities with our samples. A lack of information or taxonomic misclassifications in public DNA databases could lead to divergent results and misidentifications. In only a few samples, *trnH-psbA* was used instead of one of the other two regions for molecular identification. No dependence of the sequencing success rate on the origin of the sample (fresh or dried tissue) was observed. The combination of *ITS* and *matK* demonstrated a greater capacity to identify a species among the Greek salep samples, in agreement with previous studies concerning the family Orchidaceae [32]. Conflicting results were reported before in the literature between DNA barcoding markers within the family Orchidaceae [27]. The taxonomic identification of samples was not possible in only a few cases due to inappropriate (out-flowered) specimens collected from the wild (Tables 1, S1 and S2). While the taxonomic identification identified some specimens as *Orchis* sp. due to out-flowering appearance (Tables 1, S1 and S2), these identifications were further identified with molecular barcoding data as *O. pallens* L. (GR-1-BBGK-21,239; GR-1-BBGK-21,240; GR-1-BBGK-21,241; GR-1-BBGK-21,242; GR-1-BBGK-21,244) or *O. quadripunctata* Cirillo ex Ten. (GR-1-BBGK-21,243). In some other cases, the molecular barcoding data suggested the identification as *Dactylorhiza maculata* (L.) Soó or *Dactylorhiza sambucina* (L.) Soó for the specimen GR-1-BBGK-18,6097-26, and *D. maculata* for the specimens GR-1-BBGK-19,6097-27 and GR-1-BBGK-16,6097-29 (Tables 1 and S1). In another six cases, both the taxonomic identification and the molecular barcoding data identified the specimens only at the genus level as *Dactylorhiza* sp. (GR-1-BBGK-18,6097-1; GR-1-BBGK-18,6097-9; GR-1-BBGK-18,6097-20; GR-1-BBGK-18,6097-30; GR-1-BBGK-18,6097-32), and the specimen GR-1-BBGK-18,6097-3 as *Anacamptis* sp. (Tables 1 and S1). Except for one sample (19,402) identified as *Anacamptis pyramidalis* (L.) Rich., all other samples (n = 16) were found to be *Anacamptis morio* (L.) R.M. Bateman, Pridgeon & M.W. Chase with molecular barcoding data; the latter belong to subsp. *caucasica* (K.Koch) H. Kretzschmar, Eccarius & H. Dietr. as determined using the taxonomic identification of the collected samples (Tables 1 and S1).

Table 1. Grouping of different identification cases of the studied wild-grown salep samples (Orchidaceae) from northwestern Greece (n = 53) with consensus levels and respective reasons for the taxonomic and/or molecular approaches performed in this investigation (see Supplementary Materials Table S1).

Consensus and Explanation	Cases	Reason
Consensus at the species level (Agreement between taxonomic and molecular identification)	3	Full taxonomic identification and undoubtful matching in molecular identification
Consensus at the genus level (Agreement between taxonomic and molecular identification)	7	Out-flowered specimens; matched only at the genus level
Partial consensus (Molecular identification aided with taxonomic identification)	29	Further identification of subspecies; verification, rejection, or resolution of possible matches
Partial consensus (Taxonomic identification aided with molecular identification)	9	Out-flowered specimens
Only taxonomic identification (no molecular identification)	2	Destroyed DNA samples
Only molecular identification (no taxonomic identification)	3	Only dry specimens

The taxonomically identified *Dactylorhiza majalis* (Rchb.) P.F. Hunt & Summerh. subsp. *pythagorae* (Gölz & H.R. Reinhard) H.A. Pedersen, P.J. Cribb & Rolf Kühn (synonym *D. kalopissii* E. Nelson subsp. *pythagorae* (Gölz & H.R. Reinhard) Kreutz; sample GR-1-BBGK-21,146) and *Dactylorhiza sambucina* (samples GR-1-BBGK-22,59 and GR-1-BBGK-18,6097-7) were only identified as *Dactylorhiza* sp. with molecular barcoding data due to contradictive matching results in the NCBI database. The taxonomically identified sample GR-1-BBGK-18,6097-24 as *D. sambucina* was matched as either *D. sambucina* or *D. incarnata* (L.) Soó with molecular barcoding data; however, it is highly possible that this sample may be *D. sambucina* due to *ITS* barcoding accuracy (Tables 1, S1 and S2). The taxonomically identified *Dactylorhiza maculata* (L.) Soó subsp. *saccifera* (Brongn.) Diklic (synonym of *D. saccifera* (Brongn.) Soó subsp. *Saccifera*) was only identified at the species level with molecular barcoding data (n = 6 samples); a possible molecular match with *D. maculata* subsp. *fuchsii* (Druce) Soó for the specimen GR-1-BBGK-21,119 (Tables 1, S1 and S2) was disregarded since this subspecies is absent from Greece (https://portal.cybertaxonomy.org/flora-greece/intro, accessed on 1 May 2023) [39]. In another case, the possible matching of samples GR-1-BBGK-18,6097-2, GR-1-BBGK-18,6097-18, and GR-1-BBGK-18,6097-29 with either *D. sambucina* or *Dactylorhiza viridis* (L.) R.M.Bateman, Pridgeon & M.W.Chase (synonym *Coeloglossum viride* (L.) Hartm.) was rejected (Tables 1, S1 and S2) due to the absence of key characters in the examined samples such as yellow-green or tinged purple flowers. Finally, the taxonomically identified sample GR-1-BBGK-22,61 as *Himantoglossum calcaratum* (Beck) Schltr. subsp. *rumelicum* (H.Baumann & R.Lorenz) Niketic & Djordjevic (synonym *Himantoglossum jankae* Somlyay, Kreutz & Óvári) was only identified at the genus level with molecular barcoding data (Tables 1, S1 and S2) due to possible matching with *H. caprinum* (M.Bieb.) Spreng. (synonym *H. affine* (Boiss.) Schltr.); however, the latter is not reported as currently present in the Greek territory (https://portal.cybertaxonomy.org/flora-greece/intro, accessed on 1 May 2023) [39]. In general, the molecular barcoding data based on genetic similarity identification was in full agreement with the taxonomic classification based on morphological data (Tables 1, S1 and S2).

Using the BLAST algorithm on the NCBI database, we were able to identify all the Greek orchid samples collected at the genus level. Some samples were identified at the species level, but none were identified at the subspecies level. In general, from the 53 samples examined, half of them were classified as species of the genus *Dactylorhiza* (52.9%). Seventeen samples were classified as *Anacamptis* (33.3%) (Figure 1). In many cases, our sequences were found to be highly similar to the sequence KU931620 deposited in GenBank (Orchidaceae member AG-2017 voucher T20); according to Ghorbani et al. [35], the latter sequence comes from a tuber in Tehran, Iran, that was characterized as *Anacamptis morio*. Nonetheless, all Greek samples examined herein were taxonomically identified as *Anacamptis morio* subsp. *caucasica*. Salep samples from Iran have been reported to include mainly different members of the genus *Orchis* [35]. Surprisingly, in this investigation, only six samples were identified as *Orchis* (11.8%) (Figure 1).

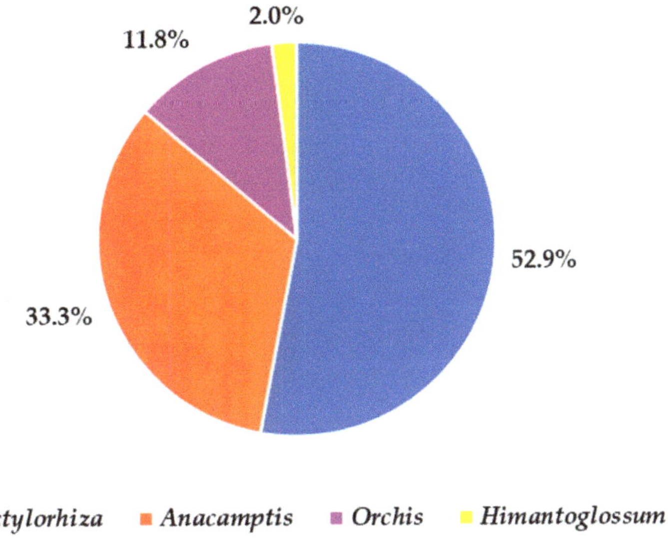

Figure 1. The proportion of different genera of Orchidaceae identified in wild-growing salep samples from northwestern Greece.

Given that superior-quality salep beverages come from tubers that belong to *Orchis* spp., the number of samples classified as such in the sample set from northwestern Greece is unexpectedly low. According to the BLAST results, the Greek samples probably belong to *Orchis pallens* or *Orchis quadripunctata*. A closer inspection of the *ITS* barcoding region nucleotide alignment analysis indicates that the Greek samples are highly similar to *O. pallens*, and thus are different from other *Orchis* spp. (indicative arrows in Figure 2).

However, according to the phylogenetic analysis of the *matK* barcoding region, the same Greek samples were placed close to *O. quadripunctata* and *O. mascula* (Figure 3). Still, there is a lack of *O. pallens matK* regions deposited in NCBI that could facilitate an accurate molecular identification. The *trnH-psbA* barcoding region was of no use for these samples as it produced completely contrasting results with the other two barcoding regions (Table S2). All these samples were collected from one specific geographical area, and their barcoding sequences were almost identical, thus proving their close taxonomic proximity.

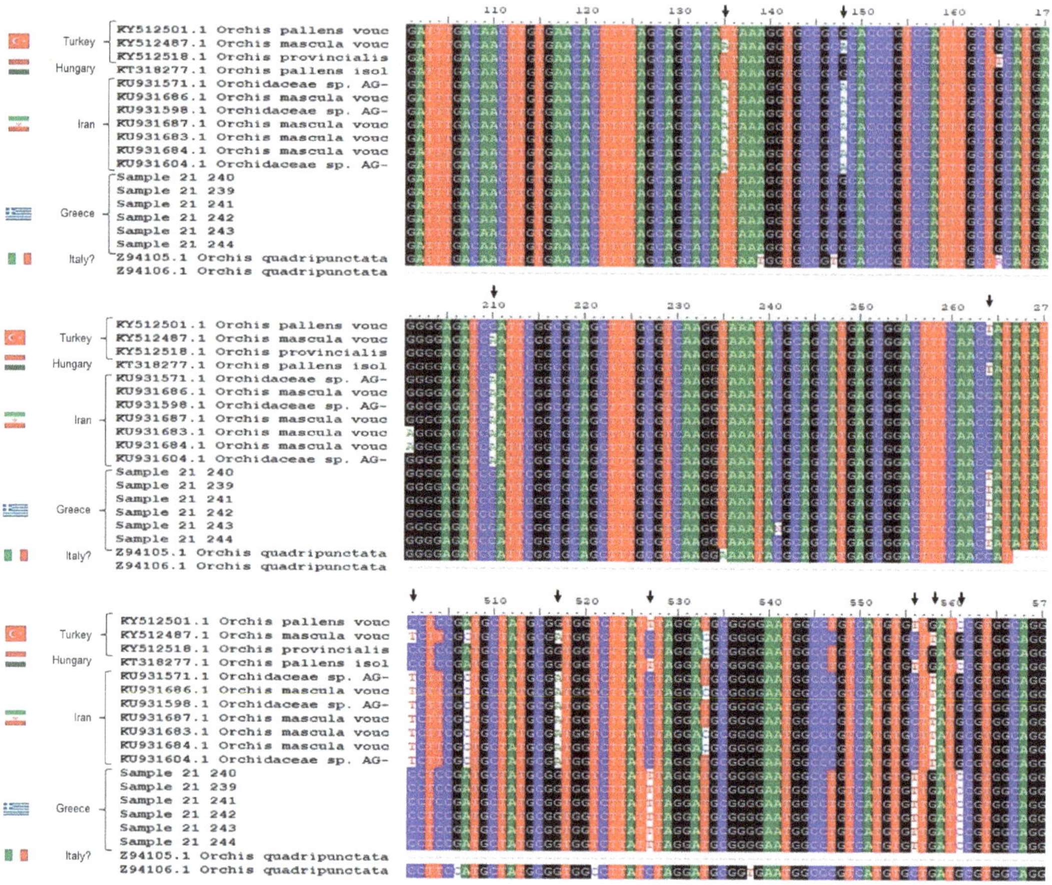

Figure 2. Multiple sequence alignment (CLUSTALW) of the *ITS* regions of the orchid samples from northwestern Greece (samples GR-1-BBGK-21,239 to GR-1-BBGK-21,244; only the last four or five digits are shown in the graphs) compared with NCBI-retrieved sequences belonging to different species of the genus *Orchis* (*Orchis* spp.). The country where the samples were collected is reported beside each of the sequences (Italian entries with no indication of the collection site in the database). Arrows indicate nucleotides that differ significantly among sequences, for example, in position 556, where the sequences of the Greek samples and *O. pallens* have thymine (T), while the sequences of *O. mascula*/*O. provincialis*/*O. quadripunctata* have cytosine (C). Alignment was edited using Bioedit software [40].

Collectively, this study generated a database of 53 original reference sequences from wild-growing samples of Greek origin, thus providing a valuable resource for the identification of other salep samples from different regions. The results of DNA barcoding using applied markers revealed that salep samples from northwestern Greece represent nine members of four different genera of Orhidaceae (Figure 1). Other review studies [8] reported as many as 46 tuberous and 1 rhizomatous orchid species used for salep in the European context, up to 90 different taxa in Turkey [10], and up to 38 species from 7 genera in Iran [12,13]. Nonetheless, it seems that members of specific genera such as *Orchis*, *Anacamptis*, and *Ophrys* in addition to *Dactylorhiza*, *Himantoglossum*, *Neotinea*, *Platanthera*, and *Serapias* are usually common between different regions [8]. In agreement with the latter observation, nine members of the genera *Dactylorhiza*, *Anacamptis*, *Orchis*, and *Himantoglossum* were also reported in the present investigation (Table S1).

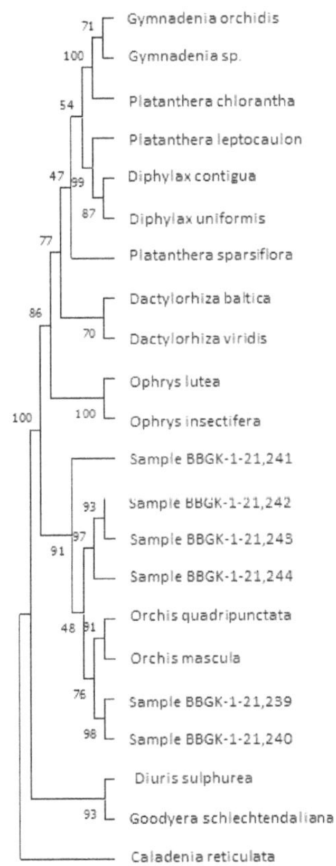

Figure 3. Phylogenetic comparison using the *matK* marker for the Greek samples of *Orchis* spp. and NCBI-retrieved data for members of other genera in Orchidaceae (Table S3).

Previous studies from Greece reported that at least 14 orchid species from 4 genera were used for salep in Greece in older times, further highlighting that up to 7 species from 3 genera were recorded in modern times [14]. With wild sourcing of salep tubers in Greece dating back to the 1800s [22], and still practiced to date [14,17], a change in the utilization of *Dactylorhiza* species in modern times has been suggested, at least in Greece [14]. The latter assumption is based on increased rarity after the overharvesting of more appreciated orchids such as the members of genera *Anacamptis* and *Ophrys*, which were considered much more abundant in Macedonia and Thessaly in older times [22]. Although perhaps true, no such decline can be found in modern studies due to the absence of monitoring programs for wild-growing orchid populations. *Dactylorhiza sambucina* is considered the most commonly over-collected salep orchid in Greece to date [17]. According to interviewees from northern Greece reported in the literature [17], the over-collection of *D. sambucina* by the local community has decreased but harvesting by people outside the local community has increased. However, this is coupled with obscure differences in orchid abundance compared with the past despite possibly variable population sizes or diverse population densities at different times [17].

Along this line and with the aim to protect orchid resources, previously published recommendations have suggested the need for long-term monitoring of wild-growing salep orchid populations in Greece and elsewhere and effective modeling regarding the response of different salep species to different harvesting pressures [17]. Other recommendations include the implementation of collection bans in over-collected areas; strengthening and monitoring current regulations and how they are enforced; the development of a straight-forward DNA barcoding-based molecular identification system to monitor and track species diversity in trade combined with controls at customs offices [13]; the establishment of specific important orchid conservation areas (in parallel with IPAs—Important Plant Areas, see https://www.plantlifeipa.org/about, accessed on 1 May 2023) [41] for in situ conservation with the monitoring of wild-growing populations; the training of local stakeholders on sustainable collection practices; and the establishment of small-scale pilot cultivation to alleviate the effect of over-harvesting [13]. With only a few of these recommendations extant in Greece to date, we developed ongoing cooperation with Greek companies interested in sustainably cultivated salep orchids of Greece, providing expertise on sustainable conservation collections as well as ex situ propagation and cultivation of the selected species reported herein in artificial environments. Conservation-wise, the Natura 2000 sites in northwestern Greece currently offer in situ protection for the wild-growing salep orchid populations. The natural conditions of the Natura 2000 network in several areas of Greece and elsewhere may create suitable circumstances for the existence of many orchid taxa [42], and many of these orchids are included in the appendices of the EU Directive 92/43/EEC as "Other Important Species", thus facilitating suitable management actions and ensuring possibilities for the survival of these plants in the future.

3. Materials and Methods

3.1. Authorized Collections of Wild-Growing Samples

Plant samples of Greek orchids used for salep were collected from mountainous areas in northwestern Greece (mainly) during 2018–2022 (Figure 4, Figure 5 and Figure S1; Table S1).

Figure 4. Collection areas of the different Orchidaceae specimens used for salep making with an emphasis on northwestern Greece (see also Supplementary Table S1).

Information regarding the localities and habitats where salep orchids naturally grow in the wild was obtained through informal conversations with (a) local residents of mountain areas in northwestern Greece who collect bulbs for personal use and (b) shepherds of the Sarakatsani ethnic Greek subpopulation group who traditionally graze their flocks at high altitudes and are aware of the local flora [14]. These people traditionally collect bulbs and consume salepi, practically discerning the species in relation to the quality of the drink they produce. Local shepherds accompanied the research team to difficult-to-access localities (Figure 5c,d), and sometimes horses were used to transport the collected plant material.

(**a**) (**b**) (**c**) (**d**) (**e**) (**f**)

Figure 5. Collection, transport, and handling of wild-growing Greek orchid samples for ex situ conservation at the premises of the Institute of Plant Breeding and Genetic Resources, Agricultural Organization Demeter (Thermi, metropolitan Thessaloniki, Greece): (**a**,**b**) Collection of wild-grown *Orchis* sp. tubers from Mt. Smolikas, northwestern Greece. (**c**,**d**) Wild-growing *Dactylorhiza sambucina* individuals, with typical yellow flowers bearing light reddish stains (**left inset**) and purple flowers speckled with darker spots on the labellum (**right inset**), and transport of collected living orchid samples in difficult-to-access areas of Mt Smolikas, northern Greece. (**e**) Plant individual collected from the wild. (**f**) Separation of individual orchid tubers.

All collections were performed between May and early July each year. When not possible to collect living plant individuals, dried inflorescences were collected (Figures 5e,f, S1c,d and S2). The wild-growing Greek orchid samples were in situ photographed to assist taxonomic identification and were collected after obtaining special permission (permits 154553/1861 and 182336/879 of 13-7-2017, 182336/879 of 16-5-2019, and 64886/2959 of 6-7-2020) issued by the national competent authority, namely, the Greek Ministry of Environment and Energy. After collection and taxonomic identification, an IPEN (International Plant Exchange Network) number was assigned to each sample maintained in the ex situ collections (salep individuals, Figures 5e,f and S1c,d) at the Balkan Botanic Garden of Kroussia, Institute of Plant Breeding and Genetic Resources, Hellenic Agricultural Organization—Demeter (ELGO-DIMITRA).

3.2. *Plant Nomenclature*

To allow comparisons across geographical scales and with other studies, all the species and subspecies are hereby cited with their currently accepted names according to the POWO (Plants of the World Online) database (https://powo.science.kew.org, accessed on 1 May 2023) [43] and not according to the Vascular Plants Checklist of Greece (https://portal.cybertaxonomy.org/flora-greece/intro, accessed on 1 May 2023) [39]. The latter online source was consulted for the local distribution of the studied orchids in the phytogeographical regions of Greece.

3.3. *DNA Extraction, Amplification of Barcoding Regions, and Sequencing*

Total genomic DNA was extracted using a modified CTAB protocol [44]. The extraction method was modified by facilitating cell lysis, extending the preparation time, and increasing the number of steps in DNA purification. Alterations and optimizations were performed regarding the added concentration of PVP-40, RNase, proteinase, and β-mercaptoethanol. Fresh tissue used for grinding under liquid nitrogen included leaves, flowers, tissue culture, and young plantlets, while dried tissue included inflorescences, whole plants, and whole tubers (Figure S2).

Four barcoding regions were tested, namely, *Internal Transcribed Spacer* (*ITS*) [45], *trnH-psbA intergenic spacer region* [46], *maturase k (matK)* [47] and *ribulose-bisphosphate carboxylase gene* (*rbcL*) [48]. The used PCR primers are reported in Supplementary Table S4. PCR reactions were prepared as follows: 1× KAPA Taq buffer (KAPA BIOSYSTEMS), 0.2 Mm Dntp mix, 0.4 Mμ of each primer, 1U/μL of KAPA Taq DNA Polymerase (KAPA BIOSYSTEMS), 20 to 40 ng DNA, and water to a final volume of 50 μL. The reaction conditions were 3 min at 95 °C, and 35 cycles of 30 s at 95 °C, 30 s at 53 °C, and 1.15 min at 72 °C followed by a final extension step of 1 min at 72 °C. All PCR amplicons were run in a standard 1.5% agarose gel where they were checked for specificity and quantity using a standard DNA ladder (markers). The amplicons were then cleaned up either from the gel or directly using a kit (Nucleospin Gel and PCR-Clean-up, Macherey-Nagel) and were sent for Sanger sequencing using a commercial provider. The same primers used for the PCR reactions were used for sequencing.

3.4. *Analysis of Data*

Sequencing data were manually checked for their quality, and contigs were assembled using Bioedit software [40] under default parameters. All generated sequences were used in Basic Local Alignment Search Tool (BLAST) [49] searches of the NCBI database. BLAST is regularly used for detecting sequence similarity in DNA barcoding projects. The identification of genera was performed using the top BLAST hit identifications provided by at least two DNA barcoding markers. The identification of species was performed when molecular data coincided and were corroborated with taxonomic identification. The further bioinformatic analysis included multiple sequence alignments (using CLUSTALW) and phylogenetic analysis using the UPGMA method [50]. The bootstrap consensus tree inferred from 500 replicates [51] was considered to represent the phylogeny of the taxa analyzed [51]. Branches corresponding to partitions reproduced in less than 50%

of bootstrap replicates were collapsed. The percentage of replicate trees in which the associated taxa clustered together in the bootstrap test (500 replicates) was shown next to the branches [51]. The phylogenetic distances were computed using the maximum composite likelihood method [52] and were in the units of the number of base substitutions per site. This analysis involved 22 nucleotide sequences. All ambiguous positions were removed for each sequence pair (pairwise deletion option). There was a total of 1821 positions in the final dataset. The phylogenetic analyses were conducted in MEGA11 [45]. Sequences from *Diuris sulphurea* R.Br., *Goodyera schlechtendaliana* Rchb. f., and *Caladenia reticulata* Fitzg. were used herein as outgroups based on previous references from the literature (Figure 3).

4. Conclusions

The investigation herein has shown that DNA barcoding markers widely used in different orchid-related studies can also be used for the identification of wild orchids collected for salep by the Sarakatsani ethnic Greek population subgroup in northwestern Greece. The combined molecular barcoding (*ITS* and *matK*) and taxonomic classification of orchids collected from different sites in northern Greece showed that 53% represent members of the genus *Dactylorhiza*, 33.3% are members of the genus *Anacamptis*, and 11.8% are members of the genus *Orchis*. Considering the reasons that sometimes create boundaries in DNA barcoding, reliable projects require holistic approaches given that biological and technical complications/issues are inevitably insurmountable. The use of such straightforward DNA barcoding protocols for salep orchids may assist in tracking, monitoring, and regulating the trade of wild-harvested products (fair and traceable salep trade) and can also facilitate the conservation of natural populations of Greek salep orchids.

Although in situ conservation of these orchid plants is ensured in various sites of the Natura 2000 network in Greece, there are still many wild-growing salep orchid populations outside the protected zones that suffer extensive over-collection directly from the wild. Therefore, long-term monitoring of wild-growing salep orchid populations is needed in Greece, which should be combined with ecological modeling regarding the response of different salep species to varied harvesting pressures. In addition, collection bans should be enforced in severely affected areas and in cases of illegal collections and exports from the country, while close monitoring of species diversity in extant salep trade should be coupled with informed controls at domestic customs offices. Training on sustainable collection practices and facilitation for the establishment of small-scale pilot cultivations of salep orchids are already in place in Greece, and these should be extended to foreign stakeholders with the aim to alleviate future collection pressure on wild-growing populations of these protected phytogenetic resources.

Supplementary Materials: The following supporting information can be downloaded at: https://www.mdpi.com/article/10.3390/plants12173038/s1, Table S1: Collection information for different Greek wild-growing species of salep orchids (Orchidaceae) assigned with International Plant Exchange Network (IPEN) accession numbers and combined taxonomic and/or molecular identification based on analysis of morphological data and BLAST tool results, respectively. Table S2: Percentage identities of BLAST results for the 51 samples used for molecular barcoding indicated with their IPEN (International Plant Exchange Network) accession numbers. Low BLAST similarities appear in red. Asterisks (*) denote divergent results (different genera). Bold letters in samples 48 to 51 are *rbcL* barcoding marker results. Table S3: NCBI GenBank-deposited accession numbers for 11 Orchidaceae species. Table S4: Sequences of the PCR primers used in the experiments. Figure S1: Collection and handling of wild-grown Greek orchid samples during ex situ conservation at the premises of the Institute of Plant Breeding and Genetic Resources, Agricultural Organization—Demeter (Thermi, metropolitan Thessaloniki, Greece). (a,b) Collection of wild-growing *Orchis* sp. individuals from Mt. Smolikas, northwestern Greece for ex situ conservation. (c) Separation of individual orchid tubers. (d) Potted individual orchid tubers. Figure S2: (a) Dried inflorescences; (b) above-ground parts; and (c) tubers used for DNA extraction.

Author Contributions: Conceptualization, K.G.; methodology, A.T., G.K., N.K. and K.G.; software, A.T., G.K., N.K., V.S. and K.G.; validation, A.T., N.K. and K.G.; formal analysis, A.T., G.K., N.K. and V.S.; investigation, A.T., G.K., N.K. and V.S.; resources, K.G. and P.B.; data curation, A.T., G.K., N.K. and V.S.; writing—original draft preparation, A.T., G.K., N.K., V.S. and K.G.; writing—review and editing, A.T., N.K., V.S. and K.G.; visualization, A.T., G.K., N.K., V.S. and K.G.; supervision, K.G.; project administration, K.G.; funding acquisition, K.G. In particular, molecular DNA extraction, amplification of barcoding regions and sequencing, and analysis of data including bioinformatics were conducted by A.T. and G.K. Taxonomic identification and botanical classification were performed by N.K. Plant material from the wild was collected by K.G. and P.B. A.T., N.K. and K.G. coordinated the data collection. P.B. is the director of a private company dealing with the trade of medicinal herbs and products including salep, co-offering the initial propagating material (whole plants) for further experimentation. All authors have read and agreed to the published version of the manuscript.

Funding: This work was co-financed by the European Regional Development Fund of the European Union and Greek national funds through the Operational Program of Western Macedonia 2014–2020 with OPS code number 2836, under the call "Cooperation and networking actions between research bodies, educational institutions and businesses in priority areas of the Region's smart specialization strategic plan—RIS3" entitled "Propagation and cultivation of Orchidaceae family species (orchids) for the production of innovative salep beverage" (Acronym: SALEBA, Grant No. DMR1-0014355, MIS 5069228).

Institutional Review Board Statement: Not applicable.

Informed Consent Statement: Not applicable.

Data Availability Statement: Data are available on request due to restrictions. The data presented in this study are available on request from the corresponding author. The data are not publicly available due to privacy restrictions.

Acknowledgments: We would like to thank Spyros Tsiftsis (domestic orchid expert) for taxonomic verification of selected specimens.

Conflicts of Interest: The authors declare no conflict of interest.

References

1. Christenhusz, M.J.M.; Byng, J.W. The number of known plants species in the world and its annual increase. *Phytotaxa* **2016**, *261*, 201–217. [CrossRef]
2. CITES—Convention on International Trade in Endangered Species of Wild Fauna and Flora. Available online: https://cites.org/eng (accessed on 7 March 2023).
3. Brinkmann, J.A. Quick Scan of Orchidaceae Species in European Commerce as Components of Cosmetic, Food and Medicinal Products. PC22 Doc. 22.1 Annex. 2014. Available online: https://cites.org/eng/com/pc/22/index.php (accessed on 15 November 2022).
4. Hinsley, A.; de Boer, H.J.; Fay, M.F.; Gale, S.W.; Gardiner, L.M.; Gunasekara, R.S.; Kumar, P.; Masters, S.; Metusala, D.; Roberts, D.L.; et al. A review of the trade in orchids and its implications for conservation. *Bot. J. Linn. Soc.* **2017**, *186*, 435–455. [CrossRef]
5. Hossain, M.M. Therapeutic orchids: Traditional uses and recent advances—An overview. *Fitoterapia* **2011**, *82*, 102–140. [CrossRef]
6. Pant, B.; Raskoti, B.B. *Medicinal Orchids of Nepal*; Himalayan Map House: Kathmandu, Nepal, 2013.
7. Teoh, E.S. *Medicinal Orchids of Asia*; Springer International Publishing: Cham, Switzerland, 2016; Available online: https://link.springer.com/book/10.1007/978-3-319-24274-3 (accessed on 1 May 2023).
8. Bazzicalupo, M.; Calevo, J.; Smeriglio, A.; Cornara, L. Traditional, therapeutic uses and phytochemistry of terrestrial European orchids and implications for conservation. *Plants* **2023**, *12*, 257. [CrossRef]
9. Sezik, E. Turkish orchids and salep. *Acta Pharm. Sci.* **2002**, *44*, 151–157. Available online: https://www.actapharmsci.com/uploads/pdf/pdf_264.pdf (accessed on 20 March 2023).
10. Tamer, E.; Karaman, C.; Utku Copur, B.O. A traditional Turkish beverage: Salep. *Food Rev. Int.* **2006**, *22*, 43–50. [CrossRef]
11. Matović, M.; Nikolić, B.; Đelić, G.; Marković, M. Natural potentials of the medicinal plants from the Orchidaceae family with mucus as the main ingredients from Zlatar mountain. *Biol. Nyssana* **2010**, *1*, 43–47. Available online: https://journal.pmf.ni.ac.rs/bionys/index.php/bionys/article/view/54 (accessed on 20 March 2023).
12. Ghorbani, A.; Gravendeel, B.; Zarre, S.; de Boer, H. Illegal wild collection and international trade in CITES-listed terrestrial orchid tubers in Iran. *TRAFFIC Bull.* **2014**, *26*, 52–58.
13. Ghorbani, A.; Gravendeel, B.; Naghibi, F.; de Boer, H. Wild orchid tuber collection in Iran: A wakeup call for conservation. *Biodivers. Conserv.* **2014**, *23*, 2749–2760. [CrossRef]

14. Kreziou, A.; de Boer, H.; Gravendeel, B. Harvesting of salep orchids in North-Western Greece continues to threaten natural populations. *Oryx* **2016**, *50*, 393–396. [CrossRef]
15. Molnár, V.A.; Nagy, T.; Löki, V.; Süveges, K.; Takács, A.; Bódis, J.; Tökölyi, J. Turkish graveyards as refuges for orchids against tuber harvest. *Ecol. Evol.* **2017**, *7*, 11257–11264. [CrossRef] [PubMed]
16. Mincheva, I.; Petrova, A.; Yordanova, M.; Kozuharova, E. Is the traditional use of "salep" in the Bulgarian Rhodopes hazardous for the wild populations of terrestrial orchids? *Flora Mediterr.* **2018**, *28*, 399–418. [CrossRef]
17. Charitonidou, M.; Stara, K.; Kougioumoutzis, K.; Halley, J.M. Implications of salep collection for the conservation of the elder-flowered orchid (*Dactylorhiza sambucina*) in Epirus, Greece. *J. Biol. Res. Thessalon.* **2019**, *26*, 18. [CrossRef]
18. Bozyel, M.E.; Merdamert-Bozyel, E. Ethnomedicinal uses of Orchidaceae taxa in Turkish traditional medicine. *Int. Res. J. Biol. Sci.* **2020**, *9*, 52–63.
19. De Boer, H.J.; Ghorbani, A.; Manzanilla, V.; Raclariu, A.C.; Kreziou, A.; Ounjai, S.; Osathanunkul, M.; Gravendeel, B. DNA metabarcoding of orchid-derived products reveals widespread illegal orchid trade. *Proc. R. Soc. B Biol. Sci.* **2017**, *284*, 20171182. [CrossRef]
20. Kasparek, M.; Grimm, U. European trade in Turkish salep with special reference to Germany. *Econ. Bot.* **1999**, *53*, 396–406. [CrossRef]
21. Masters, S.; Anthoons, B.; Madesis, P.; Saroja, S.G.; Schermer, M.; Gerritsen, W.; Karahan, A.; Verdoes, R.; Schwallier, R.; van Andel, T.; et al. Quantifying an online wildlife trade using a web crawler. *Biodivers. Conserv.* **2022**, *31*, 855–869. [CrossRef]
22. Landerer, X. Naturgeschichte und Pharmakognosie. Beiträge zur Pharmakognosie. Ueber Salep und die Salepisiden. *Arch. Pharm.* **1850**, *112*, 177–180. [CrossRef]
23. Fay, M.F. Orchid conservation: How can we meet the challenges in the twenty-first century? *Bot. Stud.* **2018**, *59*, 16. [CrossRef]
24. Al-Snafi, A.E. Pharmacological potential of *Orchis mascula*—A review. *IOSR J. Pharm.* **2020**, *10*, 1–6. Available online: http://iosrphr.org/papers/vol10-issue3/A1003010106.pdf (accessed on 20 March 2023).
25. Cameron, K.M. Utility of plastid *psaB* gene sequences for investigating intrafamilial relationships within Orchidaceae. *Mol. Phylogenet. Evol.* **2004**, *31*, 1157–1180. [CrossRef] [PubMed]
26. Kim, H.M.; Oh, S.H.; Bhandari, G.S.; Kim, C.S.; Park, C.W. DNA barcoding of Orchidaceae in Korea. *Mol. Ecol. Resour.* **2014**, *14*, 499–507. [CrossRef] [PubMed]
27. Jin, W.T.; Schuiteman, A.; Chase, M.W.; Li, J.W.; Chung, S.W.; Hsu, T.C.; Jin, X.H. Phylogenetics of subtribe Orchidinae s.l. (Orchidaceae; Orchidoideae) based on seven markers (plastid *matK*, *psaB*, *rbcL*, *trnL-F*, *trnH-psbA*, and nuclear nrITS, *Xdh*): Implications for generic delimitation. *BMC Plant Biol.* **2017**, *17*, 222. [CrossRef] [PubMed]
28. Xu, S.; Li, D.; Li, J.; Xiang, X.; Jin, W.; Huang, W.; Jin, X.; Huang, L. Evaluation of the DNA barcodes in *Dendrobium* (Orchidaceae) from mainland Asia. *PLoS ONE* **2015**, *10*, e0115168. [CrossRef] [PubMed]
29. Hebert, P.D.N.; Cywinska, A.; Ball, S.L.; DeWaard, J.R. Biological identifications through DNA barcodes. *Proc. R. Soc. B Biol. Sci.* **2003**, *270*, 313–321. [CrossRef]
30. Kress, W.J.; Erickson, D.L. DNA barcodes: Genes, genomics, and bioinformatics. *Proc. Natl. Acad. Sci. USA* **2008**, *105*, 2761–2762. [CrossRef]
31. Kehie, M.; Kumaria, S.; Devi, K.S.; Tandon, P. Genetic diversity and molecular evolution of Naga King Chili inferred from internal transcribed spacer sequence of nuclear ribosomal DNA. *Meta Gene* **2016**, *7*, 56–63. [CrossRef]
32. Xiang, X.G.; Zhang, J.B.; Lu, A.M.; Li, R.Q. Molecular identification of species in Juglandaceae: A tiered method. *J. Syst. Evol.* **2011**, *49*, 252–260. [CrossRef]
33. Inda, L.A.; Pimentel, M.; Chase, M.W. Phylogenetics of tribe Orchideae (Orchidaceae: Orchidoideae) based on combined DNA matrices: Inferences regarding timing of diversification and evolution of pollination syndromes. *Ann. Bot.* **2012**, *110*, 71–90. [CrossRef]
34. Parveen, I.; Singh, H.K.; Raghuvanshi, S.; Pradhan, U.C.; Babbar, S.B. DNA barcoding of endangered Indian *Paphiopedilum* species. *Mol. Ecol. Resour.* **2012**, *12*, 82–90. [CrossRef]
35. Ghorbani, A.; Saeedi, Y.; de Boer, H.J. Unidentifiable by morphology: DNA barcoding of plant material in local markets in Iran. *PLoS ONE* **2017**, *12*, e0175722. [CrossRef]
36. Theissinger, K.; Fernandes, C.; Formenti, G.; Bista, I.; Berg, P.R.; Bleidorn, C.; Bombarely, A.; Crottini, A.; Gallo, G.R.; Godoy, J.A.; et al. How genomics can help biodiversity conservation. *Trends Genet.* **2023**, *39*, 545–559. [CrossRef]
37. Li, Y.L.; Tong, Y.; Xing, F.W. DNA barcoding evaluation and its taxonomic implications in the recently evolved genus *Oberonia* Lindl. (Orchidaceae) in China. *Front. Plant Sci.* **2016**, *7*, 1791. [CrossRef]
38. Casiraghi, M.; Labra, M.; Ferri, E.; Galimberti, A.; De Mattia, F. DNA barcoding: A six-question tour to improve users' awareness about the method. *Brief Bioinform.* **2010**, *11*, 440–453. [CrossRef]
39. Flora of Greece Web—Vascular Plants Checklist of Greece. Available online: https://portal.cybertaxonomy.org/flora-greece/intro (accessed on 7 March 2023).
40. Hall, T.; Biociencias, I.; Carlsbad, C. BioEdit: Un software importante para la biología molecular. *GERF Bull. Biosci.* **2011**, *2*, 60–61.
41. Plantlife—Important Plant Areas (IPAs). Available online: https://www.plantlifeipa.org/about (accessed on 13 May 2023).
42. Tsiftsis, S. The role of Natura 2000 network in protecting the orchid flora of East Macedonia (NE Greece). *Eur. J. Environ. Sci.* **2021**, *11*, 71–78. [CrossRef]

43. POWO—Plants of the World Online Database. Royal Botanic Gardens, Kew. Available online: https://powo.science.kew.org (accessed on 2 February 2023).
44. Lade, B.D.; Patil, A.S.; Paikrao, H.M. Efficient genomic DNA extraction protocol from medicinal rich *Passiflora foetida* containing high level of polysaccharide and polyphenol. *SpringerPlus* **2014**, *3*, 457. [CrossRef]
45. Tamura, K.; Stecher, G.; Kumar, S. MEGA11: Molecular Evolutionary Genetics Analysis version 11. *Mol. Biol. Evol.* **2021**, *38*, 3022–3027. [CrossRef]
46. Costion, C.; Ford, A.; Cross, H.; Crayn, D.; Harrington, M.; Lowe, A. Plant DNA barcodes can accurately estimate species richness in poorly known floras. *PLoS ONE* **2011**, *6*, e26841. [CrossRef] [PubMed]
47. Cuénoud, P.; Savolainen, V.; Chatrou, L.W.; Powell, M.; Grayer, R.J.; Chase, M.W. Molecular phylogenetics of Caryophyllales based on nuclear 18S rDNA and plastid *rbcL*, *atpB*, and *matK* DNA sequences. *Am. J. Bot.* **2002**, *89*, 132–144. [CrossRef]
48. Ismail, M.; Ahmad, A.; Nadeem, M.; Javed, M.A.; Khan, S.H.; Khawaish, I.; Sthanadar, A.A.; Qari, S.H.; Alghanem, S.M.; Khan, K.A.; et al. Development of DNA barcodes for selected *Acacia* species by using *rbcL* and *matK* DNA markers. *Saudi J. Biol. Sci.* **2020**, *27*, 3735–3742. [CrossRef]
49. Altschul, S.F. *BLAST Algorithm. Encyclopedia of Life Sciences (eLS)*; John Wiley & Sons, Ltd.: Hoboken, NJ, USA, 2001. [CrossRef]
50. Tamura, K.; Nei, M.; Kumar, S. Prospects for inferring very large phylogenies by using the neighbor-joining method. *Proc. Natl. Acad. Sci. USA* **2004**, *101*, 11030–11035. [CrossRef]
51. Sneath, P.H.A.; Sokal, R.R. *Numerical Taxonomy: The Principles and Practice of Numerical Classification*; WF Freeman & Co.: San Francisco, CA, USA, 1973.
52. Felsenstein, J. Confidence limits on phylogenies: An approach using the bootstrap. *Evolution* **1985**, *39*, 783–790. [CrossRef]

plants

MDPI

Article

More than Moths: Flower Visitors of a Night-Blooming Plant in South Florida Pine Rocklands, USA

María Cleopatra Pimienta and Suzanne Koptur *

Department of Biological Sciences, Florida International University, 11200 S.W. 8th Street, Miami, FL 33199, USA
* Correspondence: kopturs@fiu.edu; Tel.: +1-305-984-0539

Abstract: Plants whose flowers open at night but remain open during the day also attract diurnal flower visitors, potentially boosting their pollination rates and providing resources that can support diverse arthropod communities. The rough-leaf velvetseed, *Guettarda scabra* (Rubiaceae), is an evergreen shrub that thrives only in the imperiled pine rockland habitat in south Florida. Its white, tubular, and fragrant flowers open during late afternoon, exhibiting traits strongly associated with the attraction of nocturnal hawkmoths (Sphingidae). Flowers of *G. scabra* remain open until the following morning, becoming available to a wider array of visitors, bringing into question the expectation that sphingophilous flowers are visited mainly by hawkmoths. To evaluate whether the flowers of *G. scabra* are mainly visited by nocturnal hawkmoths and understand the role of this plant in the pine rockland habitat, we characterized the arthropod fauna associated with its flowers during the morning, evening, and at night. We found that most flower visitors were diurnal insects of the orders Hymenoptera and Lepidoptera, although we observed other arthropod groups too. Visitation at night was dominated by two species of hawkmoths. Nectar was the main resource used by the arthropod community during this study. Legitimate visitation and nectar-robbing were the behaviors most frequently observed among the flower visitors. Our results suggest that flowers of the night-blooming *G. scabra* constitute an important food source for both diurnal and nocturnal arthropod fauna in the fire-dependent pine rocklands of southern Florida. Our study provides novel data to support efforts to conserve and protect pine rocklands and the plants and animals that inhabit them.

Keywords: butterflies; floral resources; *Guettarda scabra*; hawkmoths; insects; nectar robbing; pine rockland; pollination

Citation: Pimienta, M.C.; Koptur, S. More than Moths: Flower Visitors of a Night-Blooming Plant in South Florida Pine Rocklands, USA. *Plants* **2022**, *11*, 2799. https://doi.org/10.3390/plants11202799

Academic Editors: Brenda Molano-Flores and James Cohen

Received: 4 September 2022
Accepted: 19 October 2022
Published: 21 October 2022

1. Introduction

Although floral morphology often suggests coevolution with determined pollen vectors, flowers usually attract other visitors too [1–3]. The availability of these visitors and the reproductive success of the plant are affected by the time at which flowers open and for how long they remain available for visits [4]. As such, night-blooming plants whose flowers remain open during the day are likely to receive diurnal visitations, boosting their pollination opportunities.

The rough-leaf velvetseed (Figure 1a), *Guettarda scabra* (L.) Vent. (Rubiaceae), is a tropical evergreen shrub native to the Caribbean, ranging from the northern parts of Colombia and Venezuela to the southern portion of Florida (USA) [5–8]. In south Florida, *G. scabra* grows only in the last remnants of pine rockland (Figure 1b) and hardwood hammock habitats on the peninsular mainland, where it is abundant [9,10]. Pineland *G.scabra* plants are short in stature and allocate much more energy to flowering and fruiting than do the tall individuals persisting in hardwood hammocks [6].

Figure 1. (**a**) Recently opened flowers of *Guettarda scabra*, during late afternoon. Some individuals, such as the one in this picture, have a long pistil that raises the stigma above the deep corolla tube. Exudates from the stigma were occasionally consumed by visitors such as flies, beetles, and possibly spiders during this study; (**b**) general view of pine rockland habitat at Long Pine Key, Everglades National Park in south Florida, USA. *Guettarda scabra* plants are abundant in patches scattered among *Pinus elliottii* trees.

Guettarda scabra flowers exhibit a set of traits associated with the attraction of nocturnal lepidopterans, particularly hawkmoths (Sphingidae). Sphingophilous flowers are pale, with long-tubed corollas, and emit a strong sweet scent [11]. Anthesis in *G. scabra* happens during late afternoon [12], which led to the assumption that they were exclusively for night-time visitors [13], particularly hawkmoths [10]. Recent observations have shown that these flowers remain open through the following morning and are visited by butterflies [14], suggesting that they can be attractive to other visitors too, providing resources to a larger arthropod community. Despite its local abundance, and its presence in the disappearing pine rocklands, the structure of the community of flower visitors associated with *G. scabra* has not been studied in detail, even though *G. scabra* thrives in an imperiled habitat and allegedly depends upon pollinators whose populations may be declining [15].

To test the hypothesis that flowers of this species are mainly visited by nocturnal Lepidoptera, we observed flowering plants during day and night. Besides nocturnal lepidopterans, we expected to find many other visitors to the flowers, not only at night, but evening and morning, during times the flowers are open, but hawkmoths are not present. We thoroughly characterize the local arthropod fauna associated with flowers of *G. scabra*, their behavior, and floral resources they use. We offer insights into the role played by this native plant species in its rockland habitat and identify many *G. scabra* potential pollinators, providing the basis for a deeper understanding of its pollination biology and its role in supporting the arthropod community of this imperiled ecosystem. By learning more about the relationships *G. scabra* has with pine rockland fauna, we test the traditional view of pollination syndromes and also elucidate the multitude of interactions a single plant species may have. In this approach, our study may reach beyond its local rare habitat and be relevant to other plant species worldwide.

2. Results

Flowers of *G. scabra* were visited by 46 species of arthropods, belonging to 8 orders and 20 families (Table 1). Most visitors were insects from the orders Lepidoptera and Hymenoptera (27 species total, vs. 17 other species; Fisher's exact test $p < 0.01$), making up 63% of all species recorded. The proportions of visitors in these two orders did not differ significantly (Fisher's exact test $p > 0.05$). The remaining were arachnids of the order

Araneae, or insects belonging to the orders Coleoptera, Diptera, Hemiptera, Mantodea, and Blattodea (Figure 2).

Table 1. Array of arthropods associated with *Guettarda scabra*, their behaviors, and plant resources used at two pine rockland sites (Larry and Penny Thompson Memorial Park: LPT; and Long Pine Key, Everglades National Park: ENP) in south Florida. Observed behaviors abbreviated as follows: predation on other arthropods (pr), legitimate visitation (lv), primary nectar robbing (1nr), secondary nectar robbing (2nr), and herbivory (h). Plant resources used by visitor abbreviated as follows: arthropod prey (ap), nectar (n), pollen (p), floral tissue (f), and leaves (l). Asterisks signify caterpillar stage.

CLASS **ORDER** **Family** **Species (Author)**	**Behavior on Plant**	**Resource Used**	**Study Site**	
			LPT	**ENP**
ARACHNIDA				
ARANEAE				
Araneidae				
Acacesia hamata Hentz	pr	ap	x	
Thomisidae				
Mecaphesa sp. 1	pr	ap	x	
Mecaphesa sp. 2	pr	ap	x	
Mecaphesa sp. 3	pr	ap	x	
Mecaphesa sp. 4	pr	ap	x	
INSECTA				
BLATTODEA				
Blattodea sp.1	2nr	n	x	
COLEOPTERA				
Cerambycidae				
Eburia stigma Oliver	lv	p		x
Plectromerus dentipes Oliver	lv	p		x
Scarabaeidae				
Euphoria sepulcralis Fabricius	lv	p	x	
Phyllophaga sp.	h	f	x	
DIPTERA				
Syrphidae				
Ornidia obesa Fabricius	lv	p	x	
Tipulidae				
Tipulidae sp.1	2nr	n	x	
Tipulidae sp.2	2nr	n		x
HEMIPTERA				
Aphididae				
Aphididae sp.1	h	f	x	
Largidae				
Largus succinctus Linnaeus	2nr	n	x	x
HYMENOPTERA				
Apidae				
Apis mellifera Linnaeus	2nr	n		x
Euglosssa dilemma Bembé & Eltz	lv	n		x
Xylocopa micans Lepeletier	1nr	n		x
Crabronidae				
Cerceris rufopicta Smith	2nr	n	x	
Formicidae				
Camponotus floridanus Buckley	2nr	n	x	
Pseudomyrmex gracilis Fabricius	2nr	n	x	
Wasmannia auropunctata Roger	2nr	n	x	
Halictidae				
Augochloropsis sp.	2nr	n		x
Scoliidae				
Dielis trifasciata Fabricius	lv	p	x	
Vespidae				
Mischocyttarus mexicanus cubicola Richards	2nr	n	x	
Pachodynerus erynnis Lepeletier	2nr	n	x	
Stenodynerus sp.	1nr, 2nr	n	x	
Vespidae sp.1	1nr	n		x
Zethus slossonae Fox	1nr	n	x	x

Table 1. *Cont.*

CLASS ORDER Family Species (Author)	Behavior on Plant	Resource Used	Study Site	
			LPT	ENP
LEPIDOPTERA				
Erebidae				
Calidota laqueata Edwards *	h	l	x	x
Hypercompe scriboni Stoll *	h	l	x	
Seirarctia echo Smith *	h	l	x	
Hesperiidae				
Asbolis capucinus Lucas	lv	n	x	x
Cymaenes tripunctus Herrich-Schäffer	lv	n	x	
Ephyriades brunnea Herrich-Schäffer	lv	n	x	x
Polites baracoa Lucas	lv	n	x	
Nymphalidae				
Agraulis vanillae Linnaeus	lv	n	x	x
Heliconius charithonia Linnaeus	lv	n	x	
Papilionidae				
Papilio palamedes Drury	lv	n		x
Papilio polyxenes Fabricius	lv	n	x	
Sphingidae				
Aellopos tantalus Linnaeus	lv	n	x	
Eumorpha fasciatus Sulzer	lv	n		x
Perigonia lusca Fabricius	lv	n		x
Xylophanes tersa Linnaeus	lv	n	x	x
MANTODEA				
Mantidae				
Mantidae sp.1	pr	ap	x	
Stagmomantis floridensis Davis	pr	ap	x	

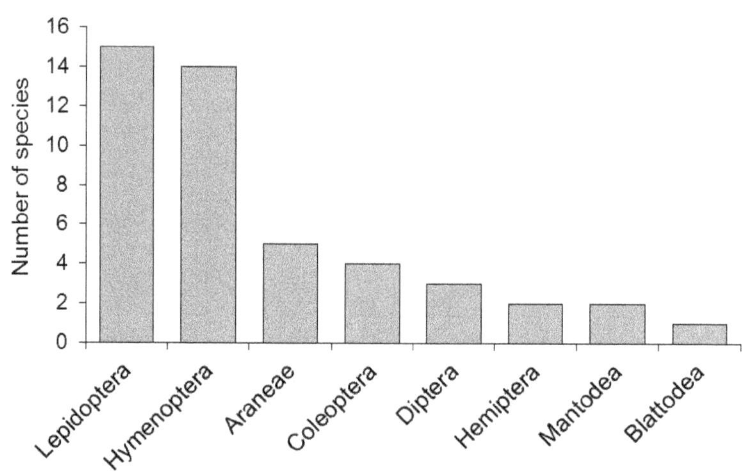

Figure 2. Arthropod orders observed on flowers of *Guettarda scabra*, sorted according to the number of species found. The large numbers of lepidopterans and hymenopterans are mostly related to diurnal activity in these two groups.

2.1. Occurrence

Most species were found only in one of the study sites: 61% of the total number of species observed at Larry and Penny Thompson Park were unique to that site; 24% of species observed at Everglades National Park were observed only there. Only a small fraction of the total species observed (15%) was common to both sites (Table 1). The proportion of unique species observed at each site (80% at LPT, 61% at ENP) did not, however, differ significantly with Fisher's Exact Test. Most arthropods registered (76%)

were seen exclusively during daytime (especially the morning hours), substantially more (Fisher's Exact test $p < 0.01$) than those observed to visit only at night (15.2%). An even smaller proportion (8.7%) of the species visited flowers both day and night (Figure 3). Overall, visitors were observed 3x more frequently in the daytime observations than in the evening observation periods, and 6x more frequently than during the night. Lepidoptera were the order most commonly observed during the morning and night; Hymenoptera most commonly in the morning and more than twice as often as Lepidoptera in the afternoon. Araneae, Diptera, and Hemiptera much more common in morning and evening; Coleoptera most often observed at night.

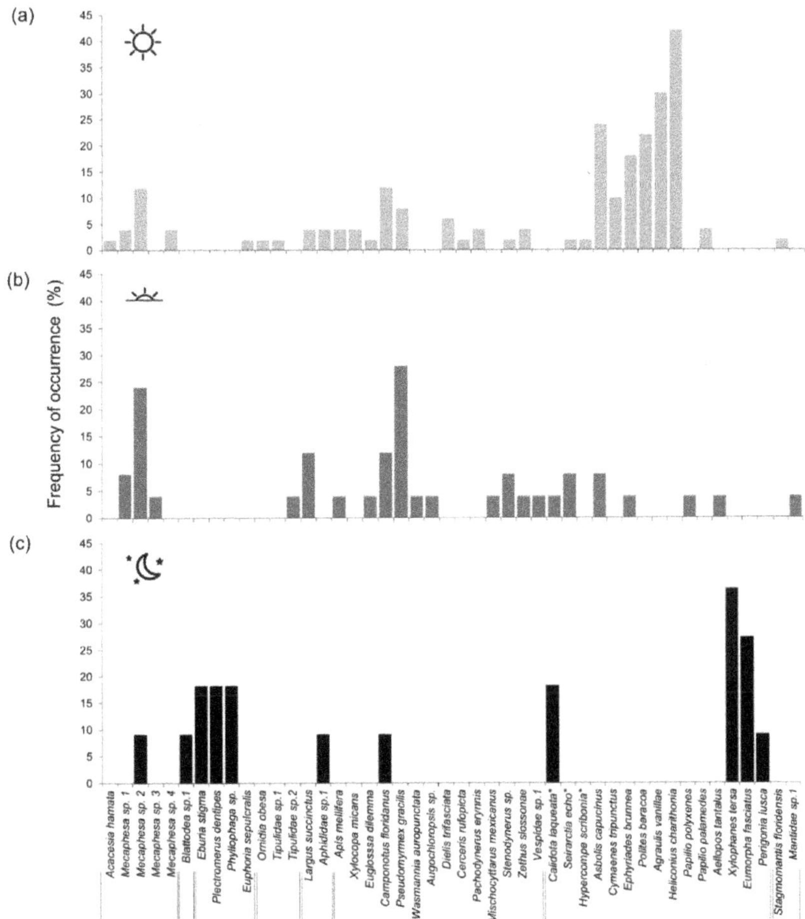

Figure 3. Frequency of occurrence of each visitor species on flowers of *Guettarda scabra*, relative to the total number of observation periods carried on during the (**a**) morning ($N = 50$), (**b**) evening ($N = 25$), or (**c**) night ($N = 11$). Ar: Araneae; Bl: Blattodea; Co: Coleoptera; Di: Diptera; He: Hemiptera; Hy: Hymenoptera; Le: Lepidoptera; Ma: Mantodea. Asterisks (*) refer to caterpillars.

Flowers of *G. scabra* were visited in the morning mainly by two butterfly species, *Heliconius charithonia* and *Agraulis vanillae*, and three species of skippers, *Asbolis capucinus*, *Polites baracoa*, and *Ephyriades brunnea* (Figure 3). Evening visits were dominated by *Pseudomyrmex gracilis* ants and crab spiders of the genus *Mecaphesa*; while at night the hawkmoths *Xylophanes tersa* and *Eumorpha fasciatus* showed the highest occurrence (Figure 3).

2.2. Visitor Behavior

We identified four behaviors among arthropods visiting *G. scabra* flowers: (a) legitimate visitation, consumption of pollen or nectar through the opening of the corolla tube involving contact with the anthers, stigma, or both and potentially resulting in pollination; (b) nectar robbing, consumption of nectar through a perforation of the corolla either made by the visitor itself (primary robber) or left by a previous visitor (secondary robber); (c) predation on other arthropods; and (d) herbivory, feeding on leaves or flowers (Table 1, Figures 4–6).

Figure 4. Overview of some species of flower visitors performing legitimate visitation and behaving as potential pollinators of *Guettarda scabra* at two pine rockland sites (Larry and Penny Thompson Memorial Park, LPT; and Long Pine Key, Everglades National Park, ENP) in south Florida, USA. Some lepidopterans such as (**a**) *Ephyriades brunnea*, (**b**) *Agraulis vanillae*, and (**c**) *Asbolis capucinus* were observed in both study sites, while (**d**) *Polites baracoa* and (**e**) *Heliconius charithonia* were seen only in LPT. Other visitors were only seen in ENP, such as (**f**) *Papilio palamedes* that feeds on nectar during daytime, and the nocturnal hawkmoths (**g**) *Perigonia lusca* and (**h**) *Eumorpha fasciatus*, represented here by an individual resting after a feeding bout. Besides lepidopterans, (**i**) the wasp *Dielis trifasciata* is seen here coming in close contact with the exposed stigma of a flower as it feeds on pollen during the morning. Beetles such as (**j**) *Eburia stigma* and (**k**) *Euphoria sepulcralis* visited flowers to feed on pollen and stigma exudates.

Figure 5. Overview of nectar robbers found on flowers of *Guettarda scabra* in pine rockland habitats in south Florida, USA. Diurnal primary nectar robbers such as the wasps (**a**) *Zethus slossonae*, (**b**) *Stenodynerus* sp., and the bee (**c**) *Xylocopa micans* use their mandibles to pierce the base of the corolla to access the nectar. Notice how the hairy underside of the abdomen in *X. micans* comes in close contact with the stigma of the flower, as the bee cuts the corolla tissue, potentially leading to pollen transfer. Secondary nectar robbers such as the honeybee (**d**) *Apis mellifera*, the ants (**e**) *Pseudomyrmex gracilis* and (**f**) *Camponotus floridanus*, the true bug (**g**) *Largus succinctus* (nymph), and (**h**) a cockroach (*Blattodea* sp.), drink nectar through holes cut at the base of the corolla tube by a previous visitor. Opportunistic ant visitors such as *P. gracilis* can also feed on nectar from postfloral nectaries (**i**) or as observed in *C. floridanus*, feed on honeydew secreted by aphids (**j**).

Overall, legitimate visitation and nectar robbing were the most common behaviors observed among the flower-visiting species (Figure 7). Fisher's exact test showed those behaviors combined were substantially greater than the others combined ($p < 0.05$), but neither was significantly different from the other, nor were predation and herbivory different from one another. The same patterns were seen at both sites separately. More than half (56%) of flower visitors at ENP and 34% at LPT visited flowers legitimately, and these were mainly lepidopterans (Table 1, Figure 4). Nectar robbing was performed by 32% of the visitors observed at LPT, and by 39% at ENP, mostly Hymenoptera (Table 1, Figure 5). Of the nectar robbers, 75% acted as secondary nectar robbers (Table 1). Herbivory was performed by different groups of insects at both study sites (Table 1, Figure 6), while predation was only witnessed at LPT and performed by spiders and mantises (Table 1, Figure 6).

Figure 6. Overview of predatory and herbivorous arthropods on *Guettarda scabra* in pine rockland habitats in south Florida, USA. (**a**) Crab spiders of the genus *Mecaphesa* in hunting position on a corolla, and (**b**) on an unopened bud. (**c**) Orbweaver spider *Acacesia hamata* sitting on an open flower almost touching the exposed stigma. (**d**) Praying mantis *Stagmomantis floridensis* exploring a branch in the morning. Caterpillars of the erebid moths (**e**) *Calidota laqueata*, (**f**) *Hypercompe scribonia* and (**g**) *Seirarctia echo*, found consuming leaves of *G. scabra*. Other herbivores found associated with flowers include (**h**) clusters of aphids sucking sap from a flower bud, and a (**i**) May beetle *Phyllophaga* sp. chewing on a flower bud at night.

2.3. Resources Consumed by Visitors

Visitors obtained five types of resources from *G. scabra* plants: nectar, pollen, floral tissue, leaves, and small insects attracted to the plant serving as prey (Table 1). Nectar was by far the main resource consumed by the arthropod community overall (Fisher's exact test, $p < 0.01$) as well as in both ENP ($p < 0.01$) and LPT ($p < 0.01$) (Figure 8), mostly Lepidoptera and Hymenoptera (Table 1). A surprising result was that some insects consumed post-floral nectar secreted after the corollas fell, the first time this has been observed in *G. scabra*. Consumption of other resources involved 43% of visitor species at LPT and only 17% of them at ENP (Figure 8). Just as with predation, we did not witness any visitors feeding on floral tissue at ENP.

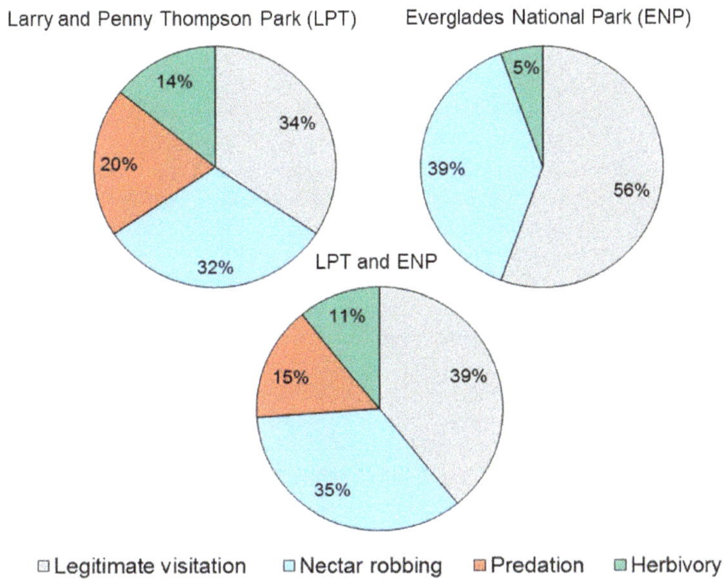

Figure 7. Relative occurrence of the four visitor behaviors observed on flowers of *Guettarda scabra*, among arthropod species in pine rockland habitats in south Florida, USA. Percentages represent the fraction of species observed engaging in a particular behavior on flowers, with respect to the total number of species found in Larry and Penny Thompson LPT (35 species), in Everglades National Park ENP (18 species), or in both sites combined (46 species).

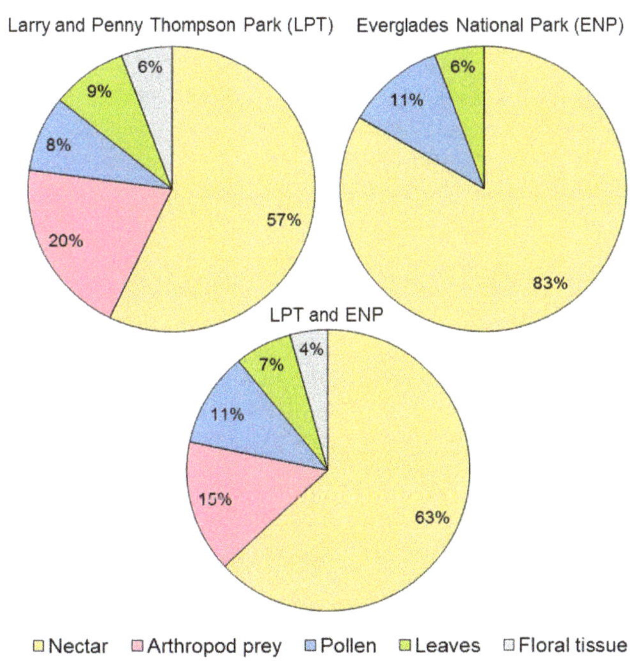

Figure 8. Relative usage of plant resources provided by *Guettarda scabra*, among arthropods in pine rockland habitats in south Florida USA. Percentages represent the fraction of species benefiting from a particular resource, with respect to the total number of species found in Larry and Penny Thompson LPT (35 species), in Everglades National Park ENP (18 species), or in both sites combined (46 species).

2.4. Taxonomic Diversity and Behaviors of Visitors

2.4.1. Lepidoptera

This order contains 15 of the 46 species found visiting flowers of *G. scabra* in both study sites, making it one of two orders of arthropods with the greatest species richness attracted to these flowers. Among Lepidoptera, 11 were skippers (Hesperiidae), butterflies (Nymphalidae, Papilionidae), and moths (Erebidae), while the remaining were hawkmoths (Sphingidae) (Table 1). Among the 15 species of Lepidoptera found, 5 were observed in both study sites (e.g., Figure 4a–c), 7 were seen only in LPT (e.g., Figure 4d–e), and 3 only in ENP (Figure 4f–h) (Table 1).

While most Lepidoptera visited flowers during the day, the hawkmoths (*Eumorpha fasciatus, Perigonia lusca*, and *Xylophanes tersa*) were observed exclusively at night (Figure 3). In general, hawkmoths approached the plants by flying fast through the vegetation and fed only on fresh flowers by hovering above the corolla with their proboscis extended. Moths tended to visit a couple of flowers per plant and then fly away, maintaining a low number of visits per night. Individuals of *E. fasciatus* were often seen hanging motionless on branches of different plants around 2130 h, after visiting flowers (Figure 4h).

In contrast to hawkmoths, butterflies and skippers were observed foraging more frequently and visiting most of the flowers available in a single plant before moving to a nearby individual. Their intensive foraging strategy often resulted in multiple individuals and species feeding simultaneously on a single plant, occasionally even on withered flowers. Butterflies and skippers fed by landing on flowers and inserting their proboscis, and at times part of their head, into the corolla tube to reach the nectar (Figure 4a–f), sometimes resulting in large amounts of pollen being deposited on their mouthparts.

Besides adult lepidopterans, caterpillars of the erebid moths *Calidota laqueata, Hypercompe scribonia*, and *Seirarctia echo* were found feeding on leaves of *G. scabra* (Figure 6e–g). None of the adults of these species were observed visiting flowers.

2.4.2. Hymenoptera

Hymenoptera was the other order with many species visiting flowers of *G. scabra* (30.4% of all recorded visitors). Over half of them (57%) were found exclusively at LPT and only one species (*Zethus slossonae*) observed in both study sites (Table 1). Most Hymenoptera observed were either wasps or bees, while ants were represented by only a few species (Table 1). The ants *Pseudomyrmex gracilis* (Figure 5e) and *Camponotus floridanus* (Figure 5f) were the most frequently found throughout this study (Figure 3), although only at LPT. Notably, both the carpenter bee *Xylocopa micans* (Figure 5c) and the honeybee *Apis mellifera* (Figure 5d) were a common sight in ENP, in contrast to the remaining species.

Hymenopterans were observed to be active exclusively during the day, except for *C. floridanus* which foraged during the night as well (Figure 3). All hymenopterans visiting flowers of *G. scabra* fed on nectar, except for *Dielis trifasciata* (Figure 4i, Table 1) that consumed only pollen by inserting its head into the corolla opening. *Euglossa dilemma* was the only hymenopteran feeding on nectar through the natural opening of the flower while hovering over it, whereas *X. micans, Z. slossonae*, and an unidentified vespid wasp (*Vespidae* sp. 1) actively pierced the base of the corolla to access nectar (Figure 5a,c). Notably, individuals of *X. micans* observed during this study landed on the flowers and positioned themselves facing the base of the corolla, with their abdomen directed toward the flower opening. As the large bee cut the corolla tube, its abdominal hairs were rubbed against the anthers and sometimes the stigma (Figure 5c).

Remaining Hymenoptera acted as secondary nectar robbers, except for the wasp *Stenodynerus* sp. (Figure 5b), the only species behaving as both primary and secondary nectar thief. Ants moved through the plants constantly, exploring flowers to feed on nectar even after corolla abscission, upon which they visited post-floral nectaries (Figure 5i). Whenever scale insects or aphids were present, ants were seen protecting them and feeding on honeydew, which led to some aggressive interactions observed in LPT between *Wasmannia auropunctata* and *P. gracilis*.

2.4.3. Coleoptera

Flowers of *G. scabra* at ENP were visited at night by two species of long-horned beetles (Cerambycidae) that fed on pollen: *Eburia stigma* (Figure 4j) and *Plectromerus dentipes*. These beetles flew through the vegetation visiting one or two flowers per plant, where they were seen feeding on secretions of the stigma and inserting their head into the corolla to reach the pollen on the anthers (Figure 4j). In contrast, plants at LPT were visited by two species of scarab beetles (Scarabaeidae) (Table 1). *Euphoria sepulcralis* (Figure 4k) fed on pollen during the morning (Figure 4k), while *Phyllophaga* sp. was observed consuming open flowers and large buds at night, after 2100 h (Figure 6i).

2.4.4. Other Insect Orders

Observations of flies visiting flowers of *G. scabra* were uncommon (Table 1). There was a single morning sighting of the flower fly *Ornidia obesa* at LPT, during which the fly hovered before landing on flowers to consume pollen through the natural opening of the corolla, contacting the exposed stigma with its mouthparts in the process. Additionally, two species of crane flies (Tipulidae) were found acting as secondary nectar robbers during the day, feeding through holes available at the base of the corolla at both study sites.

The only Hemiptera observed at both study sites were *Largus succinctus*, a secondary nectar robber (Figure 5g). Occasional observations of aphids (Aphididae) extracting sap from flowers and buds only occurred at LPT (Figure 6h). Aphids were often accompanied by *C. floridanus* ants (Figure 5j), and in one instance also by a silver fly *Leucopis* sp. (see [16]). Scale insects (Coccoidea) were also found on inflorescences at LPT, but their presence was not recorded systematically.

Finally, two groups of Orthoptera were found only at LPT: a species of cockroach (Blattodea) acting as a secondary nectar robber and also visiting post-floral nectaries at night (Figure 5h); and two species of praying mantises (Mantidae) perched at the base of the inflorescences during daytime (Figure 6d).

2.4.5. Aranae

Five species of spiders were observed during this study, all of them at LPT: the orbweaver spider *Acacesia hamata* and four crab spiders of the genus *Mecaphesa* (Table 1). All spiders were observed sitting on the corolla, close to the pistil in both fresh and withered flowers, as well as on inflorescences with unopened buds (Figure 6a–c). They were observed either capturing small insects or resting on a flower with their front legs held out to each side of their body, a characteristic pose in this group.

3. Discussion

Although it was previously assumed that *G. scabra* is a moth-pollinated plant, our findings show that their flowers are visited by a wide array of arthropods that can act as pollinators, most of them diurnal. Such diversity is not surprising, since nearly 30% of arthropods species visit flowers regularly and potentially pollinate them [17]. Likewise, differences in diversity of visitors between night and day occur in many other plants whose flowers exhibit sphingophily, particularly the diurnal dominance of Hymenoptera and Lepidoptera that we observed in *G. scabra* (e.g., [18–21]), both groups being the largest insect taxa containing important pollinators [17].

In general, night-blooming species whose flowers remain open into the morning may be attractive to diurnal visitors, especially those unvisited flowers that accumulated nectar through the night [22]. Diurnal visitation of nocturnal flowers by a variety of animals has been reported across different families of plants. Examples highlighting the taxonomic diversity of plants include species of the families Caprifoliaceae and Cactaceae that are visited by bees [19,23], Apocynaceae and Rubiaceae by bees and butterflies [18,24], and Bromeliaceae by bees, ants, and flies [25]. The availability of nectar in the morning can even attract hummingbirds, as observed in Bromeliaceae [25] and Rubiaceae [26]. In the latter family, the genus *Guettarda* contains several species with this pattern of anthesis in which sphingophylous flowers remain open through the morning making nectar and pollen available to diurnal visitors. Observations on *G. speciosa* in south China [27] and *G. clarensis* in Cuba [28] revealed that both species were visited by a single local species of nocturnal hawkmoth and some diurnal insects, mostly lepidopterans, hymenopterans, and dipterans. While these two species were visited by both nocturnal and diurnal insects, *G. platypoda* in Brazil was solely visited by nocturnal moths of three species [29]. These observations contrast with our findings, since *G. scabra* flowers are visited by a much larger number of species during the day and night. However, the frequency of visits by nocturnal hawkmoths was as low as in *G. platypoda* [29] and *G. speciosa* [27] (M.C.P. unpublished observations). Attracting a larger set of flower visitors may be advantageous for *G. scabra*, as non-hawkmoth visitors may provide some pollination when specialized hawkmoth pollinators are not available.

3.1. Occurrence

Our findings suggest that the flowers of *G. scabra* are visited by a community of arthropods whose structure differs between study sites. These differences may be linked to variations in the availability of biotic components of the ecosystem that depend on the presence of particular species of arthropods. Some of the species visiting flowers of *G. scabra* may require other resources that can vary between study sites, such as the presence of host plant species in the case of Lepidoptera, or nesting and shelter spaces for other arthropods. Carpenter bees (*X. micans*), for example, rely on the availability of dead wood they need to build their nests [30]. The scarcity of this resource might explain the absence of this species in LPT. On the other hand, our observations of the skipper *E. brunnea* in both study sites are clearly related to the availability of its host plant *Byrsonima lucida* (Malpighiaceae) [31] in both areas.

Surprisingly, the lepidopterans *Heliconius charithonia, Polites baracoa, Cymaenes tripunctus,* and *Papilio polyxenes*, which were all reported present all year round in the Long Pine Key area of ENP more than 40 years ago [31], were not observed in that area during this study, although we did observe them visiting flowers in LPT. Other notable absences in ENP include the caterpillars of three erebid moths (*Seirarctia echo*, *Spilosoma virginica,* and *Pyrrharctia isabella*), a paper wasp (*Mischocyttarus* sp.) and a species of flower fly (*Copestylum mexicanum*) seen visiting flowers of *G. scabra* over 30 years ago [14]. However, *S. echo* and a species of *Mischocyttarus* were found on *G. scabra* in LPT.

Interestingly, almost 25% of the total number of arthropod species found visiting flowers of *G. scabra* were only at ENP, a site that we undersampled with respect to LPT. While the sampling effort was different enough between both sites to prevent us from drawing any solid conclusions, the high proportion of species found only in ENP suggests that the communities of floral visitors are indeed different between study sites. It is possible that the arthropod community associated with *G. scabra* flowers in south Florida is even more taxonomically diverse than reported here.

3.2. Potential Pollinators

Guettarda scabra is visited by a wide range of potential pollinators besides lepidopterans. In fact, plants whose flowers fit a particular pollination syndrome may still receive visits from opportunistic insects capable of contributing to their fitness [3,23,32].

Due to floral morphology in *G. scabra*, most of their visitors with short mouthparts (such as bees, wasps, flies, and beetles) encounter anthers, stigma, or both while foraging, potentially serving as pollen vectors for this plant. Since anthers in flowers (of all morphs) of *G. scabra* are located at the opening of the corolla, short-tongued visitors can access pollen grains in any open flower and may then transfer them to flowers with exserted stigmas. Successful pollination of flowers with long corollas by short-tonged insects has been observed in other distylous Rubiaceae, such as *Psychotria homalosperma*. While that plant is presumably pollinated by long-tongued moths, in their absence, honeybees (*Apis mellifera*) manage to pollinate it with their short mouthparts by moving pollen unidirectionally from short- to long-styled flowers [33]. We think that a similar scenario occurs in *G. scabra*, where both short- and long-tongued visitors may promote pollination.

While short-tongued visitors could contribute to reproduction in *G. scabra*, floral traits in this species suggest the existence of a most effective pollinator with long mouthparts capable of encountering the stigma, regardless of how deep in the corolla it is located. Regarding the identity of such pollinator, previous authors have found nocturnal hawkmoths to be the main pollen vectors for other species in the genus *Guettarda*, such as *G. platypoda* [24,29], *G. speciosa* [27], and *G. clarensis* [28]. In fact, the nocturnal hawkmoth *X. tersa* was a common flower visitor of *G. scabra* during this study, and the same hawkmoth was the most frequent pollinator for *G. platypoda* in Brazil [29], suggesting a particular association between this moth and *Guettarda* plants when both are present. Besides lepidopterans, the bee *E. dilemma* was the only other visitor with a tongue long enough to reach nectar deep in the corolla of the flowers of *G. scabra*. This bee has a mutualistic relationship with orchids in its Central American native range, and was recently introduced to south Florida, where it has been reported (as *E. viridissima*) visiting a wide variety of non-orchid plants [34]; (Brittany M. Harris, personal communication). Our study provides the first record of *E. dilemma* visiting flowers of *G. scabra*.

Attracting different types of potential pollinators could enhance fruit production in areas where the most effective pollinator is absent or scarce [19], and/or when weather conditions disrupt foraging activity [35]. This plant has survived in the highly fragmented pine rockland habitat, being regularly exposed to extreme weather events such as heavy rainfall and flooding, hurricanes, and fire. Despite the low frequency of visits by nocturnal hawkmoths locally (M.C.P. unpublished observations), flowers of *G. scabra* may increase their chances of being pollinated by receiving visits from other pollen vectors observed during this study. In fact, day-active flower visitors may complement the effect of nocturnal ones in this species, as has been suggested by indirect observations [36].

3.3. Nectar Robbing: A Common Behavior

The fact that nectar robbing was a very common behavior observed among the floral visitors of *G. scabra* agrees with other instances in which more than half of the species of flower visitors are nectar robbers [37]. This behavior is known to happen in other species of the genus *Guettarda*. In *G. clarensis*, for example, nectar robbing reduces fruit production, negatively impacting reproduction [38]. Interestingly, the main robbers in *G. clarensis* (*Largus sellatus* and *Xylocopa cubaecola*) belong to the same genera as two common robbers we found in *G. scabra* (*L. succinctus* and *X. micans*) [39]. Nectar robbing has also been reported in *G. speciosa* [27], but there is no detailed account of these observations. In *G. scabra*, we did not observe damage caused by nectar robbers on sexual structures of the flowers (i.e., pistil or stamens), which could directly interfere with pollination, but it is unknown whether robbing can affect reproduction in this species.

While nectar robbing may be detrimental for plant reproduction [40], under certain conditions it may also have positive effects [41], such as contributing to pollination. Some *Xylocopa* bees for example, have been reported robbing nectar from plants with long tubular flowers [33,42–46]. In certain cases, they have been seen touching the anthers and stigma of flowers as they feed, promoting pollination [44–47]. Our observations on the foraging behavior of *X. micans* suggest that these bees may transfer pollen in *G. scabra* during nectar robbing. However, no other nectar robber observed during this study behaved or positioned its body in a way that could result in pollen transfer while they were feeding.

In addition to robbers depositing pollen, they may benefit the plants they rob in another way: by causing floral visitors to visit fewer flowers and move to other plants more quickly [48]. This is especially beneficial in plants that are self-incompatible [49] but may be important in avoiding inbreeding depression in those that are self-compatible as well by reducing geitonogamy [50]. As *G. scabra* is self-compatible, it may benefit from the actions of its numerous nectar robbers.

3.4. Guettarda scabra as Food Source for Local Arthropod Fauna

Floral resources can be a limiting factor in many habitats during a particular season. In pine rocklands, most species flower from January to April [51] and initiate fruit formation during summer [52], reducing the availability of floral resources during this time. In contrast, most *G. scabra* individuals are fully in flower in June and July, when few other species are flowering, making them a valuable source of floral rewards. Our findings suggest that *G. scabra* may be a keystone species in the pine rockland habitat of south Florida, as it is an important source of food and foraging grounds for the local arthropod fauna during its flowering season. Flowers of this plant provide highly nutritious resources in the form of pollen and nectar to visitors, as well as flower parts and leaves for herbivores, making this plant attractive to a large variety of arthropods with diverse natural histories.

In fact, *G. scabra* flower rewards are used by wasps found only in Florida, such as *Z. slossonae* [53] and *D. trifasciata* [54]. Flowers of *G. scabra* also provide nectar for adult lepidopterans with distributions restricted to the southern half of Florida, such as *Perigonia lusca* [55] and *Cymaenes tripunctus* [56], along with *Ephyriades brunnea* whose populations have declined in recent years [57]. Such a critical role in the maintenance of the local pollinator fauna was observed also in *G. platypoda* in Brazil, where hawkmoth communities rely on its nectar as an energy source [24]. Although most adult lepidopterans visit *G. scabra* to feed on nectar, *H. charithonia* probably also consumes pollen, a resource reported as part of its diet [58–60]. Besides the erebid moths reported in this study, *G. scabra* is the host plant for caterpillars of other species of moths in south Florida, such as *Spilosoma virginica* and *Pyrrharctia isabella* [14], as well as the hawkmoths *P. lusca* and *Eupyrrhoglossum sagra* [55,61].

Our observations also suggest that the pollen of *G. scabra* is an important food source for local populations of some long-horned and scarabeid beetles. In fact, scarabeids may rely on more than pollen from this plant, since at least *Phyllophaga* sp. was observed consuming its flowers during this study. It is also possible that *Euphoria sepulcralis* feeds on flower tissue of *G. scabra* as well, based on field observations of this species consuming flowers of other plants in LPT, including *Bidens* sp. (Asteraceae), *Spermacoce* sp. (Rubiaceae), and *Lantana* sp. (Verbenaceae) (M.C.P. personal observations), and occasional reports of this species as flower-damaging pest in some fruit trees in south Florida [62].

Besides insects, spiders may spend time on flowers benefiting from food sources other than prey. Spiders can feed on stigma exudates, nectar, and pollen [63–68]. While we did not witness this behavior directly, we often saw individual spiders sitting on the corolla, with their mouthparts very close to the stigma, anthers, or postfloral nectaries. Considering that the stigma of *G. scabra* remains moist throughout anthesis, and even after the corolla tube is wilted, spiders may have been feeding on stigmatic exudates. Interestingly, most of the spiders observed on flowers of *G. scabra* belong to the family Thomisidae, a group also commonly observed on flowers of *G. clarensis* [28].

The effect of predatory visitors on the reproductive success of *G. scabra* is unknown. In general, predators can harmfully disrupt pollination by consuming pollen vectors [69,70] or decreasing the frequency and duration of their visits [71–75]. Sometimes predators may benefit plants by causing pollinators to move between plants more [76], promoting outcrossing [77] as can nectar robbers [48]. At the same time, they can benefit the plant by decimating insects feeding on it [78]. In fact, some of the wasps observed during this study are known to attack phytophagous larvae, such as *Pachodynerus erynnis* that feeds on caterpillars of several families [79,80], or *D. trifasciata* which parasitizes larvae of the beetle *Phyllophaga portoricensis* [81]. Interestingly, we found a species of *Phyllophaga* consuming flowers of *G. scabra*, raising the question of whether *D. trifasciata* can control the local population of this beetle and benefit *G. scabra* in the process.

4. Materials and Methods

4.1. Plant Species

The rough-leaf velvetseed *Guettarda scabra* (Rubiaceae) is a tropical shrub usually less than 1.5 m tall when it grows in pine rockland forests in south Florida. Its blooming season begins in April and peaks between May and July [12]. Plants resprout after fire, but do not bloom the summer after burning, taking two years from fire until blooming again [36]. Flowers are white, often with a pink-flushed corolla tube, about 2 cm long that holds nectar at its base (Figure 1a). Flowers are arranged in dichasial cymose inflorescences and open sequentially over several weeks, usually one to three flowers per inflorescence per day, releasing a strong, sweet scent. Anthesis occurs during late afternoon and flowers remain fresh through the following morning [10]. Flower senescence occurs usually by noon, when the corolla turns brown and dehydrates, remaining attached to the calyx for about a day [10].

Guettarda scabra exhibits a special case of distyly, in which both the anther height and style length vary continuously in the population [10]. Plants are self-compatible, sometimes setting fruit without visitation, but pollen vectors are required for greater fruit production [10].

4.2. Study Sites

This study was conducted in two natural areas in Miami-Dade County, Florida, USA: (a) Larry and Penny Thompson Memorial Park (LPT), a county park containing the largest fragment of pine rockland habitat in the city of Miami (25°35′55″ N 80°23′55″ W); and (b) the Long Pine Key area (25°24′13.2″ N 80°39′33.2″ W), within a large, continuous pine rockland forest in Everglades National Park (ENP) (Figure 1b). The pine rockland habitat is unique to south Florida and the Caribbean and is considered critically imperiled due to a substantial loss of its original extent [82,83]. Although the objective of this study was not to compare the two sampling sites, for some aspects the data are shown separately to discuss general trends.

Rockland habitats are greatly reduced from their original extent as they have undergone extensive human development over the last century [83–85]. Pine rocklands are considered globally imperiled [86] with many endemic plant taxa in the diverse understory of more than 225 native plant species, of which 10% are considered threatened or endangered at the state level, eight of which are federally endangered [87].

4.3. Flower Visitor Observations

We surveyed arthropods visiting *G. scabra* flowers and/or feeding on the plant during the blooming seasons of 2016, 2018, and 2019 (17, 3, and 31 days respectively) at LPT, and during 2018 and 2019 (5 and 3 days respectively) at ENP. Observations were carried out on groups of plants with open flowers for 30 min at a time, three times a day. Surveys done between 0700–1200 h were considered to have been performed in the morning, 1800–2019 h in the evening, and 2020–2300 h at night. Nocturnal observations were made using red light lanterns to minimize disturbing the behavior of insect visitors. A total of 75 of these observation periods were conducted in LPT (48 mornings, 20 evenings and 7 nights) on 25 plants, and 11 in ENP (2 mornings, 5 evenings and 4 nights) on 20 plants. Additionally, visitors spotted on flowers of *G. scabra* incidentally while walking through the study sites were recorded. The data reported are the number of observation periods in which each type of visitor was observed.

All arthropods observed touching flowers were considered floral visitors. Due to the potential relevance of lepidopterans in the pollination biology of *G. scabra*, caterpillars feeding on plants were documented, collected, and reared for species determination. Flower visitors were recorded, noting their time of activity and behavior (harvesting reward, contacting sexual organs of the flower, and interacting with other species), and photographed if possible. When necessary, voucher specimens were preserved to confirm identification. These specimens will be deposited in the Florida State Collection of Arthropods (Gainesville, FL, USA).

4.4. Statistical Comparisons

To evaluate the relative importance of different groups of visitors, their behaviors, resources utilized, and activity periods, we used Fisher's exact test (which is appropriate for small sample sizes) to compare the numbers of species associated with each of those parameters. We used a significance level of $p < 0.05$ for single comparisons and $p < 0.01$ for multiple.

5. Conclusions

Although *G. scabra* flowers have traits traditionally associated with attracting nocturnal moths, they open in the evening and remain open into the morning, luring in a much wider array of floral visitors. Despite recent work on the diversity of flower-visiting arthropods in the Everglades [88–90] and pollination of plants in the pine rockland habitat [91–96], little is known about the entire array of flower visitors to any particular plant species. The maintenance of healthy pine rockland habitat requires periodic fires to prevent succession to hardwood hammock forest [84], and in the open pine rockland understory *G. scabra* grow relatively free of competition from other hardwoods, investing much energy into flowering [6]. This study constitutes the first in-depth survey of insects and arachnids associated with the abundant flowers of *G. scabra* in this habitat.

Our findings show that *G. scabra* is not only visited by nocturnal hawkmoths as expected, but many other potential pollen vectors, beyond those predicted by its pollination syndrome. Our observations also suggest that this plant provides an important foraging and food resource for the local arthropod fauna. Our research provides baseline data on the local arthropod fauna associated with a native plant species, along with insights into the complexity of trophic interactions in the pine rockland habitat. There are 147 recognized species of the genus worldwide [97], but no species of *Guettarda* are considered rare, and those that are ranked by conservation organizations are apparently secure, the habitats in which many occur are imperiled or unranked and threatened in ways similar to the pine rocklands. The richness of floral visitors to *G. scabra* and the critical role this plant may play in sustaining that community indicates that plants may host a wide array of arthropods, regardless of the presence of adaptations suggesting coevolution with a much narrower set of visitors. Our observations on the natural history of *G. scabra* offer a glimpse of how intricate plant-animal interactions can be. For threatened habitats such as the pine rocklands in south Florida, studies like this yield needed information to support efforts to conserve and protect them along with their associated diversity of plants and animals.

Author Contributions: Conceptualization, M.C.P. and S.K.; methodology, M.C.P.; formal analysis, M.C.P.; investigation, M.C.P.; resources, M.C.P. and S.K.; data curation, M.C.P.; writing—original draft preparation, M.C.P.; writing—review and editing, S.K.; visualization, M.C.P.; supervision, S.K.; project administration, S.K.; funding acquisition, M.C.P. All authors have read and agreed to the published version of the manuscript.

Funding: This research was funded by grants to M.C.P. by The Botanical Society of America, The Kelly Foundation for Tropical Botany, and FIU Tropics.

Data Availability Statement: The data that support the findings of this study are openly available in the FIU Research Data Portal at https://doi.org/10.34703/gzx1-9v95/3BRPS3 (accessed on 19 October 2022).

Acknowledgments: We thank Carlos Ruiz for assistance in the field, constructive advice during the development of this research, and reviewing the manuscript. We also thank Diego Salazar Amoretti, Jamie Theobold, Florence George, Javier Francisco Ortega, and Brittany Harris for insightful comments on the manuscript. This research was conducted under Permit # 181R from Natural Areas Management, Miami-Dade County Park and Recreation, and Everglades National Park Research Permit # EVER-2018-SCI-0012. We thank both agencies for access to the study sites and excellent management of the pine rocklands. This is contribution #1498 from the Institute of Environment at Florida International University.

Conflicts of Interest: The authors declare no conflict of interest.

References

1. Waser, N.M.; Chittka, L.; Price, M.V.; Williams, N.M.; Ollerton, J. Generalization in pollination systems, and why it matters. *Ecology* **1996**, *77*, 1043–1060. [CrossRef]
2. Fleming, T.H.; Sahley, C.T.; Holland, J.N.; Nason, J.D.; Hamrick, J.L. Sonoran Desert columnar cacti and the evolution of generalized pollination systems. *Ecol. Monogr.* **2001**, *71*, 511–530. [CrossRef]
3. De Merxem, D.G.; Borremans, B.; De Jäger, M.L.; Johnson, T.; Jooste, M.; Ros, P.; Zenni, R.D.; Ellis, A.G.; Anderson, B. The importance of flower visitors not predicted by floral syndromes. *S. Afr. J. Bot.* **2009**, *75*, 660–667. [CrossRef]
4. Miyake, T.; Yahara, T. Theoretical evaluation of pollen transfer by nocturnal and diurnal pollinators: When should a flower open? *Oikos* **1999**, *86*, 233–240. [CrossRef]
5. Acevedo-Rodríguez, P.; Strong, M.T. *Catalogue of Seed Plants of the West Indies*; Smithsonian Institution: Washington, DC, USA, 2012. [CrossRef]
6. Koptur, S.; Garcia, D. Habitat differences in morphology and reproductive allocation in *Guettarda scabra* (Rubiaceae). *Castanea* **2017**, *82*, 51–57. [CrossRef]
7. Roberts, A. *Guettarda scabra*. The IUCN Red List of Threatened Species. 2014. e.T56503696A56503850. Available online: https://www.iucnredlist.org/species/56503696/56503850 (accessed on 19 October 2022).
8. WCSP: World Checklist of Selected Plant Families. Facilitated by the Royal Botanic Gardens, Kew. 2022. Available online: https://wcsp.science.kew.org (accessed on 19 October 2022).
9. Davis, A.P.; Govaerts, R.; Bridson, D.M.; Ruhsam, M.; Moat, J.; Brummitt, N.A. A global assessment of distribution, diversity, endemism, and taxonomic effort in the Rubiaceae 1. *Ann. Mo. Bot. Gard.* **2009**, *96*, 68–78. [CrossRef]
10. Richards, J.H.; Koptur, S. Floral variation and distyly in *Guettarda scabra* (Rubiaceae). *Am. J. Bot.* **1993**, *80*, 31–40. [CrossRef]
11. Faegri, K.; van der Pijl, L. *The Principles of Pollination Ecology*; Pergamon Press: Oxford, UK, 1979.
12. Tomlinson, P.B. *The Biology of Trees Native to Tropical Florida*; Harvard University Printing Office: Allston, MA, USA, 1980.
13. Austin, D.F. *Florida Ethnobotany*; CRC Press: Boca Raton, FL, USA, 2004.
14. Koptur, S. Scientific Note: Insects associated with *Guettarda scabra* in Everglades National Park, Florida. *Castanea* **2020**, *85*, 155–158. [CrossRef]
15. Wagner, D.L.; Grames, E.M.; Forister, M.L.; Berenbaum, M.R.; Stopak, D. Insect decline in the Anthropocene: Death by a thousand cuts. *Proc. Natl. Acad. Sci. USA* **2021**, *118*, e2023989118. [CrossRef] [PubMed]
16. Ruiz, C.; Pimienta, M.C. Behavior of adult *Leucopis* sp. (Chamaemyiidae) associated with aphids feeding on flowers of the rough-leaf velvetseed (*Guettarda scabra*: Rubiaceae) in south Florida. *Fly Times* **2019**, *62*, 8–10. Available online: https://www.nadsdiptera.org/News/FlyTimes/issue62.pdf (accessed on 19 October 2022).
17. Wardhaugh, C.W. How many species of arthropods visit flowers? *Arthropod-Plant Interact.* **2015**, *9*, 547–565. [CrossRef]
18. Maruyama, P.K.; Amorim, F.W.; Oliveira, P.E. Night and day service: Distyly and mixed pollination system in *Faramea cyanea* (Rubiaceae). *Flora Morphol. Distrib. Funct. Ecol. Plants* **2010**, *205*, 818–824. [CrossRef]
19. Locatelli, E.; Machado, I.C.S. Floral biology of *Cereus fernambucensis*: A sphingophilous cactus of restinga. *Bradleya* **1999**, *17*, 86–94. [CrossRef]
20. Walter, H.E. Floral biology of *Echinopsis chiloensis* ssp. *chiloensis* (Cactaceae): Evidence for a mixed pollination syndrome. *Flora Morphol. Distrib. Funct. Ecol. Plants* **2010**, *205*, 757–763. [CrossRef]
21. Young, H.J. Diurnal and nocturnal pollination of *Silene alba* (Caryophyllaceae). *Am. J. Bot.* **2002**, *89*, 433–440. [CrossRef]
22. Haber, W.A.; Frankie, G.W. A tropical hawkmoth community: Costa Rican dry forest Sphingidae. *Biotropica* **1989**, *21*, 155–172. [CrossRef]
23. Miyake, T.; Yahara, T. Why does the flower of *Lonicera japonica* open at dusk? *Can. J. Bot.* **1998**, *76*, 1806–1811. [CrossRef]
24. Darrault, R.O.; Schlindwein, C. Esfingídeos (Lepidoptera, Sphingidae) no Tabuleiro Paraibano, nordeste do Brasil: Abundância, riqueza e relação com plantas esfingófilas. *Rev. Bras. Zool.* **2002**, *19*, 429–443. [CrossRef]
25. Aguilar-Rodríguez, P.A.; Krömer, T.; García-Franco, J.G.; MacSwiney, G.M.C. From dusk till dawn: Nocturnal and diurnal pollination in the epiphyte *Tillandsia heterophylla* (Bromeliaceae). *Plant Biol.* **2016**, *18*, 37–45. [CrossRef]
26. Wolff, D.; Braun, M.; Liede, S. Nocturnal versus diurnal pollination success in *Isertia laevis* (Rubiaceae): A sphingophilous plant visited by hummingbirds. *Plant Biol.* **2003**, *5*, 71–78. [CrossRef]
27. Xu, Y.; Luo, Z.; Gao, S.; Zhang, D. Pollination niche availability facilitates colonization of *Guettarda speciosa* with heteromorphic self-incompatibility on oceanic islands. *Sci. Rep.* **2018**, *8*, 13765. [CrossRef]
28. Martínez, L. Fenología Reproductiva y Efecto del Robo de Néctar en el Éxito Reproductivo de *Guettarda clarensis*, en. Bachelor's Thesis, Universidad Central "Marta Abreu" de Las Villas, Santa Clara, Villa Clara, Cuba, 2013. Available online: https://dspace.uclv.edu.cu/handle/123456789/1723 (accessed on 19 October 2022).
29. Novo, R.R.; Consolaro, H.; Almeida, N.M.; Castro, C.C. Floral biology of the velvetseed *Guettarda platypoda* DC. (Rubiaceae): Atypical distyly or style dimorphism? *Flora Morphol. Distrib. Funct. Ecol. Plants* **2018**, *239*, 62–70. [CrossRef]
30. Warriner, M.D. A range extension for the large carpenter bee *Xylocopa micans* (Hymenoptera: Apidae) with notes on floral and habitat associations. *J. Kans. Entomol. Soc.* **2010**, *83*, 267–269. [CrossRef]
31. Lenczewski, B. *Butterflies of Everglades National Park*; Report T-588; South Florida Research Center: Homestead, FL, USA, 1980; 110p.
32. Fishbein, M.; Venable, D.L. Diversity and temporal change in the effective pollinators of *Asclepias tuberosa*. *Ecology* **1996**, *77*, 1061–1073. [CrossRef]

33. Watanabe, K.; Kato, H.; Kuraya, E.; Sugawara, T. Pollination and reproduction of *Psychotria homalosperma*, an endangered distylous tree endemic to the oceanic Bonin (Ogasawara) Islands, Japan. *Plant Species Biol.* **2018**, *33*, 16–27. [CrossRef]
34. Pemberton, R.W.; Wheeler, G.S. Orchid bees don't need orchids: Evidence from the naturalization of an orchid bee in Florida. *Ecology* **2006**, *87*, 1995–2001. [CrossRef]
35. Koptur, S. Flowering phenology and floral biology of *Inga* (Fabaceae: Mimosoideae). *Syst. Bot.* **1983**, *8*, 354–368. [CrossRef]
36. Koptur, S.; Peña, S.; Barrios, R.B. Do morning butterfly visitors benefit a night-flowering hawkmoth pollinated plant? *Castanea* **2021**, *86*, 100–111. [CrossRef]
37. Genini, J.; Morellato, L.P.C.; Guimarães, P.R., Jr.; Olesen, J.M. Cheaters in mutualism networks. *Biol. Lett.* **2010**, *6*, 494–497. [CrossRef]
38. Martínez-Pérez, L.; Faife-Cabrera, M. Nectar Robbing in *Guettarda clarensis* (Rubiaceae): Does Floral Neighborhood Matter? *Rev. Del. Jardín Botánico Nac.* **2019**, *40*, 47–57. Available online: https://www.rjbn.uh.cu/index.php/RJBN/article/view/443/465 (accessed on 19 October 2022).
39. Martínez, L. Relación del Vecindario Floral Con la Frecuencia de Robo de Néctar en *Guettarda clarensis* (Rubiaceae). Master's Thesis, Universidad Central "Marta Abreu" de Las Villas, Santa Clara, Villa Clara, Cuba, 2017. Available online: https://dspace.uclv.edu.cu/handle/123456789/10813 (accessed on 19 October 2022).
40. Schlindwein, C.; Westerkamp, C.; Carvalho, A.T.; Milet-Pinheiro, P. Visual signalling of nectar-offering flowers and specific morphological traits favour robust bee pollinators in the mass-flowering tree Handro-anthus impetiginosus (Bignoniaceae). *Bot. J. Linn. Soc.* **2014**, *176*, 396–407. [CrossRef]
41. Irwin, R.E.; Bronstein, J.L.; Manson, J.S.; Richardson, L. Nectar robbing: Ecological and evolutionary perspectives. *Annu. Rev. Ecol. Evol. Syst.* **2010**, *41*, 271–292. [CrossRef]
42. Barrows, E.M. Robbing of exotic plants by introduced carpenters and honeybees in Hawaii, with comparative notes. *Biotropica* **1980**, *12*, 23–29. [CrossRef]
43. Dedej, S.; Delaplane, K.S. Nectar-robbing carpenter bees reduce seed-setting capability of honey bees (Hymenoptera: Apidae) in rabbiteye blueberry, *Vaccinium ashei*, 'Climax'. *Environ. Entomol.* **2004**, *33*, 100–106. [CrossRef]
44. Guitián, J.; Sánchez, J.M.; Guitián, P. Pollination ecology of *Petrocoptis grandiflora* Rothm. (Caryophyllaceae); a species endemic to the north-west part of the Iberian Peninsula. *Bot. J. Linn. Soc.* **1994**, *115*, 19–27. [CrossRef]
45. Schmid, S.; Schmid, V.S.; Zillikens, A.; Steiner, J. Diversity of flower visitors and their role for pollination in the ornithophilous bromeliad *Vriesea friburgensis* in two different habitats in southern Brazil. *Ecotropica* **2011**, *17*, 91–102.
46. Zhang, Y.W.; Robert, G.W.; Wang, Y.; Guo, Y.H. Nectar robbing of a carpenter bee and its effects on the reproductive fitness of *Glechoma longituba* (Lamiaceae). *Plant Ecol.* **2007**, *193*, 1–13. [CrossRef]
47. Scott, P.E.; Buchmann, S.L.; O'Rourke, M.K. Evidence for mutualism between a flower-piercing carpenter bee and ocotillo: Use of pollen and nectar by nesting bees. *Ecol. Entomol.* **1993**, *18*, 234–240. [CrossRef]
48. Gottsberger, G. Some pollination strategies in neotropical savannas and forests. *Plant Syst. Evol.* **2004**, *152*, 29–45. [CrossRef]
49. Singh, V.K.; Barman, C.; Tandon, R. Nectar Robbing Positively Influences the Reproductive Success of *Tecomella undulata* (Bignoniaceae). *PLoS ONE* **2014**, *9*, e102607. [CrossRef]
50. Maloof, J. The effects of a bumble bee nectar robber on plant reproductive success and pollinator behavior. *Am. J. Bot.* **2001**, *88*, 1960–1965. [CrossRef]
51. Loope, L.L. *Phenology of Flowering and Fruiting in Plant Communities of Everglades National Park and Biscayne National Monument, Florida*; Report T-593; South Florida Research Center: Homestead, FL, USA, 1980; 50p.
52. Gunderson, L.; Taylor, D.; Craig, J. *Fire Effects on Flowering and Fruiting Patterns of Understory Plants in Pinelands of Everglades National Park*; Report SFRC-83/04; South Florida Research Center: Homestead, FL, USA, 1983; 36p.
53. Grissell, E.E. *Mason Wasps of Florida, Zethus spp. (Insecta: Hymenoptera: Vespidae: Eumeninae)*; Entomology Circular 153; Florida Department of Agriculture and Consumer Services, Division of Plant Industry, UF/IFAS Extension: Gainesville, FL, USA, 2021; pp. 1–3.
54. Grissell, E.E. *Scoliid Wasps of Florida, Campsomeris, Scolia and Trieliss pp. (Insecta: Hymenoptera: Scoliidae)*; Entomology Circular 179 and 185; Florida Department of Agriculture and Consumer Services, Division of Plant Industry, UF/IFAS Extension: Gainesville, FL, USA, 2017; pp. 1–14.
55. Tuttle, J.P. *The Hawk Moths of North America: A Natural History Study of the Sphingidae of the United States and Canada*; Wedge Entomological Research Foundation: Washington, DC, USA, 2007.
56. Warren, A.D.; Davis, K.J.; Grishin, N.V.; Pelham, J.P.; Stangeland, E.M. Interactive Listing of American Butterflies. 2012. Available online: https://www.butterfliesofamerica.com/ (accessed on 19 October 2022).
57. Daniels, J. Florida butterflies. *Wings: Essays Invertebr. Conserv.* **2010**, *33*, 18–21. Available online: https://xerces.org/wings (accessed on 19 October 2022).
58. Boggs, C.L. Nutritional and life-history determinants of resource allocation in holometabolous insects. *Am. Nat.* **1981**, *117*, 692–709. [CrossRef]
59. Gilbert, L.E. Pollen feeding and reproductive biology of *Heliconius* butterflies. *Proc. Natl. Acad. Sci. USA* **1972**, *69*, 1403–1407. [CrossRef] [PubMed]
60. O'Brien, D.M.; Boggs, C.L.; Fogel, M.L. Pollen feeding in the butterfly *Heliconius charitonia*: Isotopic evidence for essential amino acid transfer from pollen to eggs. *Proc. R. Soc. London. Ser. B Biol. Sci.* **2003**, *270*, 2631–2636. [CrossRef] [PubMed]

61. Slotten, J.R.; Miller, W. Occurrence of *Eupyrrhoglossum sagra* and *Perigonia lusca* in Florida (Lepidoptera: Sphingidae). *Holarct. Lepid.* **2000**, *7*, 59–63.
62. Thomas, M.C. *A Flower Beetle, Euphoria Sepulcralis (Fabricius) (Insecta: Coleoptera: Scarabaeidae)*; Entomology Circular 386; Florida Department of Agriculture and Consumer Services, Division of Plant Industry, UF/IFAS Extension: Gainesville, FL, USA, 2019; pp. 1–6.
63. Jackson, R.R.; Pollard, S.D.; Nelson, X.J.; Edwards, G.B.; Barrion, A.T. Jumping spiders (Araneae: Salticidae) that feed on nectar. *J. Zool.* **2001**, *255*, 25–29. [CrossRef]
64. Marquínez, X.; Cepeda, J.; Lara, K.; Sarmiento, R. Arañas asociadas a la floración de *Drimys granadensis* (Winteraceae). *Rev. Colomb. Entomol.* **2010**, *36*, 172–175. [CrossRef]
65. Nahas, L.; Gonzaga, M.O.; Del-Claro, K. Wandering and web spiders feeding on the nectar from extrafloral nectaries in neotropical savanna. *J. Zool.* **2017**, *301*, 125–132. [CrossRef]
66. Nyffeler, M.; Olson, E.J.; Symondson, W.O. Plant-eating by spiders. *J. Arachnol.* **2016**, *44*, 15–27. [CrossRef]
67. Taylor, R.M.; Foster, W.A. Spider nectarivory. *Am. Entomol.* **1996**, *42*, 82–86. [CrossRef]
68. Vogelei, A.; Greissl, R. Survival strategies of the crab spider *Thomisus onustus* Walckenaer 1806 (Chelicerata, Arachnida, Thomisidae). *Oecologia* **1989**, *80*, 513–515. [CrossRef] [PubMed]
69. Dukas, R. Effects of perceived danger on flower choice by bees. *Ecol. Lett.* **2001**, *4*, 327–333. [CrossRef]
70. Morse, D.H. Choice of hunting site as a consequence of experience in late-instar crab spiders. *Oecologia* **1999**, *120*, 252–257. [CrossRef]
71. Dukas, R.; Morse, D.H. Crab spiders affect flower visitation by bees. *Oikos* **2003**, *101*, 157–163. [CrossRef]
72. Llandres, A.L.; De Mas, E.; Rodriguez-Girones, M.A. Response of pollinators to the tradeoff between resource acquisition and predator avoidance. *Oikos* **2012**, *121*, 687–696. [CrossRef]
73. Robertson, I.C.; Maguire, D.K. Crab spiders deter insect visitations to slickspot peppergrass flowers. *Oikos* **2005**, *109*, 577–582. [CrossRef]
74. Romero, G.Q.; Antiqueira, P.A.; Koricheva, J. A meta-analysis of predation risk effects on pollinator behaviour. *PLoS ONE* **2011**, *6*, e20689. [CrossRef] [PubMed]
75. Suttle, K.B. Pollinators as mediators of top-down effects on plants. *Ecol. Lett.* **2003**, *6*, 688–694. [CrossRef]
76. Gentry, A.H. Anti-Pollinators for Mass-Flowering Plants? *Biotropica* **1978**, *10*, 68–69. [CrossRef]
77. Hopkins, H.C.; Hopkins, M.J. Predation by a snake of a flower-visiting bat at *Parkia nitida* (Leguminosae: Mimosoideae). *Brittonia* **1982**, *34*, 225–227. [CrossRef]
78. Romero, G.Q.; Vasconcellos-Neto, J. Beneficial effects of flower-dwelling predators on their host plant. *Ecology* **2004**, *85*, 446–457. [CrossRef]
79. Carpenter, J.M. The genus *Pachodynerus* in North America (Hymenoptera: Vespidae: Eumeninae). *Proc. Entomol. Soc. Wash.* **1986**, *88*, 572–577.
80. Krombein, K.V. *Trap-Nesting Wasps and Bees: Life Histories, Nests and Associates*; Smithsonian Press: Washington, DC, USA, 1967.
81. Bradley, J.C. The species of *Campsomeris* (Hymenoptera-Scoliidae) of the Plumipes Group, inhabiting the United States, the Greater Antilles, and the Bahama Islands. *Proc. Acad. Sci. Phila.* **1928**, *80*, 313–337.
82. Florida Natural Areas Inventory (FNAI). Pine Rockland. In *Guide to the Natural Communities of Florida*; Florida Natural Areas Inventory: Tallahassee, FL, USA, 2010; pp. 69–72.
83. Koptur, S. The conservation of specialized and generalized pollination systems in subtropical ecosystems: A case study. In *Plant–Pollinator Interactions: From Specialization to Generalization*; Waser, N., Ollerton, J., Eds.; University of Chicago Press: Chicago, IL, USA, 2006; pp. 341–361.
84. Snyder, J.R.; Herndon, A.; Robertson, W.B., Jr. South Florida Rockland. In *Ecosystems of Florida*; Myers, R.L., Ewel, J.J., Eds.; University of Central Florida Press: Orlando, FL, USA, 1990; pp. 230–274.
85. Peña, A.L.; Koptur, S. A Historical Floristic Inventory of Pine Rockland Fabaceae (Leguminosae). *Nat. Areas J.* **2021**, *41*, 258–272. [CrossRef]
86. Possley, J.E.; Maschinski, J.M.; Maguire, J.; Guerra, C. Vegetation Monitoring to Guide Management Decisions in Miami's Urban Pine Rockland Preserves. *Nat. Areas J.* **2014**, *34*, 154–165. [CrossRef]
87. Natureserve. 2022. Available online: https://explorer.natureserve.org/Taxon/ELEMENT_GLOBAL.2.723149/South_Florida_Pine_Rockland (accessed on 19 October 2022).
88. Artz, D.R.; Waddington, K.D. The Effects of Neighbouring Tree Islands on Pollinator Density and Diversity, and on Pollination of a Wet Prairie Species, *Asclepias lanceolata* (Apocynaceae). *J. Ecol.* **2006**, *94*, 597–608. [CrossRef]
89. Pascarella, J.B.; Waddington, K.D.; Neal, P.R. The Bee Fauna (Hymenoptera: Apoidea) of Everglades National Park, Florida and Adjacent Areas: Distribution, Phenology, and Biogeography. *J. Kans. Entomol. Soc.* **1999**, *72*, 32–45.
90. Pascarella, J.B.; Waddington, K.D.; Neal, P.R. Non-apoid flower-visiting fauna of Everglades National Park, Florida. *Biodivers. Conserv.* **2001**, *10*, 551–566. [CrossRef]
91. Pascarella, J.B. Pollination Ecology of *Ardisia escallonioides* (Myrsinaceae). *Castanea* **1997**, *62*, 1–7.
92. Pascarella, J.B. Hurricane Disturbance, Plant-Animal Interactions, and the Reproductive Success of a Tropical Shrub. *Biotropica* **1998**, *30*, 416–424. [CrossRef]
93. Cardel, Y.; Koptur, S. Effects of Florivory on the Pollination of Flowers: An Experimental Field Study with a Perennial Plant. *Int. J. Plant Sci.* **2010**, *171*, 283–292. [CrossRef]

94. Harris, B.M.; Koptur, S. Facilitated fecundity in sand flax: Pollination in an endangered herb of pine rocklands. *Flora* **2022**, *289*, 152041. [CrossRef]
95. Linares, L.J.; Koptur, S. Floral Biology and Breeding System of the Crenulate Leadplant, *Amorpha herbacea* var. *crenulata*, an Endangered South Florida Pine Rockland Endemic. *Nat. Areas J.* **2010**, *30*, 138–147. [CrossRef]
96. Liu, H.; Koptur, S. Breeding System and Pollination of a Narrowly Endemic Herb of the Lower Florida Keys: Impacts of the Urban Wildland Interface. *Am. J. Bot.* **2003**, *90*, 1180–1187. [CrossRef]
97. Royal Botanic Gardens, Kew 2021. The World Checklist of Vascular Plants (WCVP). Checklist Dataset. Available online: https://www.gbif.org/dataset/f382f0ce-323a-4091-bb9f-add557f3a9a2 (accessed on 19 October 2022).

plants

MDPI

Article

San Diego Thornmint (*Acanthomintha ilicifolia*) Populations Differ in Growth and Reproductive Responses to Differential Water Availability: Evidence from a Common Garden Experiment

Katherine D. Heineman [1], Stacy M. Anderson [2], Joseph M. Davitt [2], Laurie Lippitt [2], Bryan A. Endress [3] and Christa M. Horn [2,*]

[1] Center for Plant Conservation, 15600 San Pasqual Valley Rd., Escondido, CA 92027, USA; kheineman@saveplants.org
[2] San Diego Zoo Wildlife Alliance, 15600 San Pasqual Valley Rd., Escondido, CA 92027, USA; sanderson@sdzwa.org (S.M.A.); jdavitt@sdzwa.org (J.M.D.); lauriel@infostations.com (L.L.)
[3] Eastern Oregon Agriculture Research Center, Oregon State University, 372 S. 10th Street, Union, OR 97883, USA; bryan.endress@oregonstate.edu
* Correspondence: chorn@sdzwa.org

Abstract: The responses of rare plants to environmental stressors will determine their potential to adapt to a rapidly changing climate. We used a common garden approach to evaluate how six populations of the annual San Diego thornmint (*Acanthomintha ilicifolia* Lamiaceae; listed as endangered in the state of California and as threatened by the US Fish and Wildlife Service) from across the species range respond in terms of growth (biomass, height, and width) and reproduction (seed production, floral production, and next generation seed viability) to experimental differences in water availability. We found a significant irrigation-by-population interaction on the aboveground growth, wherein the differences in the magnitude and direction of treatment did not correlate directly with climate variables in natural populations. With respect to reproduction, the low-irrigation treatment produced more seeds per plant, more reproductive individuals, and a larger proportion of viable seed in most, but not all, populations. The seed production and the effect of irrigation on seed production correlated positively with rainfall at wild source populations. These results suggest that *Acanthomintha ilicifolia* responds to water limitation by creating more and higher-quality seed, and that plants locally adapted to a higher annual rainfall show a greater plasticity to differences in water availability than plants adapted to a lower annual rainfall, a finding that can inform the in situ demographic management and ex situ collection strategy for *Acanthomintha ilicifolia* and other rare California annuals.

Keywords: annual plants; common garden experiment; drought responses; endangered species; germination; threatened species; viability testing

Citation: Heineman, K.D.; Anderson, S.M.; Davitt, J.M.; Lippitt, L.; Endress, B.A.; Horn, C.M. San Diego Thornmint (*Acanthomintha ilicifolia*) Populations Differ in Growth and Reproductive Responses to Differential Water Availability: Evidence from a Common Garden Experiment. *Plants* **2023**, *12*, 3439. https://doi.org/10.3390/plants12193439

Academic Editors: Brenda Molano-Flores and James Cohen

Received: 16 June 2023
Revised: 19 September 2023
Accepted: 19 September 2023
Published: 29 September 2023

1. Introduction

Currently, nearly 40% of the world's plants are threatened with extinction [1], and climate change is increasingly being recognized as a threat to these plants. Rare species are disproportionately threatened by climate change, as well as other human impacts [2–5]. Thus, predicting how rare and threatened plants will react to climate change is a high and important priority for the conservation of species and the ecosystems that they occupy.

There are three possible outcomes for plant populations in a rapidly changing climate: in situ adaptation, migration, or extirpation [6,7]. Increased knowledge about the species, especially in terms of their ecological needs and their adaptive potential, can help us to identify appropriate conservation strategies in the face of climate change. Different constraints on species, including their phenotypic plasticity and adaptive potential, impact which of the three outcomes is most likely, and what strategies to undertake: assisted migration, the management of biodiversity corridors, the augmentation of the potential climate, the protection of genetic refugia, and/or ex situ conservation [7].

Ex situ conservation, largely in the form of conservation seed banking, plays a supporting role in many of these strategies, in addition to serving as a long-term safeguard against loss of genetic diversity due to extirpation and extinction [8]. There is an increasing interest in being more strategic about conservation seed collecting [9,10], and knowing more about potential local adaptation can help inform such strategies. Beyond long-term ex situ conservation, the use of the collected seed as a source for restoration efforts should consider the potential of genetic differences between populations and the potential of both locally adapted traits and phenotypic plasticity [11,12].

Common garden experiments help researchers and conservationists plan safeguards for a species, both in situ and ex situ, by examining how the local environment is driving the expression of intraspecific variation, how species or populations respond to changing climatic conditions, and more [13–16]. Many studies examining species population or species responses to climate change have established common gardens with different water availability, sometimes in conjunction with other variables, such as warming or competition [15,17]. However, the impact of drought on resource allocation for plants, specifically annuals, is not always predictable [18]. Nonetheless, common garden experiments are useful tools for gauging species- and population-level responses to environmental stress in species of a high conservation value, such as the San Diego thornmint (*Acanthomintha ilicifolia* (A.Gray) A.Gray, Lamiaceae), an herbaceous annual.

Responses to climate change may be more important to the long-term persistence of edaphic specialist plant species than to that of other rare plants, as migration to a more suitable habitat is limited by the availability of the edaphic habitat [19]. In California, climate change is predicted to shrink the edaphic habitat of rare annual herbs specialized to the hydrological and edaphic environment of vernal pools and similar habitats [20]. *Acanthomintha ilicifolia* is one such annual—a mint that is listed as endangered in the state of California and as threatened by the US Fish and Wildlife service due to the rapid loss of its native clay lens habitat spanning San Diego County, to northwest Baja California, Mexico [21]. Although an edaphic specialist, *A. ilicifolia* spans relatively broad elevation and precipitation gradients. *A. ilicifolia* produces bisexual flowers and is an outcrosser that is insect pollinated; however, there is limited information regarding its breeding system. A genetic analysis of 21 *A. ilicifolia* populations found a strong genetic structure among populations and at least two cytotypes [22]. A more recent genomic study found that these populations make up at least five unique genetic clusters within San Diego County [23], making *A. ilicifolia* a promising study species for examining the effects of local adaptation on genetic and phenotypic diversity in rare plant species.

In this study, we evaluate how experimental variation in water availability affects aspects of the growth and reproductive output of *A. ilicifolia* relevant to its long-term persistence in the wild and in ex situ conservation. We used a common garden approach to test the interactive effects of source population and irrigation treatments on the aboveground growth (biomass, height, and width) and reproductive output (flower number, seed number, and seed viability) of six populations of *A. ilicifolia* spanning a regional precipitation gradient. We hypothesized that, if plants allocate more resources to structures supporting light and root competition when water is an abundant resource, then plants supplied with ample water will invest proportionally more in aboveground growth and less in reproductive output compared to plants grown under drought stress. We further hypothesized that plants grown from seed sourced from populations with a lower average annual precipitation in nature should perform relatively better in the low-water treatment than populations with a higher average annual precipitation. Uncovering how threatened and endangered annual plants vary in key fitness traits across time and space is critical to developing an informed conservation strategy.

2. Results

2.1. Common Garden Experiment

The response of *Acanthomintha ilicifolia* to experimental watering treatments differed among source populations at all stages of plant growth and reproduction. The germination rate of wild collected seeds planted in common garden pots was 42% across all the source populations and treatments. We observed a significant treatment-by-population interaction on the germination success ($X^2 = 35.2$, $p < 0.001$), wherein the high-irrigation treatment had a greater germination success than the low-irrigation treatment in three of the six source populations, but this did not differ significantly between treatments in the two Carlsbad populations or the Mission Trails population (Figure 1a). One treatment in particular, the low-irrigation treatment for the McGinty Mountain population, showed only 20% germination and did not yield a single germinant in 24 of the 55 experimental pots. After thinning, 91% of the plants survived to harvest, and a population-by-irrigation interaction on survival ($X^2 = 15.4$, $p = 0.009$) was primarily driven by a higher proportion of surviving individuals in the low-irrigation treatment compared to the high-irrigation treatment in the Alpine and Carlsbad North 1 populations, whereas the reverse trend was true for the Carlsbad North 2 population (Figure 1b).

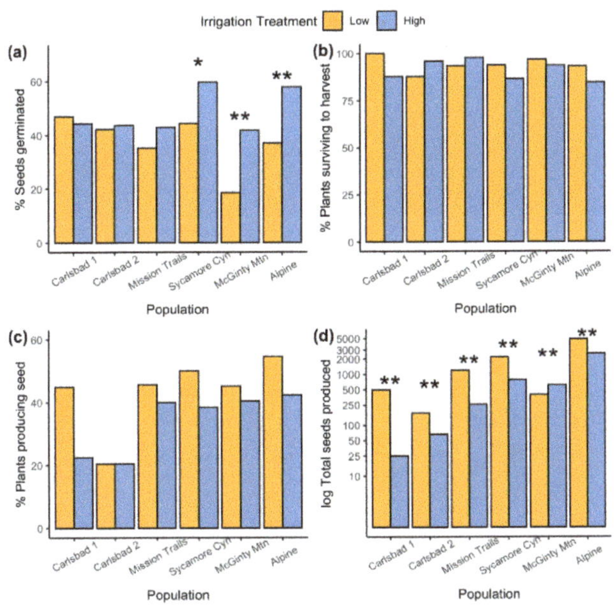

Figure 1. Comparison of the pot-based germination and reproductive performance between treatments for each population in the common garden experiment, including (**a**) the percentage of wild collected seed that germinated in pots, (**b**) the percentage of plants that survived to harvest, (**c**) the percentage of plants that produced one or more seeds, and (**d**) the total seed produced across all plants. Stars above population-by-treatment pairs (* and **) represent significant differences at alpha = 0.05 and 0.001, respectively, between treatment groups, determined via post hoc contrasts of generalized linear models evaluating the interaction effect of populations and treatments on the binomial and count-based metrics of seed performance. The populations are ordered on the x-axis by increasing mean annual rainfall.

The source population explained a greater proportion of variance than the irrigation in the average plant allocation to aboveground metrics such as biomass, height, width, flower number, and seed production (see X^2 values, Table A1). In general, the Carlsbad 1 population produced the largest plants, and the Alpine population produced the smallest plants.

All the linear models evaluating aboveground allocation had a significant population-by-treatment interaction due to the response of the southernmost McGinty Mountain source population, which had taller and wider plants, with a higher biomass and more flowers, in the high-irrigation treatment than in the low-irrigation treatment—the reverse of what was found for the other populations (Figure 2).

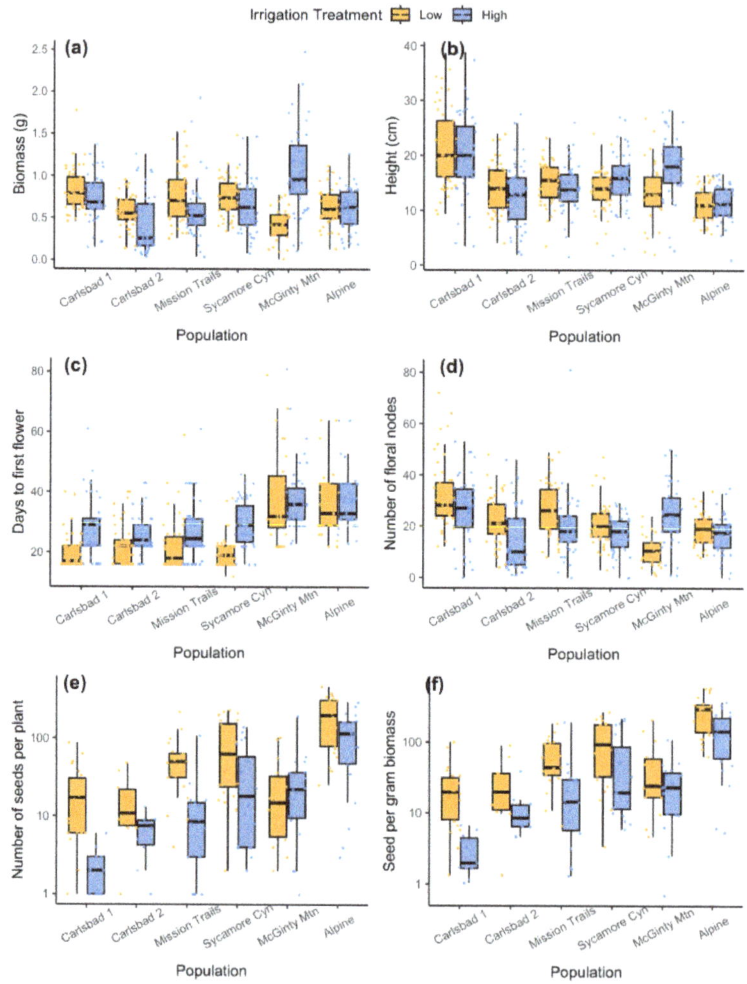

Figure 2. Box plots comparing individual plant measures of performance in the common garden experiment among populations and treatments, as measured when the experiment plants were harvested, including (**a**) aboveground biomass, (**b**) height, (**c**) width, (**d**) number of floral nodes, (**e**) number of seeds per plant, and (**f**) seeds per gram biomass (only for plants producing seeds). The populations are ordered on the x-axis by increasing mean annual rainfall.

The effect of the watering treatment on seed production was more consistent in terms of the qualitative direction of the response across populations. Across all populations and treatments, 42% of plants produced at least one seed. There was a significant main effect of both the population ($X^2 = 21.7$, $p < 0.001$) and the treatment ($X^2 = 6.01$, $p = 0.014$) on the probability of seed production. Plants under the low-irrigation treatment were 65% more likely to produce seed than plants under the high-irrigation treatment (Figure 1c). Plants in the low-irrigation treatment also produced more than double the number of seeds on average per plant (Table 1) in all populations except for McGinty Mountain (Figure 2). This resulted in a significant treatment-by-population interaction effect on both the seeds produced per plant ($X^2 = 13.0$, $p = 0.024$) and the seeds produced per unit of biomass ($X^2 = 14.5$, $p = 0.013$). Taken together, the higher proportion of seed-producing plants and the higher average seed production per plant resulted in the lower-irrigation treatment having an order-of-magnitude-higher combined seed production compared to the high-irrigation treatment in all populations except McGinty Mountain, for which the reverse was true (Figure 1d). The populations that produced more seed under low-watering treatments also flowered more quickly than plants under the high-watering treatment, but there was no phenological difference among treatments for the McGinty Mountain or Alpine populations (Figure 1c, $X^2 = 3.8$ $p = 0.002$).

Table 1. Common garden experiment results comparison for the plants' aboveground and reproductive attributes after harvest according to population and irrigation treatment.

Population	Irrigation Treatment	Sample Size at Harvest	Biomass (g)		Height (cm)		Width (cm)		Floral Nodes		Seeds Produced per Plant	
			Mean	SE	Mean	SE	Mean	SE	Mean	SE	Mean	SE
Carlsbad 1	High	43	0.74	0.04	20.2	1.04	13.7	0.7	26.0	1.7	2.3	0.5
	Low	49	0.83	0.03	21.6	0.98	14.9	0.5	32.2	1.7	22.5	4.7
Carlsbad 2	High	49	0.42	0.05	12.6	0.76	9.8	0.7	14.4	1.6	6.6	1.1
	Low	47	0.57	0.03	14.0	0.62	13.0	0.5	22.4	1.3	17.3	4.7
Alpine	High	44	0.61	0.03	11.4	0.44	10.7	0.4	16.7	1.0	115.2	17.5
	Low	41	0.63	0.03	11.1	0.42	9.61	0.3	19.1	0.8	202.8	25.4
Mission Trails	High	44	0.59	0.05	14.0	0.60	9.8	0.5	19.6	1.7	14.3	5.7
	Low	43	0.76	0.04	14.9	0.52	11.2	0.4	26.6	1.4	56.5	10.3
McGinty Mountain	High	44	1.05	0.07	17.9	0.80	12.9	0.8	25.0	1.6	32.3	9.8
	Low	30	0.42	0.03	13.3	0.78	7.7	0.4	10.6	0.7	28.4	8.9
Sycamore Canyon	High	45	0.64	0.04	15.5	0.71	13.4	0.7	16.9	0.9	39.0	9.9
	Low	45	0.76	0.03	13.9	0.50	14.4	0.4	20.8	1.1	90.5	15.6

2.2. Ex Situ Germination Trials

The germination rates of the wild seed collected in 2013 and stored at room temperature/humidity until 2019 ranged from 85 to 100% (Table A3). The results of more recent viability tests of wild seed collected from these and other *A. ilicifolia* populations tested within a year of collection by the San Diego Zoo Botanical Conservation Center range from 82 to 95%, indicating that very little viability was lost due to storage conditions. However, the ex situ germination rates of the seed produced from the common garden experiment were significantly lower than the wild collected seed in every population.

Nonetheless, we found a significant irrigation-by-population interaction on the probability of germination in the seed harvested from the common garden experiment (likelihood ratio test = 749.5, df = 11, $p < 0.001$; Figure A1). While there was no difference in the germination success among treatments in the McGinty Mountain population (Figure A1d), a significantly higher proportion of the seed harvested under the low-irrigation treatment germinated by the end of the trial compared to seed from the high-irrigation treatment in five of the six populations (Figure A1b). The relationship between seed viability and seed production across the treatment level was positive, but only marginally significant ($X^2 = 2.9$, $p = 0.084$). However, in five of the six populations, the low-irrigation treatment produced more seeds in total, as well as a higher proportion of viable seed (Figure 3a), indicating that the low-water treatment enhanced both the quantity and quality of the seed produced.

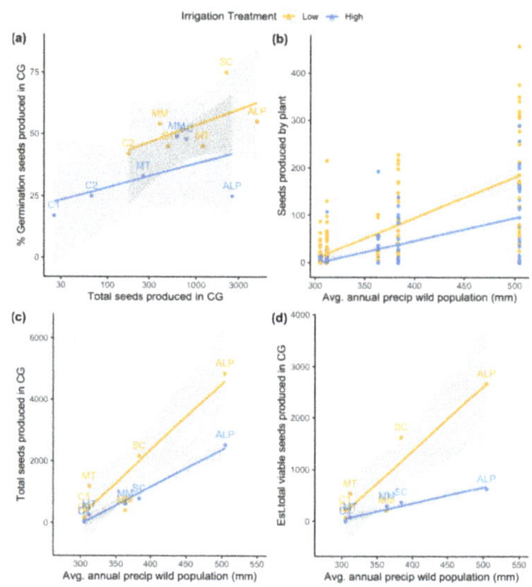

Figure 3. The relationship between (**a**) the seed viability and seed production in each common garden (CG) experiment treatment by population group (see Table 1 for population code definitions), and the relationship between seed production and annual wild source population rainfall, in terms of (**b**) the individual common garden plants, (**c**) the total seeds produced, and (**d**) the estimated viable seeds produced in each treatment by population group. Grey confidence envelopes represent one standard error.

2.3. Relationship between Common Garden Reponses and Rainfall in Wild Source Populations

The wild population climate, specifically the average annual rainfall, explained a significant proportion of the variation observed among populations in the common garden experiment with respect to the reproductive output, but not the aboveground growth (Table A2, Figure 3). Plants from wetter source populations produced more seed on average than those from drier populations, and a significant treatment-by-population interaction (X^2 = 8.2, p = 0.004) suggests that wetter source populations demonstrated a stronger response to this experimental water stress than drier populations (Figure 3b). When the seed totals were aggregated by treatment and population group, this interaction persisted, with the total seed production increasing with the average annual wild population rainfall. The positive effect of low irrigation was more pronounced at the wetter end of the rainfall gradient (X^2 = 6.3, p = 0.010, Figure 3c). While the proportion of viable seed produced from the experiment did not scale significantly with the wild source population rain availability (X^2 = 0.45, p = 0.509), the estimate of the total viable seeds produced in each treatment by population group increased sharply with the wild population in the low-irrigation treatment, whereas the viable seed production increased more slowly with the rainfall in the high-irrigation treatment (interaction effect, X^2 = 13.0, p < 0.001, Figure 3d). Other geographic variables related to the source population (elevation, latitude, longitude) either had no effect, or had a weak effect, on the treatment-wide reproductive output (Figure A2, Table A4) compared to the strong effect of the annual rainfall.

3. Discussion

3.1. Variable Growth Responses among Acanthomintha ilicifolia Populations in Common Garden

A rare plant with a climatically heterogeneous but edaphically restricted native range, *A. ilicifolia* displays a strong genetic structure among populations and significant differences in growth and reproduction among populations in a common garden setting [22,23]. Our findings add that *A. ilicifolia* populations vary in their responses to experimental water variability and exhibit directional patterns in reproductive performance along a regional climate gradient. The direction of the relationship between the wild population rainfall and common garden reproductive responses was counter to our expectations. We hypothesized that plants adapted to drier environments, where the aboveground competition is lower, should invest proportionally more resources in reproduction and be better able to respond to low water availability than plants from wetter environments. However, we found the opposite to be true. One possible explanation could be that the wet population plants experienced a higher degree of water stress (in both treatments) than the plants from dry environments at our common garden site, which is lower, drier, and warmer than that of the source populations. This relative increase in stress may have triggered wet population plants to invest a greater proportion of resources in reproduction compared to plants adapted to dry populations. Reports of correlations between the source population climate and plant responses, including reproductive phenology [24], physiology [25], and seed and leaf traits [26], have been observed in other plant species and can be interpreted as products of local adaption to climatic conditions, which may be partially responsible for the strong genetic structure among *A. ilicifolia* populations.

The non-reproductive measures of plant production measured (biomass, height, and width) did not show predictable patterns along the regional rainfall gradient, but there were striking differences among populations. For instance, the Carlsbad subpopulations, located within a kilometer of each other, differed drastically in aboveground biomass and seed production. The most idiosyncratic population was the McGinty Mountain population, which responded more favorably to the high-irrigation treatment in terms of germination, flower production, and biomass. In contrast, the Alpine population, McGinty Mountain's most similar neighbor in terms of geography (Figure 4), elevation, and rainfall (Table 2), did not show strong differences in response to aboveground allocation between treatments, and produced higher-quality, more viable seed under the low-water treatment. Past genomic work has demonstrated that, while only a few miles apart, these populations are from different genetic clusters [23]. While the belowground biomass was not evaluated in the present study, we acknowledge that the belowground allocation would likely be affected by the manipulation of belowground resource availability (e.g., water), as has been seen in prior studies on another southern California native, *Artemisia californica*, in which experimental drought stress increased the root-to-shoot ratio across a variety of treatments [27].

Maternal effects were not directly measured in the present study but are known to have important effects in common garden experiments [28] and on seed-based traits in particular [29]. While it is unclear how many maternal plants were represented in each population by treatment group at the end of the study, an effort was made to ensure that the seeds from each "stem" provided for seed processing were evenly distributed across the experiment. Based on the biology of the species, stems were likely roughly equivalent to maternal lines in this experiment, but future studies should endeavor to track and measure the maternal effect directly.

Table 2. The *Acanthomintha ilicifolia* populations represented in the common garden experiment.

Population Name	EO [3] (DeWoody et al. [22])	Genetic Cluster (Milano et al. [23])	Lat	Long	Elevation (m)	Wild Seed Collected	Mean Annual Rainfall (mm) [1]	CV Ann. Precip. [2]
Carlsbad1 (C1)	EO70A	Orange	33.14	−117.26	53	685	305	0.41
Carlsbad2 (C2)	EO70B	Orange	33.13	−117.26	53	2757	305	0.41
Sycamore Canyon (SC)	EO32-2	Green/mixed	32.93	−116.98	341	975	383	0.32
Mission Trails (MT)	EO33	Green	32.83	−117.07	153	588	312	0.40
McGinty Mountain (MM)	EO871	Purple	32.75	−116.87	655	1369	363	0.34
Alpine (ALP)	EO75	Pink	32.86	−116.74	770	1705	504	0.25

[1] 30 year normal annual rainfall values extracted from the Prism dataset 1990–2020 at 800 m resolution. [2] Coefficient of variation calculated from annual rainfall values extracted from the PRISM dataset 1980–2020 at 4 km resolution. [3] An elemental occurrence (EO) is an area of land in which a species or natural community is, or was, present. These numbers were assigned by the California Natural Diversity Database (CNDDB).

3.2. *Enhanced Reproductive Performance in Low-Irrigation Treatment*

Plants in the low-irrigation treatment tended to produce more seeds and higher-quality seeds, as measured via ex situ germination trials of the seed generated. It is unclear how the water availability in the experiment compares with natural responses to water resources, especially in the idiosyncratic clay soils that define *A. ilicifolia*'s range, but the lack of a clearly negative response in the reproductive output to what we believe is mild water stress is a positive indication for the species, as it experiences drought throughout its range. A similar common garden approach evaluating the watering treatment effects on 18 populations of *Lupinus angustifolius* in Spain found similar patterns, wherein low watering treatments produced plants with larger, higher-quality seeds, and the strength of this treatment effect varied among spatial, climatic gradients [26]. Further, our study aligns with findings across multiple species showing that plants increased their allocation to reproductive structures when grown under low-water conditions in a common garden setting (i.e., drought) [15,30].

Differing responses to water treatments across populations, including the trend toward an increased reproductive performance, suggest phenotypic plasticity (or epigenetic factors) influencing *A. ilicifolia*. Such plasticity may aid in short-term adaptation for climate change [7]. In the longer term, further study is needed to evaluate whether multiple generations of water stress would result in a qualitative trend of increased seed production in *A. ilicifolia* populations through epigenetic or biochemical effects, as has been seen in model systems [31], or selection for more fecund plants. Christmas et al. [7] note that competition from incoming species may be an ecological constraint to this adaptive potential. *A. ilicifolia* is threatened by invasive grasses, and previous work has indicated that seed production is negatively impacted by competition [32]. Further, we also observed that plants under the lower-watering treatment also flowered earlier than plants under the high-watering treatment. Phenological shifts in response to climate change can have negative consequences for the availability of pollinators [33] and the demographic dynamics of plant populations [34].

3.3. *Conservation Strategy Implications of Results*

For many conservation strategies, seed is required, to maintain or augment populations, or establish new populations within the existing range, or extend it. This study raises several interesting considerations for the strategy of maintaining a robust ex situ seed bank for *A. ilicifolia*. The results demonstrating the variety of responses to resource availability across populations emphasize the importance of conserving seed from across a species range to help preserve a variety of plant genotypes that may respond differently to climate change. The effect of watering treatments on plant performance and seed viability also indicates the importance of collecting seed in multiple years. In alpine species, seed collected in drier, warmer years has been shown to have a longer ex situ viability than seed produced in cooler, wetter years [35]. For annual species, the climate of the collection season may also influence the genetic makeup of the plants represented, especially following cases of severe drought [36].

Wild seed maintained a viability of >90% when stored at room temperature for seven years. This species appears to be extremely durable ex situ, which is good news for the effectiveness of seed banking as a safeguard against population extirpation. We do caution, however, that the conditions under which these seeds were stored are not ideal; it is always advised to store known "orthodox" seeds at −18 °C after seeds have been dried to a constant relative humidity for the best ex situ viability results [37]. It is difficult to know if the lack of appropriate conditions here altered the results for the seeds resulting from the common garden experiment, although we believe that the very high viability in the original wild collected seed helps assuage that concern.

Beyond long-term storage, our research also speaks to the curation and use of ex situ seed collections. Seed augmentations for restoration, or to replenish collections for this species, may be more successful when growing the plants under some stress (e.g., a low-water environment) to produce a higher quantity and quality of seed, an idea that warrants further research. Our observation of a much lower viability in seed produced from common garden treatments compared to wild collected seed stored in the same manner for the same amount of time indicates that pollinator limitation or environmental conditions less favorable to high-quality seed production may have reduced the seed set in the ex situ common garden. However, as all populations and treatments experienced common conditions, we do not believe that this lower production impacted the comparison results of the study. We recommend the hand pollination of this species for future experiments in an ex situ setting to prevent such a large viability loss, especially for those seeking to bulk seed for conservation translocations.

4. Materials and Methods

4.1. Population Selection

For this study, we selected six *A. ilicifolia* populations that would maximize the contrast across the species range in terms of proximity to the coast, elevation, and annual rainfall (Table 2). Two sub-populations from Carlsbad were selected to examine the intrapopulation variability across different microsites. The populations included in the common garden were later found to include four of the five genetic clusters identified via a regional analysis of 24 *A. ilicifolia* populations [23]. For each source population location, we extracted the 30 year normal annual rainfall (mm) and monthly historical rainfall from 1980 to 2020 from the PRISM climate explorer at 800 m and km resolution, respectively [38] (Table 2).

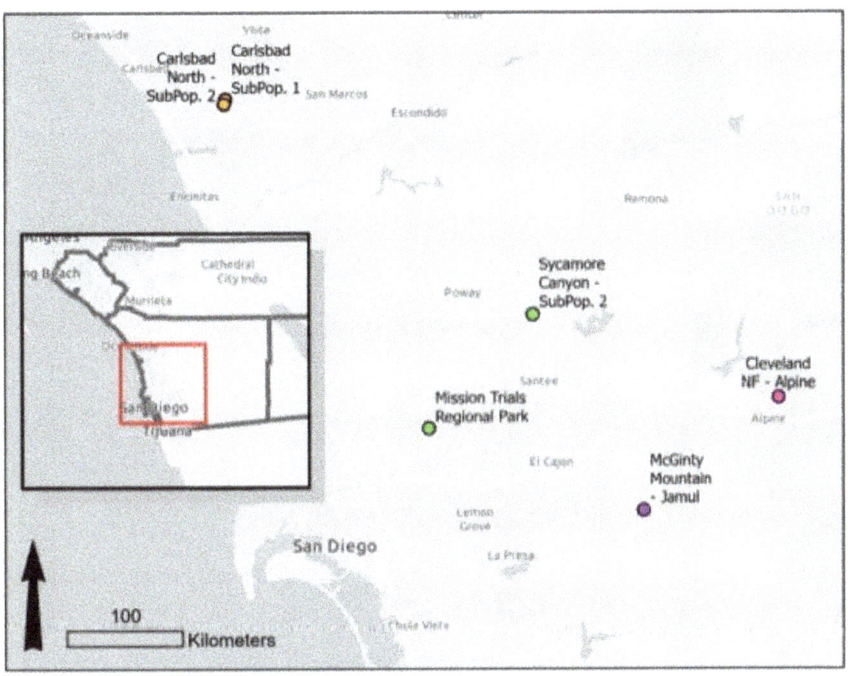

Figure 4. Map of the *Acanthomintha ilicifolia* populations represented in the common garden experiment. The colors of the population points represent the genetic clusters identified via a previous genetic analysis of this species [30], see Table 2.

4.2. Common Garden Study

In 2012, the Center for Natural Lands Management collected stems with attached spiny nodes containing seeds from nine wild *A. ilicifolia* populations and provided material to the San Diego Zoo Wildlife Alliance. To access the seed, we broke open nodes with a rubber bung or a block of wood to release seeds, a process yielding between 170 seeds and

2757 per population (Table 2). CalFIRE's L.A. Moran Reforestation Center performed X-ray analysis to determine whether the filled content of seed in each population sample was less than the 50-seed threshold. After X-ray analysis, three populations (La Costa Greens, Sycamore Canyon subpopulation 1, and Manchester) did not meet the filled seed threshold and were eliminated from the study. The common garden study began in February 2013 using six populations and two watering treatments (high and low) to observe the response of populations to variation in water availability. There were 55 pots per population per water treatment (660 total), which were then sown with 5 seeds per pot. The pots contained a substrate of 3:1 Sunshine#3 potting mix to washed sand in Anderson plant bands AB58 (5″ wide by 8″ tall). The potting mix was a fine soilless mix, which, when combined with sand, had a water-holding capacity similar to that of the native clay soil, and matched other garden studies conducted on the species [33].

The plants were grown in full sun on a single potting bench in the fenced horticultural growing area at the San Diego Zoo Safari Park (Escondido, CA, USA; see Figure 5) located within the species' natural range. The experiment consisted of 37 trays, with nine pots per tray, and both the pots within the trays, and the tray location on the potting bench, were randomized.

During the germination phase, we watered the high-water pots deeply with 500 mL water biweekly and the low-water treatment with 500 mL water monthly, while keeping the soil moist with mist in both treatments between waterings. The mean annual precipitation was not considered for the nursery-grown plants due to container plants drying out more than plants in the ground and the variance in the mean precipitation across the range of the species. Rather than mimic natural precipitation levels, the goal was to provide common conditions and two distinct watering treatments. The number of germinants per pot was recorded each weekday for four weeks, and then thinned to one plant per pot. Each pot did not necessary represent the progeny of a separate maternal plant, because the wild seed was not separated by maternal line upon collection. After thinning, plants received 1000 mL per pot every two weeks (high treatment) or every four weeks (low treatment). The choice of 1000 mL per pot was not based on precipitation; rather, 1000 mL was enough to drip through the bottom of the pot, ensuring that the entire root system received water. The rate of high-water treatment being twice as frequent as that of the low-water treatment was likely to be different enough that variances in source population responses might be observed. However, we observed water stress in the growing plants, which prompted an increase in the watering frequency for both treatments, to 1000 mL weekly in the high-water treatment and 1000 mL biweekly in the low-water treatment.

Monitoring continued five days a week, with the recording of the date when each plant produced its first flower and the date of senescence. For the cross-pollination of the experimental plants, we relied on ambient pollinator activity in the well-vegetated area of the Safari Park, which has been successful at generating ample fertilization in numerous seed bulking and propagation trials in the vicinity. After senescence, we harvested plants and measured the following attributes: height, as the distance from the potting soil to the tallest vertical point on the plant; width, as the widest distance across the plant when viewed from the top down; the number of inflorescence whorls per individual; and the F1 seed quantity. As a measure of the aboveground biomass, we then dried all the non-seed aboveground material from each pot in a drying oven until the weight was constant. An attempt to collect the root biomass was abandoned due to difficulty in separating the fine roots from the soil.

The F1 seeds collected at harvest from the common garden experiment were stored under the ambient conditions of the seedbank from 2013 to 2019 in stable ~20–22 °C temperatures year-round, in sealed vials. The humidity within the seedbank was ~30–50%. However, the eRH (estimated relative humidity) of the seed within the vials during this period is unknown as, though sealed, it was not in flux.

Figure 5. San Diego thornmint plants in the common garden experiment. Label markers enabled the tracking of individual plant flowering and reproduction, and flags indicated watering treatments.

4.3. Germination Protocols

For the germination tests, we plated 50–100 seeds (10 seeds per plate) for each of the different seed lots (wild and F1) involved in the study. The seeds were treated with a 1% bleach solution for 10 min to minimize contamination by mold, and then imbibed in R.O. water for 24 h prior to plating on a 0.5% agar medium. We labeled plate lids with the experimental treatment number and repetition number for data collection. We used ethanol to clean fine-tipped forceps while plating and when checking tests. The germination tests were placed into a germination chamber with an alternating baseline photoperiod day/night cycle of an 11 h day at 22 °C, and a 13 h night at 10 °C. We monitored tests and recorded the number of germinants every other day in the initial stages of the trials, then weekly as the seed germination rate plateaued. A seed was scored as germinated once the radicle emerged, and we removed germinants from the plate immediately to avoid the contamination of the remaining seeds.

4.4. Statistical Methods

4.4.1. The Effect of Population and Watering Treatment on Plant Growth and Reproduction

We used generalized linear mixed-effect models to evaluate the interactive effects of the source population, the source population's average annual rainfall, and the irrigation treatment on aspects of the aboveground growth, survivorship, and reproduction of *A. ilicifolia*. We evaluated the effect of the source population on the plant performance directly as a fixed effect in one series of models, and the effect of the average annual rainfall at the source population with the population as a random effect in a separate series of models. For the models evaluating the interactive effects of the source population and watering treatment, we specified the common garden bench position (east or west) as the random effect. For the models evaluating the interactive effects of the annual precipitation and the treatment as fixed effects, we specified the source population as a random effect. For binary dependent variables (germination, survivorship, and the presence of seed and flowers at harvest), we used a binomial error distribution; for count data (the total seed number across all plants in a treatment), we used a Poisson error distribution; and for all other continuous dependent variables (biomass, height, width, flower number, days to flower, seed number per plant and per unit biomass), we used a Gaussian error distribution. We natural-log-transformed the dependent variable seed number to meet the distributional assumptions of normality of errors. The models evaluating the average number of seed and inflorescences produced per plant included only plants that produced at least one flower or fruit, respectively. For each dependent variable, we present the Wald Type II test of fixed effects for the simplest model that did not improve the model AIC values by more than two points (Tables A1 and A2). We used post hoc linear model contrasts to interpret the significance of the difference between high- and low-watering treatments within populations. Generalized linear models were implemented in the lme4 package in R [39].

4.4.2. The Effect of Watering Treatments on Ex Situ Seed Viability

We statistically examined differences in ex situ germination rates among generations, populations, and treatments using cox proportional hazard regression models, a time-to-event method accounting for the number of germinants at each time point monitored. We fit one model with a generation-by-population interaction to test whether the seed produced from the experiment differed from the wild collected seed stored the same way for the same amount of time. We fit a second model with the treatment-by-population interaction evaluating the effect of the irrigation on the seed viability. Previous reviews of germination data analysis have advocated the use of Cox proportional hazard models in cases where researchers are contrasting the effect of treatments on germination outcomes [40]. We used the R package survival and function coxaph to implement the analysis [41].

4.4.3. The Effect of Source Population Climate and Geographic on ex Situ Reproductive Output

To evaluate the ecological importance of the common garden and viability testing results at the population level, we calculated two aggregate measures of the reproductive output for the common garden treatment groups: the total seeds produced, summed over all plants in the treatment by population group, and the estimated viable seed produced (the total number of seeds produced multiplied by the population-by-treatment specific ex situ viability). We then used linear mixed models to test the interaction effect of the irrigation treatment by the source population's average annual precipitation, elevation, latitude, and longitude on the following dependent variables: the total seeds produced, the ex situ viability, and the estimated viable seed produced. For each model, the population was specified as a random effect, and we used a Wald type II test of fixed effects to evaluate the significance of the fixed effects. We interpreted the simplest model that did not improve the model AIC values by more than two points.

5. Conclusions

Through our common garden approach, we learned more about how a rare species with a climatically heterogenous, but edaphically restricted, range responds phenotypically to differences in water availability. A previous report [22] exploring plants within just one of these watering treatments supported the hypothesis that genetic differences between populations exist and, thus, the differences seen across the species' range are not solely due to plasticity expressed across environmental variability. However, adding the high- and low-watering treatments allowed us to identify differences in plasticity as a type of intraspecific variation. In particular, the added knowledge that water limitation may positively impact seed production and seed viability has implications for our understanding of the demography of *Acathomintha ilicifolia* under different climate scenarios, and merits further evaluation in situ.

Our results, in tandem with other work on the species [22,23], suggest that, across the range of the species, there is both adaptive genetic potential and phenotypic plasticity along the gradients of resource availability. As an edaphic endemic, the species' migration potential is limited by the clay soil availability. Thus, using the framework set forth by Christmas et al. [7], we support the belief that the species conservation priority lies in the protection of genetic refugia and remnant populations. Specifically, we recommend land management strategies in key populations of each recognized genetic cluster [23]. Assisted migration would need to be explored if there are indicators that the adaptive capacity has been reached.

Author Contributions: Conceptualization, B.A.E., L.L. and S.M.A.; methodology, S.M.A. and J.M.D.; formal analysis, K.D.H.; data curation, S.M.A. and J.M.D.; writing—original draft preparation, C.M.H. and K.D.H.; writing—review and editing, visualization, K.D.H., C.M.H., S.M.A. and B.A.E.; investigation, L.L. and S.M.A.; funding acquisition through subcontract, B.A.E. All authors have read and agreed to the published version of the manuscript.

Funding: This research was partially funded through the Center for Land Management's grant from the San Diego Association of Governments' Environmental Mitigation Program (Grant Agreement #5001964) and a subcontract with the Zoological Society of San Diego. Additional support from the Schlum Foundation and the J.W. Sefton Foundation.

Data Availability Statement: Common garden datasets used for the manuscript were uploaded to the DRYAD repository (https://doi.org/10.5061/dryad.q573n5tpm).

Acknowledgments: We wish to sincerely thank Kaitlyn Coleman for her ex situ germination trials of *Acanthomintha ilicifolia* during her summer fellowship. We acknowledge and thank Deborah Rogers and the Center for Natural Lands Management staff for developing the SANDAG subcontract and acquiring *Acanthomintha ilicifolia* materials for the common garden study. We thank Erin Conlisk, Kelly Anderson, and Burak Pekin for providing initial data analysis for the common garden study. We thank the San Diego Zoo Wildlife Alliance Safari Park Horticulture department for providing bench space for growing plants for the common garden study.

Conflicts of Interest: The authors declare no conflict of interest.

Appendix A

Table A1. The results of Wald type II chi-squared (X^2) tests of fixed effects for generalized linear mixed-effect models evaluating the interactive effects of the source population and the irrigation treatment on aspects of San Diego thornmint growth and reproduction in the common garden experiment. For each dependent variable, we present the terms of the best model resulting from stepwise AIC model selection. Trt × pop = treatment-by-population interaction.

Dependent Variable	Model Terms	X^2	df	p
Biomass at harvest	Treatment	0.1	1	0.807
	Population	12.2	5	<0.001
	Trt × pop	19.1	5	<0.001
Height at harvest	Treatment	0.5	1	0.484
	Population	175.0	5	<0.001
	Trt × pop	19.8	5	0.001
Width at harvest	Treatment	0.5	1	0.475
	Population	74.0	5	<0.001
	Trt × pop	43.1	5	<0.001
Number of Floral nodes	Treatment	9.2	1	0.002
	Population	86.7	5	<0.001
	Trt × pop	65.0	5	<0.001
Days to flower	Treatment	31.0	1	<0.001
	Population	41.2	5	<0.001
	Trt × pop	3.8	5	0.002
Number of Seed produced	Treatment	22.8	1	<0.001
	Population	165.5	5	<0.001
	Trt × pop	13.0	5	0.024
Number of Seed/biomass	Treatment	24.0	1	<0.001
	Population	199.8	5	<0.001
	Trt × pop	14.5	5	0.013
% Germination	Treatment	66.5	1	<0.001
	Population	41.4	5	<0.001
	Trt × pop	35.2	5	<0.001
% Survival	Treatment	3.1	1	0.651
	Population	3.3	5	0.08
	Trt × pop	15.3	5	0.009
% Plants producing seeds	Treatment	6.1	5	0.014
	Population	21.7	1	<0.001

Table A2. The results of the Wald type II chi-squared (X^2) tests of fixed effects for linear mixed-effect models evaluating the interactive effects of the source population and the annual rainfall (mm) at the source population on aspects of San Diego thornmint growth and reproduction in the common garden experiment. For each dependent variable, we present the terms of the best model resulting from stepwise AIC model selection. Trt × ann. rain = treatment-by-annual-rainfall-at-wild-population interaction effect.

Dependent Variable	Model Terms	X^2	df	*p*
Biomass at harvest	None retained	-	-	-
Height at harvest	None retained	-	-	-
Width at harvest	None retained	-	-	-
Number of floral nodes	Treatment	6.5	1	0.011
Days to flower	Treatment	30.3	1	<0.001
Number of Seed produced	Treatment	23.8	1	<0.001
	Ann. rain wild pop	74.4	1	<0.001
	Trt × ann. rain	8.2	1	0.004
Number of Seed/biomass	Treatment	24.5	1	<0.001
	Ann. rain wild pop	65.7	1	<0.001
	Trt × ann. rain	13.4	1	<0.001
% Germination	Treatment	40.3	1	<0.001
	Ann. rain wild pop	0.5	1	0.49
	Trt × ann. rain	16.3	1	<0.001
% Survival	None retained	-	-	-
% Plants producing seed	Treatment	5.9	1	0.024
	Ann. rain wild pop	5.1	1	0.016

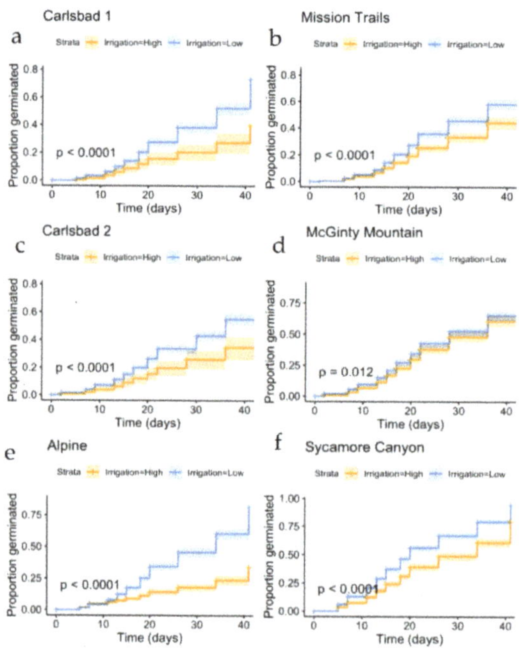

Figure A1. The germination over time (days) by population and treatment in the germination tests on the F1 seed produced from the common garden experiment and tested in 2019 at the San Diego Zoo Botanical Conservation Center, seven years after the common garden experiment. *p* values are associated with Cox proportional hazard regression models testing the difference among treatments for each population.

Table A3. The results of the ex situ germination testing of seeds stored at the San Diego Zoo Botanical Conservation Center in sealed vials at room temperature for seven years after the completion of the common garden experiment. Wild seed indicates seed collected from the natural population used in the common garden experiment. CG indicates seed that was harvested from plants grown in the respective common garden treatments.

Population	Generation–Treatment	Ex Situ Germination (2019)
Carlsbad 1	Wild	100%
	CG-Hi	17%
	CG-Low	45%
Carlsbad 2	Wild	95%
	CG-Hi	25%
	CG-Low	42%
Alpine	Wild	N/A
	CG-Hi	25%
	CG-Low	55%
Mission Trails	Wild	85%
	CG-Hi	33%
	CG-Low	45%
McGinty Mountain	Wild	96%
	CG-Hi	49%
	CG-Low	54%
Sycamore	Wild	95%
	CG-Hi	48%
	CG-Low	75%

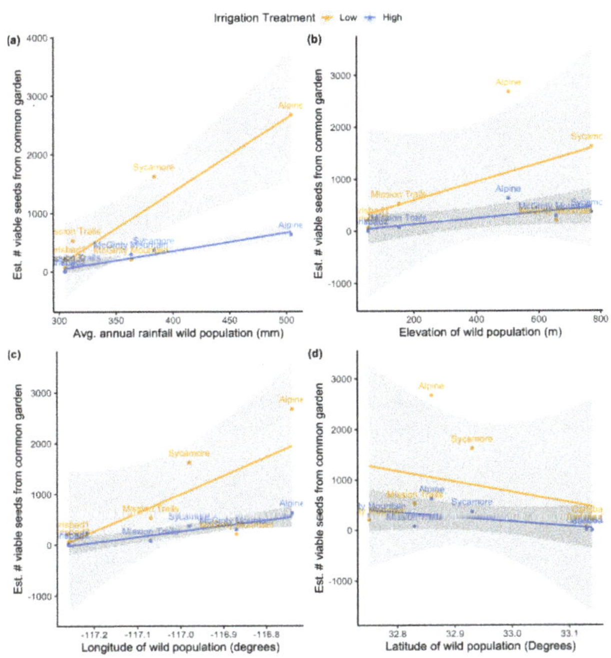

Figure A2. The relationship between the estimated total number of viable seeds produced by each population by treatment group (see Table 1 for population code definitions) in the common garden versus the measure of the source population climate and geography, including (**a**) the average annual rainfall in mm, (**b**) the elevation, (**c**) the longitude, and (**d**) the latitude. Grey confidence envelopes represent one standard error.

Table A4. The results of the Wald type II chi-squared (X^2) tests of fixed effects for linear mixed-effect models evaluating the interactive effects of the source population and several climatic and geographic variables at the source population on the treatment-wide measures for reproductive output. For each dependent variable, we present the terms of the best model resulting from stepwise AIC model selection.

Dependent Variable	Model Terms	X^2	df	*p*
	ANNUAL RAINFALL			
Total seeds	Ann. rain wild pop	52.1	1	<0.001
	Treatment	10.1	1	0.002
	Trt × ann. rain	6.4	1	0.011
Ex situ germ	Treatment	23.1	1	<0.001
Total viable seeds	Ann. rain wild pop	36.2	1	<0.001
	Treatment	12.9	1	<0.001
	Trt × ann. rain	13	1	<0.001
	ELEVATION			
Total seeds	Treatment	4.7	1	0.027
Ex situ germ	Elevation	23.1	1	<0.001
	Treatment	23.6	1	<0.001
Total viable seeds	Treatment	3.8	1	0.052
	LATITUDE			
Total seeds	Treatment	4.8	1	0.027
Ex situ germ	Treatment	23.1	1	<0.001
Total viable seeds	Treatment	3.8	1	0.052
	LONGITUDE			
Total seeds	Treatment	5.6	1	0.017
	Longitude	6.6	1	0.01
	Trt × longitude	1.8	1	0.181
Ex situ germ	Treatment	23.1	1	<0.001
Total viable seeds	Treatment	5.2	1	0.022
	Longitude	6.5	1	0.011
	Trt × longitude	2.9	1	0.091

References

1. Nic Lughadha, E.; Bachman, S.P.; Leão, T.C.C.; Forest, F.; Halley, J.M.; Moat, J.; Acedo, C.; Bacon, K.L.; Brewer, R.F.A.; Gâteblé, G.; et al. Extinction Risk and Threats to Plants and Fungi. *Plants People Planet* **2020**, *2*, 389–408. [CrossRef]
2. Vincent, H.; Bornand, C.N.; Kempel, A.; Fischer, M. Rare Species Perform Worse than Widespread Species under Changed Climate. *Biol. Conserv.* **2020**, *246*, 108586. [CrossRef]
3. Thuiller, W.; Albert, C.; Araújo, M.B.; Berry, P.M.; Cabeza, M.; Guisan, A.; Hickler, T.; Midgley, G.F.; Paterson, J.; Schurr, F.M.; et al. Predicting Global Change Impacts on Plant Species' Distributions: Future Challenges. *Perspect. Plant Ecol. Evol. Syst.* **2008**, *9*, 137–152. [CrossRef]
4. Enquist, B.J.; Feng, X.; Boyle, B.; Maitner, B.; Newman, E.A.; Jørgensen, P.M.; Roehrdanz, P.R.; Thiers, B.M.; Burger, J.R.; Corlett, R.T. The Commonness of Rarity: Global and Future Distribution of Rarity across Land Plants. *Sci. Adv.* **2019**, *5*, eaaz0414. [CrossRef] [PubMed]
5. Bartholomeus, R.P.; Witte, J.-P.M.; van Bodegom, P.M.; van Dam, J.C.; Aerts, R. Climate Change Threatens Endangered Plant Species by Stronger and Interacting Water-Related Stresses. *J. Geophys. Res. Biogeosci.* **2011**, *116*, G4. [CrossRef]
6. Aitken, S.N.; Yeaman, S.; Holliday, J.A.; Wang, T.; Curtis-McLane, S. Adaptation, Migration or Extirpation: Climate Change Outcomes for Tree Populations. *Evol. Appl.* **2008**, *1*, 95–111. [CrossRef]
7. Christmas, M.J.; Breed, M.F.; Lowe, A.J. Constraints to and Conservation Implications for Climate Change Adaptation in Plants. *Conserv. Genet.* **2016**, *17*, 305–320. [CrossRef]
8. Guerrant, E.O.; Havens, K.; Vitt, P. Sampling for Effective Ex Situ Plant Conservation. *Int. J. Plant Sci.* **2014**, *175*, 11–20. [CrossRef]
9. Hoban, S.; Volk, G.; Routson, K.J.; Walters, C.; Richards, C. Sampling Wild Species to Conserve Genetic Diversity. In *North American Crop Wild Relatives, Volume 1: Conservation Strategies*; Greene, S.L., Williams, K.A., Khoury, C.K., Kantar, M.B., Marek, L.F., Eds.; Springer International Publishing: Cham, Switzerland, 2018; pp. 209–228. ISBN 978-3-319-95101-0.
10. McGlaughlin, M.E.; Riley, L.; Brandsrud, M.; Arcibal, E.; Helenurm, M.K.; Helenurm, K. How Much Is Enough? Minimum Sampling Intensity Required to Capture Extant Genetic Diversity in Ex Situ Seed Collections: Examples from the Endangered Plant Sibara Filifolia (Brassicaceae). *Conserv. Genet.* **2015**, *16*, 253–266. [CrossRef]

11. Havens, K.; Vitt, P.; Still, S.; Kramer, A.T.; Fant, J.B.; Schatz, K. Seed Sourcing for Restoration in an Era of Climate Change. *Nat. Areas J.* **2015**, *35*, 122–133. [CrossRef]
12. Neale, J.R. Genetic Considerations in Rare Plant Reintroduction: Practical Applications (or How Are We Doing?). In *Plant Reintroduction in a Changing Climate: Promises and Perils*; Maschinski, J., Haskins, K.E., Raven, P.H., Eds.; The Science and Practice of Ecological Restoration; Island Press/Center for Resource Economics: Washington, DC, USA, 2012; pp. 71–88. ISBN 978-1-61091-183-2.
13. Christie, K.; Strauss, S. Along the Speciation Continuum: Quantifying Intrinsic and Extrinsic Isolating Barriers across Five Million Years of Evolutionary Divergence in California Jewelflowers. *Evolution* **2018**, *72*, 1063–1079. [CrossRef] [PubMed]
14. Berend, K.; Haynes, K.; MacKenzie, C.M. Common Garden Experiments as a Dynamic Tool for Ecological Studies of Alpine Plants and Communities in Northeastern North America. *Rhodora* **2019**, *121*, 174–212. [CrossRef]
15. Hamann, E.; Kesselring, H.; Stöcklin, J. Plant Responses to Simulated Warming and Drought: A Comparative Study of Functional Plasticity between Congeneric Mid and High Elevation Species. *J. Plant Ecol.* **2018**, *11*, 364–374. [CrossRef]
16. Huxman, T.E.; Winkler, D.E.; Mooney, K.A. A Common Garden Super-Experiment: An Impossible Dream to Inspire Possible Synthesis. *J. Ecol.* **2022**, *110*, 997–1004. [CrossRef]
17. Aronson, J.; Kigel, J.; Shmida, A. Reproductive Allocation Strategies in Desert and Mediterranean Populations of Annual Plants Grown with and without Water Stress. *Oecologia* **1993**, *93*, 336–342. [CrossRef] [PubMed]
18. Schwinning, S.; Lortie, C.J.; Esque, T.C.; DeFalco, L.A. What Common-Garden Experiments Tell Us about Climate Responses in Plants. *J. Ecol.* **2022**, *110*, 986–996. [CrossRef]
19. Corlett, R.T.; Tomlinson, K.W. Climate Change and Edaphic Specialists: Irresistible Force Meets Immovable Object? *Trends Ecol. Evol.* **2020**, *35*, 367–376. [CrossRef]
20. Montrone, A.; Saito, L.; Weisberg, P.J.; Gosejohan, M.; Merriam, K.; Mejia, J.F. Climate Change Impacts on Vernal Pool Hydrology and Vegetation in Northern California. *J. Hydrol.* **2019**, *574*, 1003–1013. [CrossRef]
21. US Fish and Wildlife Service Acanthomintha Ilicifolia (San Diego Thornmint). 5-Year Review: Summary and Evaluation. In *Acanthomintha Ilicifolia (San Diego thornmint). 5-Year Review: Summary and Evaluation*; U.S. Fish and Wildlife Service Carlsbad Fish and Wildlife Office: Carlsbad, CA, USA, 2009.
22. DeWoody, J.; Rogers, D.L.; Hipkins, V.D.; Endress, B.A. Spatially Explicit and Multi-Sourced Genetic Information Is Critical for Conservation of an Endangered Plant Species, San Diego Thornmint (*Acanthomintha Ilicifolia*). *Conserv. Genet.* **2018**, *19*, 893–907. [CrossRef]
23. Milano, E.R.; Vandergast, A.G. *Population Genomic Surveys for Six Rare Plant Species in San Diego County, California*; US Geological Survey: Reston, VA, USA, 2018; Volume 2018-1175, p. 72.
24. Leiblein-Wild, M.C.; Tackenberg, O. Phenotypic Variation of 38 European Ambrosia Artemisiifolia Populations Measured in a Common Garden Experiment. *Biol. Invasions* **2014**, *16*, 2003–2015. [CrossRef]
25. McKay, J.K.; Bishop, J.G.; Lin, J.-Z.; Richards, J.H.; Sala, A.; Mitchell-Olds, T. Local Adaptation across a Climatic Gradient despite Small Effective Population Size in the Rare Sapphire Rockcress. *Proc. R. Soc. Lond. Ser. B Biol. Sci.* **2001**, *268*, 1715–1721. [CrossRef] [PubMed]
26. Matesanz, S.; Ramos-Muñoz, M.; Moncalvillo, B.; Rubio Teso, M.L.; García de Dionisio, S.L.; Romero, J.; Iriondo, J.M. Plasticity to Drought and Ecotypic Differentiation in Populations of a Crop Wild Relative. *AoB Plants* **2020**, *12*, plaa006. [CrossRef] [PubMed]
27. Valliere, J.M.; Allen, E.B. Interactive Effects of Nitrogen Deposition and Drought-Stress on Plant-Soil Feedbacks of Artemisia Californica Seedlings. *Plant Soil* **2016**, *403*, 277–290. [CrossRef]
28. Roach, D.A.; Wulff, R.D. Maternal Effects in Plants. *Annu. Rev. Ecol. Syst.* **1987**, *18*, 209–235. [CrossRef]
29. Li, N.; Li, Y. Maternal Control of Seed Size in Plants. *J. Exp. Bot.* **2015**, *66*, 1087–1097. [CrossRef] [PubMed]
30. Gremer, J.R.; Kimball, S.; Keck, K.R.; Huxman, T.E.; Angert, A.L.; Venable, D.L. Water-Use Efficiency and Relative Growth Rate Mediate Competitive Interactions in Sonoran Desert Winter Annual Plants. *Am. J. Bot.* **2013**, *100*, 2009–2015. [CrossRef] [PubMed]
31. Herman, J.J.; Sultan, S.E.; Horgan-Kobelski, T.; Riggs, C. Adaptive Transgenerational Plasticity in an Annual Plant: Grandparental and Parental Drought Stress Enhance Performance of Seedlings in Dry Soil. *Integr. Comp. Biol.* **2012**, *52*, 77–88. [CrossRef]
32. Bauder, E.T.; Sakrison, J.A. *Mechanisms of Persistence of San Diego Thornmint (Acanthomintha Ilicifolia)*; San Diego State Department of Biology: San Diego, CA, USA, 1999; p. 54.
33. Memmott, J.; Craze, P.G.; Waser, N.M.; Price, M.V. Global Warming and the Disruption of Plant–Pollinator Interactions. *Ecol. Lett.* **2007**, *10*, 710–717. [CrossRef]
34. Iler, A.M.; CaraDonna, P.J.; Forrest, J.R.K.; Post, E. Demographic Consequences of Phenological Shifts in Response to Climate Change. *Annu. Rev. Ecol. Evol. Syst.* **2021**, *52*, 221–245. [CrossRef]
35. White, F.J.; Hay, F.R.; Abeli, T.; Mondoni, A. Two Decades of Climate Change Alters Seed Longevity in an Alpine Herb: Implications for Ex Situ Seed Conservation. *Alp Bot.* **2023**, *133*, 11–20. [CrossRef]
36. Welt, R.S.; Litt, A.; Franks, S.J. Analysis of Population Genetic Structure and Gene Flow in an Annual Plant before and after a Rapid Evolutionary Response to Drought. *AoB Plants* **2015**, *7*, plv026. [CrossRef] [PubMed]
37. *Center for Plant Conservation CPC Best Plant Conservation Practices to Support Species Survival in the Wild*; Center for Plant Conservation: Escondido, CA, USA, 2019.
38. PRISM Climate Group. Oregon State University PRISM Climate Explorer. Available online: https://prism.oregonstate.edu/terms/ (accessed on 28 February 2023).

39. Bates, D.; Maechler, M.; Bolker, B.; Walker, S. Lme4: Linear Mixed-Effects Models Using Eigen and S4. R Package Version 1.1-34. CRAN. 2014. Available online: https://cran.r-project.org/web/packages/lme4/lme4.pdf (accessed on 28 February 2023).
40. Romano, A.; Stevanato, P. Germination Data Analysis by Time-to-Event Approaches. *Plants* **2020**, *9*, 617. [CrossRef] [PubMed]
41. Therneau, T.M.; Elizabeth, A.; Cynthia, C.; Lumley, T. *Survival: Survival Analysis*, R Package Version 3.5-7. CRAN; 2023. Available online: https://cran.r-project.org/web/packages/survival/survival.pdf/ (accessed on 28 February 2023).

plants

MDPI

Article

Population Assessments of Federally Threatened Everglades Bully in Big Cypress National Preserve, Florida, USA, Using Habitat Suitability Modeling and Micromorphology

James J. Lange [1,2], Courtney L. Angelo [3], Erick Revuelta [2,4] and Jennifer Possley [2,*]

1 Smart-Sciences, Inc., 330 SW 27th Ave STE 504, Miami, FL 33135, USA; jlange@smart-sciences.com
2 Fairchild Tropical Botanic Garden, 10901 Old Cutler Rd., Miami, FL 33156, USA
3 Big Cypress National Preserve, 33100 Tamiami Trail E, Ochopee, FL 34141, USA
4 St. John's River Water Management District, 4049 Reid Street, Palatka, FL 32177, USA
* Correspondence: jpossley@fairchildgarden.org

Abstract: In Big Cypress National Preserve, the federally threatened Everglades bully (*Sideroxylon reclinatum* subsp. *austrofloridense*) is sympatric with its conspecific, more widespread relative, the Florida bully (*Sideroxylon reclinatum* subsp. *reclinatum*). In this area of overlap, the only reliable characters to distinguish the two are cryptic, micromorphological traits of the abaxial laminar surface. In order to better understand the distribution of the federally threatened taxon, we used a combination of habitat suitability modeling (HSM), field surveys, and microscopy. Using models to inform initial surveys, we collected leaf material of 96 individuals in the field, 86 of which we were able to identify to subspecies. Of these, 73 (85%) were identified as the threatened taxon, expanding both the known range and population size within Big Cypress. We used these 73 new occurrences to rerun HSMs to create a more accurate picture of where the taxon is likely to occur. A total of 15,015 hectares were predicted to be suitable habitat within Big Cypress, with 34,069 hectares across the entire study area. These model results could be used to inform the critical habitat designation for this taxon. For at-risk, cryptic taxa, such as the Everglades bully, multiple approaches are needed to inform management and conservation priorities, including the consideration of a hybridization zone.

Keywords: habitat suitability modeling; Everglades bully (*Sideroxylon reclinatum* subsp. *austrofloridense*); Big Cypress National Preserve scanning electron microscope; plant conservation; cryptic speciation

Citation: Lange, J.J.; Angelo, C.L.; Revuelta, E.; Possley, J. Population Assessments of Federally Threatened Everglades Bully in Big Cypress National Preserve, Florida, USA, Using Habitat Suitability Modeling and Micromorphology. *Plants* **2023**, *12*, 1430. https://doi.org/10.3390/plants12071430

Academic Editors: Brenda Molano-Flores and James Cohen

Received: 29 January 2023
Revised: 11 March 2023
Accepted: 13 March 2023
Published: 23 March 2023

1. Introduction

A foundation of organismal conservation is an understanding of the rarity of a given taxon. Rarity generally refers to a taxon's distribution and abundance [1], but can include other factors such as level of habitat specificity [2]. The process of quantifying rarity is inextricably linked to the discipline of taxonomy, which seeks to first define the entity of concern, be it species or intraspecific taxon, as only then can rarity be assessed [3]. Unique taxa are not always discernible using morphological characters alone, despite being evolutionarily distinct based on other criteria, a phenomenon known as cryptic speciation [4]. Advancements in genetic techniques along with a focus on behavioral ecology and various micromorphologies are increasingly identifying new taxa which do not have a clear morphological distinction, i.e., cryptic species [5–7]. These advancements in taxonomic research are providing new tools for understanding rarity, thus enabling a more complete assessment of threats and conservation priorities that can improve our ability to save the most at-risk taxa.

Morphology of laminar surfaces can be particularly informative, as these organ characters can be highly polymorphic and can generate distinguishable features between taxa. Venation and trichome characters are recognized widely as taxonomic tools, but less emphasis has been placed on micromorphological characters such as stomata and epidermal cell walls, likely due to the difficulty and cost of examination. However, this approach has been found to be instrumental in determination of taxa within several widespread genera such as *Solanum* L., *Persicaria* Mill., and *Crotalaria* L. [8–10], to name a few.

The purpose of this study was to determine the extent of an at-risk, cryptic taxon within Big Cypress National Preserve (BICY) in South Florida, USA. The Everglades bully (*Sideroxylon reclinatum* Michx. subsp. *austrofloridense* (Whetstone) Kartesz and Gandhi) was first described in 1985 [11] and was listed as federally threatened in 2017 [12]. *S. reclinatum* sensu lato (s.l.) is a woody shrub in the Sapotaceae found throughout Florida and portions of the Southeastern Coastal Plain. The subspecies was recognized on the basis of abaxial laminar surfaces, pedicels, and calyx being consistently rufous-tomentose, while these on the wider-ranging subsp. *reclinatum* were glabrous or with scattered trichomes along the abaxial midvein [11] (Figure 1). At the time of listing, subsp. *austrofloridense* was known only from Miami-Dade County, FL, chiefly within Everglades National Park (EVER) but also in a limited number of Miami-Dade County and South Florida Water Management District preserves, in habitats including marl prairie, pine rockland, and prairie/pine rockland ecotone. Subsp. *austrofloridense* was not documented in BICY until 2002, when it was discovered during a plant inventory by The Institute for Regional Conservation (IRC) [13]. IRC did not provide detailed population numbers for the taxon in BICY at the time since individual taxa were not the focus of the study. However, subsequent surveys by IRC in 2013 within the Lostmans Pines region of BICY provided a baseline for population estimates [14]. The authors discussed difficulties in identifying *S. reclinatum* s.l. to subspecies during their surveys. Many individuals displayed laminar pubescence characters intermediate between the widespread subsp. *reclinatum* and the South Florida endemic subsp. *austrofloridense*. IRC conservatively determined individuals to be subsp. *austrofloridense* only if mature leaves displayed conspicuous pubescence throughout the abaxial surface. By this standard, they documented 17 individuals of subsp. *austrofloridense* in the Lostmans Pines region.

(a)

(b)

Figure 1. In situ specimens of *S. reclinatum* (**a**) superficially resembling subsp. *reclinatum* with glabrous surfaces from Big Cypress National Preserve, and (**b**) characteristic subsp. *austrofloridense* with rufous-tomentose abaxial surfaces, and pubescent pedicels and calyx from Everglades National Park.

Shortly thereafter, Corogin and Judd [15] published a detailed analysis of the two subspecies using micromorphological characters of abaxial leaf surfaces, which they considered to be more reliable than pubescence. This approach had previously been applied by Anderson [16] in distinguishing the rare *S. thornei* (Cronquist) T.D.Penn. from superficially similar congeners. Corogin and Judd's analysis revealed that both subspecies are cryptically sympatric in BICY. Of five (5) specimens examined from the southern portion of BICY, three

(3) were determined through micromorphology to be subsp. *austrofloridense*, confirming the need for a more detailed and precise survey of the species complex in BICY.

Given the relatively recent discovery of subsp. *austrofloridense* within BICY, along with the vast unsurveyed marl prairie/pineland ecotones found within the Preserve, a more thorough evaluation of the taxon's distribution needed to be explored. For this purpose, we developed habitat suitability models (HSMs) for the taxon, which were used to inform surveys of potential suitable habitat in early 2022. HSMs generated locations for novel areas to survey under this study, both to expand our knowledge of subsp. *austrofloridense* in BICY, but also to survey for optimal translocation sites for taxon restoration, if deemed necessary. We then followed protocols developed by Corogin and Judd [15] to identify a subset of individuals to subspecies and used newly expanded occurrence data to update the HSMs. This expanded spatial dataset will help to inform resource management strategies based on the current condition and rarity of subsp. *austrofloridense*, along with the amount of potential suitable habitat for the taxon.

2. Results

2.1. Model Results

For reference, area under the curve (AUC) scores above 0.9 indicate high accuracy, scores between 0.7 and 0.9 indicate useful applications, and values of 0.5 to 0.7 indicate low accuracy [17].

2.1.1. Pre-Survey Models

The pre-survey model had high accuracy based on test data [(mean AUC = 0.965), standard deviation 0.016]. A jackknife test revealed that the environmental variable with the highest gain when used in isolation was annual maximum water depth, which therefore appeared to have the most useful information by itself. The environmental variable that decreased the gain the most when omitted was vegetation, which therefore appears to have the most information that is not present in the other variables. The pre-survey model generated 28,665 hectares of potential suitable habitat across the study area, 5562 of which were in BICY (Figure 2).

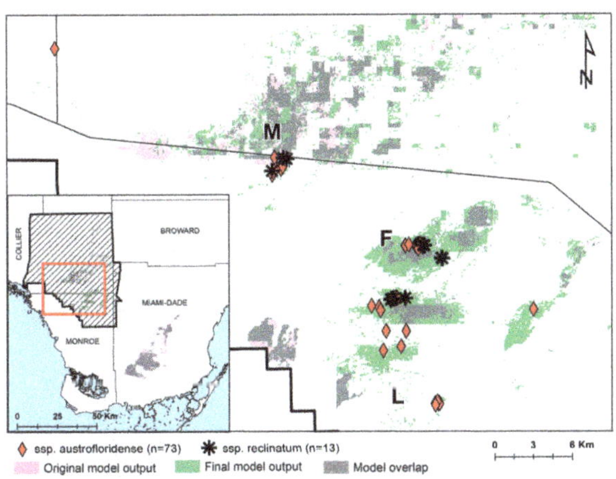

Figure 2. Approximate locations of *Sideroxylon reclinatum* s.l. found during the study that were identified to subspecies. Subsp. *austrofloridense* is represented by red diamonds and subsp. *reclinatum* is represented by black asterisks. Green polygons represent the final HSM model output for focal areas of BICY (map, extent represented by red square in inset) and for the entire study area (inset). BICY boundary is represented as black crosshatch in the inset. Landmarks discussed in results are displayed on the map: M = Monument Lake, F = Frog Hammock, L = Lostmans Pines.

2.1.2. Post-Survey Modeling

The post-survey model had a high accuracy based on test data [(mean AUC = 0.953) standard deviation 0.028]. A jackknife test revealed that the environmental variable with the highest gain when used in isolation was mean annual hydroperiod, as opposed to annual maximum water depth in the pre-survey model. The environmental variable that decreased the gain the most when omitted was again vegetation. The post-survey model generated 29,757.39 hectares of potential suitable habitat across the study area, 11,651.61 of which were in BICY.

The final model output, which included the average plus the standard deviation to allow for a more liberal estimate of potential habitat generated 34,069 hectares of potential suitable habitat across the study area, 15,015 of which were in BICY (Figure 2). In the post-survey model, 64 of the 73 subsp. *austrofloridense* points were within the model output, a vast improvement over only six individuals in the pre-survey models. Based on the large spatial area classified as suitable habitat, along with the fact that 85% of individuals identified to subspecies revealed to be subsp. *austrofloridense*, we estimate the subsp. *austrofloridense* population in BICY to be between 1000–10,000 individuals.

2.2. Surveys

Over the course of the project, we surveyed over 215,000 m or 133.5 total miles based on track log data. Based on this, our spatial extent of detailed rare plant surveys was just over 80 hectares, representing less than 1% of the potential suitable habitat. We recorded 245 separate points of *S. reclinatum* s.l. in the areas we surveyed. From these, we subsampled the leaves of 96 individuals.

2.3. Microscopy

Evaluation of the leaf samples revealed that at least 73 (85% of specimens) of the individuals we sampled were subsp. *austrofloridense*, and 13 (15% of specimens) were subsp. *reclinatum* (see Figure 2 for collection locations, and Figure 3 for images).

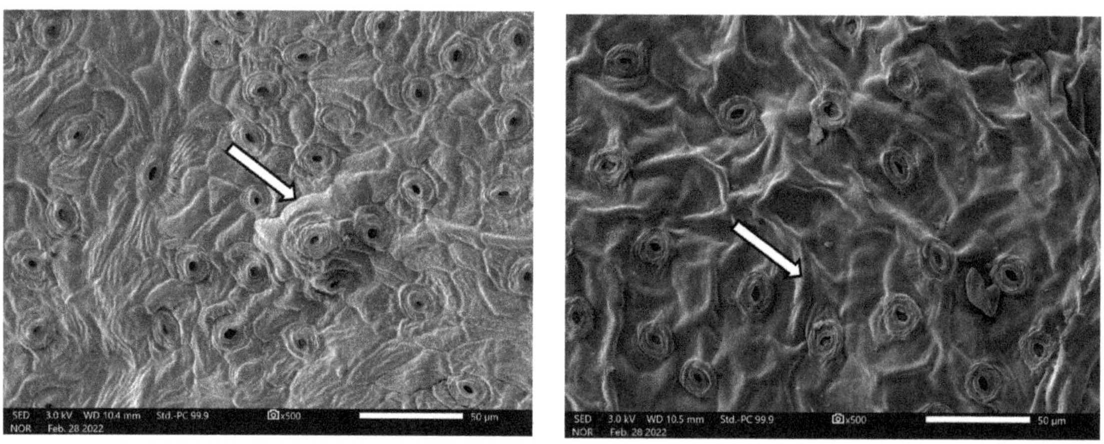

(a) (b)

Figure 3. Comparative scanning electron microscope (SEM) imagery of abaxial laminar surfaces of separate specimens of (**a**) *S. reclinatum* subsp. *austrofloridense* and (**b**) *S. reclinatum* subsp. *reclinatum*. Note the specimens on the left display distinct epidermal cell outlines marked by impressed grooves (arrows), and that specimens on the right display less-distinct cell outlines and exhibit ridges that lack impressed grooves (arrows). See Supplementary Materials for digital microscopy images.

While abaxial laminar pubescence varied with individuals and leaf maturity, we did not observe a single specimen displaying the characteristic rufous, tomentose pubescence of subsp. *austrofloridense* plants in rockland habitats of Miami-Dade County. It is worth noting that every specimen we sampled south of Loop Road was identified as subsp. *austrofloridense*. However, *S. reclinatum* s.l. for unknown reasons was far less frequently encountered in the Lostmans Pines region.

Plants were particularly abundant along the rock reef formation in the middle of the Loop Unit, running east from Frog Hammock Camp. Generally, plants were found in "thickets", or areas characterized by high coverage of other shrubs or small trees relative to the broader landscape. These conditions were found in a variety of cypress, pineland, prairie, and ecotonal habitats. By habitat, *S. reclinatum* s.l. was found most commonly in pineland/prairie ecotone (n = 96), followed by: cypress/prairie ecotone (n = 76), pineland (n = 43), pine/cypress/shrub mix (n = 15), marl prairie (n = 12), and cypress dome (n = 1). With the exception of marl prairie, where only subsp. *austrofloridense* was found, at least one of each subspecies was documented in each of the above listed habitat types, making definitive statements about habitat preference challenging. However, there was a trend toward subsp. *austrofloridense* in BICY being found in habitats with longer hydroperiods when compared to subsp. *reclinatum* (Table 1). It is important to note that 77.0% of subsp. *reclinatum* specimens were found in pine habitat or pineland/prairie ecotones, while this was only 39.7% for subsp. *austrofloridense*. Similarly, 57.5% of subsp. *austrofloridense* specimens were found associated with pond cypress (*Taxodium ascendens* Brongn.) habitat (pine/cypress or cypress/prairie), while this was only 23.1% for subsp. *reclinatum*. The specimen found in the cypress dome was not identified to a subspecies.

Table 1. Habitats in which *Sideroxylon reclinatum* s.l. specimens that were identified to subspecies were found using the percentage of each taxon's conclusive occurrences. Habitats with a "/" represent ecotonal, or mixed associations. The table represents a rough gradient of increasing hydroperiod from left to right.

Subspecies	Habitat				
	Pine	Pine/Prairie	Prairie	Pine/Cypress	Cypress/Prairie
austrofloridense	26	13.7	2.7	16.4	41.1
reclinatum	46.2	30.8	0	7.7	15.4

3. Discussion

Our results represent a significant range extension for the federally threatened subsp. *austrofloridense*. A specimen from Monument Lake was formerly the northernmost known station (Sadle, 630; Bradley, 1547), yet we collected a sample that was identified as subsp. *austrofloridense* outside of the Big Cypress Institute, roughly 16.9 km to the northwest of Monument Lake. Perhaps the most compelling (and confounding) finding from this study was the confirmation of many new occurrences of subsp. *austrofloridense* in BICY, while also confirming population-level sympatry with subsp. *reclinatum*. We found that several populations were mixed, at times with plants just meters from one another (see Figure 2), and despite all plants superficially resembling subsp. *reclinatum* (i.e., lacking abaxial laminar pubescence), the vast majority of individuals sampled matched the micromorphological character of subsp. *austrofloridense*.

We found a tendency toward subsp. *austrofloridense* being more commonly found in longer hydroperiod microsites associated with pond cypress relative to subsp. *reclinatum*, but recognize that the low sample size of the latter restricts this analysis. This result is counterintuitive, since a more sculptured laminar surface, such as found in subsp. *austrofloridense*, is more typically an adaptation to hotter, drier climates [18]. Evolving in rockland soils, such as present in southern BICY and Everglades National Park where subsp. *austrofloridense* occurs, that lack of the capillary capacity to remain saturated during the dry season would favor such adaptations, along with increased pubescence to prevent water loss.

We recognize several limitations in our modeling approach that should be considered in interpretation thereof. For one, our environmental variables were highly correlated. Factors such as hydrology, fire-return intervals, and soil type greatly influence vegetation type, making interpretation of individual variable contributions challenging [19,20]. Secondly, the geographic scope of our models was limited by the geographic extent of long-term hydrologic data provided by the United States Geological Survey's (USGS) Everglades Depth Estimation Network (EDEN). Additionally, the resolution of our models (50 × 50 m) fails to capture factors such as microtopography (e.g., solution holes or small outcrops) and higher resolution vegetation associations such as "thickets" and ecotonal habitats that are likely to influence the occurrence of subsp. *austrofloridense*. Despite these limitations, we believe that our post-survey HSM represents the best available estimate of potentially suitable habitat for subsp. *austrofloridense* and can be used as a foundation for further, more detailed spatial research as well as a baseline tool for conservation biologists.

This study also highlights the importance of the use of micromorphology in taxonomy, specifically in the identification of cryptic taxa and divergent traits. In our study, this attention to detail provided improved clarity into the spatial distribution of a federally listed taxon. We found that while SEM provides far-superior imagery, at least in this case, the relative cost and speed of processing samples should be considered when choosing a methodology. In this case, we found that a trained observer using the high-powered dissecting microscope could identify a leaf sample with confidence in a matter of seconds, despite the relatively low quality of the captured image, and so can quickly abandon the expensive and time-intensive SEM method. Most field biologists likely can get access to a high-powered dissecting microscope through a local university or research institution for a reasonable expense and thus should not be intimidated by the prospects of this level of detail in their work, if deemed necessary.

Our work has determined that these two otherwise geographically isolated taxa overlap in BICY, at the edge of their respective ranges in southwest Florida, and may in fact be actively hybridizing, with intraspecific introgression driving trait expression throughout the hybrid zone. Harrison and Larson [21] discuss the "semi-permeability" of the species boundary and outline analyses of the extent of introgression and interpretation of observed patterns in hybrid zones. Generally, these analyses take place on a geographical or ecological "cline", e.g., latitude, precipitation, etc., through which selective pressures shape allele and genotypic frequencies. These clines can be narrow, as has been documented in *Artemisia tridentata* Nutt. subspecies in Utah where the "basin" and "mountain" taxa generate distinct hybrids across a range roughly 40 m in elevation that occurs rather abruptly in the landscape [22]. However, in the case of *S. reclinatum* s.l. the range of pineland and marl prairies (at times overlying exposed limestone) broadly spans multiple kilometers across the BICY landscape, and thus a broad "hybrid zone" where taxa express superficially similar macro traits should be expected. Natural hybridization can increase intraspecific genetic diversity, and lead to increased potential for adaptation to environmental change [23], and thus it is important to protect this natural hybrid zone. Furthermore, it would not be surprising to find subsp. *austrofloridense* even further to the north, perhaps even moving north with climate change over time if it is in fact more adapted to warmer temperatures and more pronounced dry seasons.

4. Materials and Methods

4.1. Study Site

Big Cypress National Preserve (BICY) consists of 295,000 hectares made up primarily of cypress swamp, pinelands, and marsh communities located in Collier, Monroe, and Miami-Dade Counties in southwest Florida. Topography is relatively flat, gently sloping in a generally southwest direction toward sea-level [24]. The climate has been classified as tropical savanna, with hot, humid summers characterized by relatively high precipitation, and mild, dry winters [25]. The pronounced seasonal variation in precipitation leads to periods of shallow sheet flow across the landscape during the summer and fall. Sheetflow subsides following the rainy season and standing water is found only in deeper slough habitats. Despite a low-relief landscape, a patchwork of habitats is expressed largely based on subtle changes in elevation that determine the hydroperiod [24]. Generally speaking, the lowest areas contain cypress swamps, which transition to marsh habitats at moderate hydroperiods, with pinelands at the highest elevations.

In the southeast portion of BICY where most of our surveys took place, the Pliocene, quartz-rich limestone bedrock is very close to the surface and sometimes exposed, particularly in pinelands, earning them the moniker of pine rocklands. These pine rocklands have a characteristic savannah-like canopy of slash pine (*Pinus elliottii* Engelm.) with understories dominated by saw palmetto (*Serenoa repens* (W.Bartram) Small) with a diverse suite of graminoids and tropical and temperate forbs [26]. These pinelands in BICY typically flood for a short portion of the year [27]. Most of the marshes in this area are marl prairies, diverse, low-stature graminoid communities with short hydroperiods and calcareous marl soils [26]. Similar habitats exist in the Everglades National Park, yet the pine rocklands there are slightly higher, and thus rarely flood and are on a ridge of younger limestone from the Pleistocene called the Miami Rock Ridge [27]. These pine rockland/marl prairie communities are unique to South Florida and boast a high degree of endemism [28].

4.2. MaxEnt Modeling

For the initial habitat suitability model, we used subsp. *austrofloridense* occurrence data from IRC and Corogin and Judd [14,15] within BICY, and occurrence data generated by IRC from EVER. When occurrence data were in the form of a polygon, we used a 25 × 25 m fishnet in ArcMap to generate points within the polygons. For the post-survey model, we included the new occurrences documented by Fairchild Tropical Botanic Garden (FTBG) in this study.

We created raster layers for environmental variables from polygon layers of vegetation, soils, and fire frequency (see Table 2). We also worked with Brian McCloskey from the United States Geological Survey's Everglades Depth Estimation Network (EDEN) to generate raster layers of decadal means (2012–2021) of annual discontinuous hydroperiod (count of all the days in the climatic year that have water depth > 0 cm above ground surface), wet season depth (mean depth 1 June–31 October), and dry season depth (1 November–31 May). All raster layers were assimilated to cells of 50 × 50 m within the study area. The limiting factor of the study area extent was the EDEN network footprint, which is intended to cover freshwater habitats of the Everglades region. Note that the footprint of the EDEN network does not cover the entirety of BICY or EVER, but does cover all relevant areas of pine rockland and marl prairie in which subsp. *austrofloridense* has ever been known to occur, including the Lostmans Pines area, the Loop Unit and limited areas north of Tamiami Trail in BICY. The EDEN network covered all known subsp. *austrofloridense* populations in the Everglades National Park. Several small, isolated populations of subsp. *austrofloridense* occur in urban preserves of Miami-Dade County, but were not included here due to the inability to model hydrology.

Table 2. Data layers used for MaxEnt modeling with geographic extent, source, year(s), file type, and additional comments. * The spatial extent of these layers was limited to that of the USGS EDEN Network.

Data Layer	Geographic Extent	Source(s)	Year(s)	File Type	Description
Sideroxylon occurrences	BICY and EVER	IRC Corogin and Judd FTBG	2012–2014 2014 2022	Point	Includes field data points as well as XY points on 25-m grid that intersect ENP occurrence polygons. One point near Monument Lake Campground was obtained from verified specimen record in Corogin and Judd 2018.
Mean Annual Hydroperiod	BICY and EVER *	EDEN	2012–2021	Raster	Annual Discontinuous Hydroperiod (1 May–30 April climatic year; count of all the days in the climatic year that have water depth > 0 cm above ground surface; averaged across one decade (2012–2021).
Mean Wet Season Depths	BICY and EVER *	EDEN	2012–2021	Raster	Average Wet Season Water Depth (1 June–31 October) over one decade (2012–2021)
Mean Dry Season Depth	BICY and EVER *	EDEN	2012–2021	Raster	Average Dry Season Water Depth (1 November–31 May) over one decade (2012–2021).
Annual Maximum water depth	BICY and EVER *	EDEN	2012–2021	Polygon	Average Annual Maximum Water Depth (1 November–31 May) over one decade (2012–2021).
Vegetation	BICY and EVER	NPS	2017, 2020	Polygon	Level 6 Classification was selected for this model.
Soils	BICY and EVER	USGS	1948, 2012	Polygon	Soils were dissolved based on soil unit name.
Fire Frequency	BICY and EVER	NPS	1978–2020	Polygon	This raster was created by overlapping all fire polygons in the two parks and obtaining the number of fires for each specific area and dividing it by the number of years (42, 1978–2020).

We generated models using a maximum entropy approach in MaxEnt (version 3.4.4; http://www.cs.princeton.edu/~schapire/maxent/ accessed on 8 December 2021 [19,20]). For each run we used 5000 maximum iterations with 10,000 maximum number of background points and a convergence threshold of 0.00001. We subsampled 25% of the occurrence data as test data, leaving the other 75% for training the model. For each final run we used 25 replicates. Since the output of MaxEnt is a continuous probability field, we determined the suitable habitat threshold from each model using maximum training sensitivity plus specificity as suggested by Jiménez-Valverde et al. [29] and utilized by Oleas et al. [30]. Maximum training sensitivity minimizes false negatives and specificity targets a reduction in false positives. We assessed the performance of each model by a receiver operating characteristic analysis (ROC), averaged over replicate runs, using the area under the curve (AUC) of test data (i.e., Test AUC), with X axis as 1-specificity, and sensitivity (1-omission rate) on the Y axis. We assessed the contribution of each environmental variable with a jackknife analysis generated by MaxEnt, wherein the test gain for each variable is given both without said variable and with only said variable, which can be compared to the test gain where all variables are used.

4.3. *Surveys*

We conducted field surveys between January and March of 2022 on eight separate days. Biologists from FTBG were assisted by experienced FTBG volunteers as well as BICY staff, totaling approximately 300 person hours of surveys across the eight days. We prioritized areas where models predicted habitat was suitable, but which had not yet been surveyed by previous researchers. Each surveyor was equipped with a smart phone with Avenza Maps software (Avenza Systems, Toronto, ON, Canada) and a series of georeferenced HSMs in PDF format, at various scales to be able navigate and record track and point data offline in remote parts of BICY. A schema was generated within Avenza with dropdown tabs for taxon, number of individuals, habitat, etc., to maintain consistency in data collection between surveyors. When surveying in teams, individuals would spread out by a minimum of 10 m. To estimate a range of total spatial coverage of surveys, we applied a conservative five-meter buffer to all track logs to account for detectability in a variety of habitats, deleting all duplicate tracks along more frequently used access roads and trails.

When a *S. reclinatum* s.l. individual was located, surveyors collected a point, along with species name, number of individuals, habitat, observer name, presence of standing water, sample number (if collection was made), along with any additional notes or photographs. Since reliable identification characters of subspecies were known to be microscopic, we collected multiple leaves from a subset of individuals (n = 96) across the range of the survey area for later determination. Leaves were placed in small coin envelopes and given a sample number. At the close of each day, envelopes were placed into a plant press and were later dried before analysis.

4.4. *Microscopy*

A combination of scanning electron microscopy (SEM) and high-powered dissecting scopes (1000×) was employed for examination of abaxial leaf surface micromorphology. All leaf material collected during our field surveys was pressed and dried before examination. Specimens were identified to subspecies following [12] wherein: subsp. *austrofloridense* "epidermal cell outlines always clearly visible and marked by an impressed groove. Surface elaborately ornamented with reticulate pattern in strong relief", and subsp. *reclinatum* "epidermal cell outlines not always clearly visible. Surface generally smooth and irregularly undulating". Microscopy imagery from that publication were referred to as a guide.

4.4.1. SEM

Leaf material was mounted on carbon adhesive tabs on aluminum specimen mounts. Samples were rendered conductive by coating them with a gold-palladium alloy in argon vacuum for 90 s using a SPI-MODULE sputter coater (Structure Probe, Inc., West Chester,

PA, USA). Samples were examined with a JEOL JSM 5900LV low vacuum SEM (SEMTech Solutions, Inc., North Ballerica, MA, USA), which includes X-Stream Imaging System to acquire digital micrographs. SEM work was conducted at the Florida Center for Analytical Microscopy at Florida International University. A total of thirteen (13) specimens were examined using this method, and images were processed at 500× magnification.

4.4.2. Digital Microscopy

A total of 83 specimens were examined using this method, and images were processed at 1000× magnification using a Keyence VHX 1000 (Keyence Corporation of America, Itasca, IL, USA) digital microscope with integrated charged-coupled device (CCD) camera. At this high level of magnification, the operator could pan and use the auto-focus function to view multiple sections of leaf, but could often only get a portion of photos in clear focus. Since the goal was to examine as many leaves as possible and the focus was on identification and not final images, we did not stack images for a clearer final product, admittedly leaving portions of photos in poor focus. This work was conducted at the Florida International University Trace Evidence Analysis Facility.

5. Conclusions

This study has revealed the geographic range of subsp. *austrofloridense* to be far more extensive than previously known. As a result, federally designated critical habitat may need to expand. Sympatry of both subspecies, along with seemingly intermediate forms differing merely by cryptic morphological differences suggests potential for an active hybridization zone. Thus, drawing clear boundaries for each subspecies, i.e., putative parents versus hybrids, etc., will remain a challenge. Despite this, the hybridization zone warrants protection as hybridization is an important mechanism in plant evolution [21,23]. Future efforts in defining and conserving these taxa should include additional field surveys and genetic analyses to determine the degree to which the micromorphological differences correspond with genetic differences, and if in fact the genetic differences are significant enough to warrant taxonomic recognition for cryptic BICY populations of subsp. *austrofloridense*. Consideration of how to deal with potential intraspecific hybrids or introgressed populations should be considered from both a taxonomic and regulatory perspective. Genotypic studies within BICY could make for a fascinating story in active evolution taking place in South Florida and should be pursued to better understand gene flow between the two taxa and overall rarity of subsp. *austrofloridense* to inform management and conservation priorities.

Supplementary Materials: The following supporting information can be downloaded at: https://www.mdpi.com/article/10.3390/plants12071430/s1, Figure S1: Comparative digital microscope imagery of abaxial laminar surfaces of separate specimens of (**a**) *S. reclinatum* subsp. *austrofloridense* and (**b**) *S. reclinatum* subsp. *reclinatum*.

Author Contributions: Conceptualization, J.J.L. and C.L.A.; methodology, J.J.L., E.R. and C.L.A.; modeling, J.J.L. and E.R.; analysis, J.J.L.; investigation, J.J.L. and E.R; resources, J.P. and C.L.A.; data curation, J.J.L.; writing—original draft preparation, J.J.L.; writing—review and editing, J.P., E.R. and C.L.A.; project administration, C.L.A. and J.P.; funding acquisition, C.L.A. and J.J.L. All authors have read and agreed to the published version of the manuscript.

Funding: This research was funded by National Park Service, Natural Resource Condition Assessment Program, grant number P21AC10099-00.

Institutional Review Board Statement: Not applicable.

Informed Consent Statement: Not applicable.

Data Availability Statement: All data generated for this project can be made available upon request to Big Cypress National Preserve. Please contact Courtney Angelo at courtney_angelo@nps.gov.

Acknowledgments: Thank you to Brian McCloskey from the United States Geological Survey's Everglades Depth Estimation Network (EDEN) for assistance in compiling hydrologic layers used in our modeling. Steve Schulze of BICY provided soil data for modeling. A special thanks to Thomas H.

Beasley and Ping Jiang from Florida International University's Florida Center for Analytical Electron Microscopy and Trace Evidence Analysis Facility, respectively, for their assistance in obtaining high-quality microscopy data for our *Sideroxylon* analysis. The Big Cypress Institute provided affordable housing during our field trips that made our work both more enjoyable and productive. The Institute for Regional Conservation kindly shared survey data from their previous work in Big Cypress National Preserve. Finally, thank you to the excellent field biologists that contributed their time, effort, and expertise in the field: Ray Morris, Lydia Cuni, Kara Driscoll, Noah Frade, Emily Guinan, and Joshua McBride. Thank you to Meike de Vringer for data organization early in the process. This truly was a group effort. Finally, many thanks to Tony Pernas for encouraging and supporting the funding application for this project.

Conflicts of Interest: The authors declare no conflict of interest.

References

1. Gaston, K.J. *Rarity*; Chapman and Hall: London, UK, 1994; ISBN 978-04-124-7500-9.
2. Coates, D.; Dixon, K. Current perspectives in plant conservation biology. *Aust. J. Bot.* **2007**, *55*, 187–193. [CrossRef]
3. Mace, G.M. The role of taxonomy in species conservation. *Phil. Trans. R. Soc. Lond. B* **2004**, *359*, 711–719. [CrossRef] [PubMed]
4. Bickford, D.; Lohman, D.J.; Sodhi, N.S.; Ng, P.K.; Meier, R.; Winker, K.; Ingram, K.K.; Das, I. Cryptic species as a window on diversity and conservation. *Trends Ecol. Evol.* **2007**, *22*, 148–155. [CrossRef] [PubMed]
5. Edwards, C.E.; Tessier, B.C.; Swift, J.F.; Bassüner, B.; Linan, A.G.; Albrecht, M.A.; Yatskievych, G.A. Conservation genetics of the threatened plant species *Physaria filiformis* (Missouri bladderpod) reveals strong genetic structure and a possible cryptic species. *PLoS ONE* **2021**, *16*, e0247586. [CrossRef] [PubMed]
6. Casacci, L.P.; Barbero, F.; Balletto, E. The "Evolutionarily Significant Unit" concept and its applicability in biological conservation. *Ital. J. Zoo.* **2014**, *81*, 182–193. [CrossRef]
7. Sokoloff, D.D.; Marques, I.; Macfarlane, T.D.; Remizowa, M.V.; Lam, V.K.Y.; Pellicer, J.; Hildalgo, O.; Rudall, P.J.; Graham, S.W. Cryptic species in an ancient flowering-plant lineage (Hydatellaceae, Nymphaeales) revealed by molecular and micromorphological data. *TAXON* **2019**, *68*, 1–19. [CrossRef]
8. Kumar, V.S.A.; Murugan, K. Taxonomic Significance of Foliar Micromorphology and Their Systematic Relevance in the Genus *Solanum* (Solanaceae). In *Prospects in Bioscience: Addressing the Issues*; Sabu, A., Augustine, A., Eds.; Springer: Delhi, India, 2012; pp. 343–349. ISBN 978-81-322-0810-5.
9. Payel, P.; Monoranjan, C. Foliar micromorphology as a tool for identification of Indian taxa of Polygonaceae. *J. Asia Pac. Biodivers.* **2021**, *14*, 569–589. [CrossRef]
10. Parveen, S.N.; Murthy, K.S.; Pullaiah, T. Leaf epidermal characters in *Crotalaria* species (Papilionoideae) from Eastern Ghats. *Phytomorphology* **2000**, *50*, 205–212.
11. Whetstone, R.D. *Bumelia reclinata* var. *austrofloridensis* (Sapotaceae), a new variety from South Florida, USA. *Ann. Mo. Bot. Gard.* **1985**, *72*, 544–547.
12. U.S. Fish and Wildlife Service. Endangered Species Status for *Dalea carthagenensis* var. *floridana* (Florida Prairie-clover), and Threatened Species Status for *Sideroxylon reclinatum* subsp. *austrofloridense* (Everglades Bully), *Digitaria pauciflora* (Florida Pineland Crabgrass), and *Chamaesyce deltoidea* subsp. *pinetorum* (Pineland Sandmat). 82 FR 46691-46715. 2017. Available online: https://www.federalregister.gov/documents/2017/10/06/2017-21617/endangered-and-threatened-wildlife-and-plants-endangered-species-status-for-dalea-carthagenensis-var (accessed on 15 December 2022).
13. Bradley, K.A.; Woodmansee, S.W.; Sadle, J.L.; Gann, G.D. *A Quantitative Plant Inventory of the Big Cypress National Preserve. Report Submitted to the National Park Service Inventory and Monitoring Program*; The Institute for Regional Conservation: Delray Beach, FL, USA, 2005; Available online: https://www.regionalconservation.org/ircs/pdf/publications/2005_01.pdf (accessed on 12 December 2022).
14. Bradley, K.A.; Gann, G.D.; van der Heiden, C.; Green, S. *Status Survey of Everglades Bully and Florida Pineland Crabgrass in the Big Cypress National Preserve, Florida*; Report Submitted to U.S. Fish and Wildlife Service; Institute for Regional Conservation with Mitigation Resources: Vero Beach, FL, USA, 2013.
15. Corogin, P.T.; Judd, W.S. New geographical and morphological data for *Sideroxylon reclinatum* subspecies *austrofloridense* (Sapotaceae), a taxon endemic to southeastern peninsular Florida, USA. *J. Bot. Res. Inst. Texas* **2014**, *8*, 403–417.
16. Anderson, L.C. New geographical and morphological data for *Sideroxylon thornei* (Sapotaceae). *Sida* **1996**, *17*, 343–348.
17. Swets, J.A. Measuring the accuracy of diagnostic systems. *Science* **1998**, *240*, 1285–1293. [CrossRef] [PubMed]
18. Barthlott, W. Epidermal and seed surface characters of plants; systematic applicability and some evolutionary aspects. *Nord. J. Bot.* **1981**, *1*, 345–355. [CrossRef]
19. Phillips, S.J.; Anderson, R.P.; Schapire, R.E. Maximum entropy modeling of species geographic distributions. *Ecol. Modell.* **2006**, *190*, 231–259. [CrossRef]
20. Phillips, S.J.; Dudík, M.; Schapire, R.E. Maxent software for modeling species niches and distributions (Version 3.4.1). Available online: http://biodiversityinformatics.amnh.org/open_source/maxent/ (accessed on 8 December 2021).

21. Harrison, R.G.; Larson, E.L. Hybridization, introgression, and the nature of species boundaries. *J. Hered.* **2014**, *105*, 795–809. [CrossRef] [PubMed]
22. Graham, J.H.; Freeman, D.C.; McArthur, E.D. Narrow hybrid zone between two subspecies of big sagebrush (*Artemisia tridentata*: Asteraceae). II. Selection gradients and hybrid fitness. *Am. J. Bot.* **1995**, *82*, 709–716. [CrossRef]
23. Rieseberg, L.H. Hybrid Origins of Plant Species. *Annu. Rev. Ecol. Evol. Syst.* **1997**, *28*, 359–389. [CrossRef]
24. Duever, M.J.; Carlson, J.E.; Meeder, J.F.; Duever, L.C.; Gunderson, L.H.; Riopelle, L.A.; Alexander, T.R.; Myers, R.L.; Spangler, D.P. *The Big Cypress National Preserve*; Research Report No. 8 of the Audubon Society; National Audubon Society: New York, NY, USA, 1979; Available online: https://docslib.org/doc/8609159/the-big-cypress-national-preserve (accessed on 15 December 2022).
25. Hela, J. Remarks on the Climate of Southern Florida. *Bull. Mar. Sci. Gulf Carib.* **1952**, *2*, 438–447.
26. Florida Natural Areas Inventory. *Guide to the Natural Communities of Florida*; Florida Natural Areas Inventory: Tallahassee, FL, USA, 2010; Available online: https://www.fnai.org/species-communities/natcom-guide (accessed on 15 December 2022).
27. Ross, M.S.; Meeder, J.F.; Sah, J.P.; Herndon, A.; Ruiz, P.L.; Telesnicki, G. *Windthrow in South Florida Pine Rocklands: Pit-and-Mound Features and Plant Microhabitats Associations Following Hurricane Andrew*; Report Submitted to Southeast Environmental Research Program; Florida International University: Miami, FL, USA, 1997; Available online: https://www.researchgate.net/publication/331386100_Windthrow_in_South_Florida_Pine_Rocklands_Pit-and-Mound_Features_and_Plant_Microhabitat_Associations_Following_Hurricane_Andrew/link/5c771472299bf1268d2b0ac7/download (accessed on 15 December 2022).
28. Gann, G.D.; Stocking, C.G. *Floristic Inventory of South Florida Database Online. 2001–2022*; The Institute for Regional Conservation: Delray Beach, FL, USA, 2017; Available online: https://regionalconservation.org/ircs/database/database.asp (accessed on 15 December 2022).
29. Jiménez-Valverde, A.; Lobo, J.M.; Hortal, J. Not as good as they seem: The importance of concepts in species distribution modelling. *Divers. Distrib.* **2008**, *14*, 885–890. [CrossRef]
30. Oleas, N.H.; Meerow, A.W.; Feeley, K.J.; Gebelein, J.; Francisco-Ortega, J. Using species distribution models as a tool to discover new populations of *Phaedranassa brevifolia* Meerow, 1987 (Liliopsida: Amaryllidaceae) in Northern Ecuador. *Check List* **2014**, *10*, 689–691. [CrossRef]

plants

MDPI

Article

Assessing the Reproductive Ecology of a Rare Mint, *Macbridea alba*, an Endangered Species Act Protected Species

Sara A. Johnson [1,2,*], Janice Coons [2,3], David N. Zaya [2] and Brenda Molano-Flores [2]

1 Department of Natural Resources and Environmental Sciences, University of Illinois Urbana-Champaign, 1102 S Goodwin Ave, Urbana, IL 61801, USA
2 Illinois Natural History Survey, 1816 South Oak Street, Champaign, IL 61820, USA; molano1@illinois.edu (B.M.-F.)
3 Department of Biological Sciences, Eastern Illinois University, 600 Lincoln Ave, Charleston, IL 61920, USA
* Correspondence: saraaj@illinois.edu

Abstract: Many rare plant species lack up-to-date research about their reproductive ecology, which challenges effective in situ and ex situ conservation, particularly in the face of ongoing environmental and anthropogenic changes. For protected species, outdated and incomplete information also creates barriers to successful recovery planning and delisting. In this study, we gathered a range of reproductive metrics for the federally threatened and state endangered Florida endemic mint, *Macbridea alba* Chapman (Lamiaceae). We collected data at seven populations within Apalachicola National Forest (Florida, USA) and conducted germination trials to estimate reproductive potential. Additionally, we observed a previously undocumented lepidopteran seed predator for the species and confirmed the occurrence of vivipary. The seed set was low with less than 20% of flowers per inflorescence producing seed across populations; however, germination was high with more than 60% of seeds germinating in five of seven populations. When comparing our results to previous research conducted more than 20 years ago, the results were similar overall (i.e., germination, vivipary); however, new information emerged (i.e., herbivore pressure). As *M. alba* undergoes reassessment as a potential candidate for delisting from the Endangered Species Act (ESA) list, this information is critical for assessing recovery goals and decisions regarding the species' protected status. For recovery needs related to propagation and reintroduction, these results can inform future seed collection and propagation efforts for the species.

Keywords: endemic plants; ex situ conservation; Lamiaceae; *Macbridea alba*; rare plants; recruitment; seed ecology; seed predator; vivipary; herbivory

Citation: Johnson, S.A.; Coons, J.; Zaya, D.N.; Molano-Flores, B. Assessing the Reproductive Ecology of a Rare Mint, *Macbridea alba*, an Endangered Species Act Protected Species. *Plants* **2023**, *12*, 1485. https://doi.org/10.3390/plants12071485

Academic Editor: Luis Echeverria Navarro

Received: 20 February 2023
Revised: 24 March 2023
Accepted: 26 March 2023
Published: 28 March 2023

1. Introduction

Rare and endemic plants are at an increased risk of decline and extinction [1,2] as many of these species exhibit narrow distributions, specialized ecologies, and are low in abundance [3,4]. These characteristics, exacerbated by anthropogenic and natural causes like fragmentation of natural populations, have led to the decline of plant species globally [5,6]. Conservation biologists emphasize that a barrier to effective conservation and recovery planning for rare species is the lack of data regarding species' reproductive biology and ecology [7]. This type of data can improve conservation efforts for at-risk species by documenting limits to recruitment that can inform in situ and ex situ safeguarding efforts [7–9], particularly regarding reintroduction and habitat management plans [10].

The Florida Panhandle is a region within the northwestern part of the state of Florida within the United States of America. This region includes the 10 counties west of the Apalachicola River. It is considered within the United States to be a biodiversity hotspot and harbors many state and federally listed plant species [11], including *Macbridea alba* Chapman (Lamiaceae, *M. alba*, from hereon). *Macbridea alba* is both federally threatened and state endangered. While the species is considered locally abundant, it is geographically restricted to a range of specific habitats in four counties of the Florida panhandle region, or the northwestern portion of Florida in the United States of America [12]. During the 1990s and early 2000s, several *M. alba* studies provided information about pollinators, population genetics, reproductive output, and germination to assist with its conservation [13–17]. *M. alba* is currently a candidate for delisting from the Endangered Species Protection Act list, yet recovery plans are based on outdated information about the current population size, distribution, and reproductive potential of this species, presenting persistent obstacles to conservation efforts [18].

Much has changed throughout the species' range in the decades since those earlier studies. With increased pressure on native populations from current and projected environmental change [19], increased stochastic weather events [20], land use change and development [21], threats to pollinator services [22], and the encroachment of exotic and invasive species [23], it is important to provide an updated account of the species' status. In the case of *M. alba*, few populations are known outside of protected areas and public lands, and those that persist are at a high risk of further decline [12]. In addition, few ex situ collections exist to supplement natural populations. Efforts to propagate, maintain ex situ collections, or reintroduce species to new habitats should be informed by a thorough review of the current life history, habitat specificities, and reproductive ecology. If delisting is the final objective for U.S. rare species, the most up-to-date information regarding their biology and ecology should be a priority to support such delisting.

The goals of this study were to (1) document population size and the number of total flowers per stem among seven populations during one sampling period within Apalachicola National Forest (Tallahassee, FL, USA), (2) document fruit and seed production and germination success for each population, (3) investigate potential variables (i.e., vivipary, herbivory) correlated with *M. alba* seed ecology and possible recruitment, and (4) investigate how reproductive output, germination, and herbivory are correlated to population size and total number of flowers. We collected infructescences from each population to document reproductive output (e.g., flowering, fruit set, and seed set) and conducted germination studies to compare our findings to previous research [14,16]. Based on previous work conducted with the species [14,16], we expected *M. alba* seeds to exhibit high germination and seed viability across populations; however, we expected reproductive output to vary by population. These results will aid in the safeguarding and management of *M. alba* populations in the face of current and future change to advise its protected status.

2. Results

2.1. Fruit Set, Seed Set, Vivipary, and Herbivory

Overall population size ranged from 67 to 800 individuals with a total number of flowers ranging from 177 to 865 across populations (Table 1). The fruit set was generally low across populations with a maximum documented fruit set of 18% (Table 2). Seed set was also low with a maximum documented seed set of 8% (Table 2). Significant differences were observed in fruit set ($\chi 2 = 31.8$, $p < 0.001$, df = 6) and seed set ($\chi 2 = 35.7$, $p < 0.001$, df = 6) among populations (Table 2).

As many as 33% of calyces across individuals within a population contained a pre-germinated seed (Tables 1 and 2). Vivipary (i.e., pre-germinated seed within the calyx) was documented in all populations, and significant differences were found among populations ($\chi 2 = 59.1$, $p < 0.001$, df = 6). In the case of the transplanted viviparous seeds, about 40% of the original 135 transplants survived after ~14 months. However, after ~20 months, all plants perished possibly due to pest damage from mealybugs and growing conditions (i.e., sensitivity to fertilizer application and possible drainage issues with soil mixture).

The proportion of fruits showing signs of herbivory varied significantly among populations ($\chi 2 = 53.3$, $p < 0.001$, df = 6) with up to 37% of calyces per stem across individuals within a population showing signs of herbivore damage (Table 2). Four of the seven sites surveyed exhibited herbivore damage on stems in at least 50% of individuals, with the most herbivore damage documented in 77% of sampled individuals in one population (Table 2).

Micro-lepidoptera specimens, as well as pupal cases, were identified as *Endothenia hebesana* (Walker) (Tortricidae; Verbena Bud Moth).

Table 1. Summary of the number of sampled individual stems, estimated population size, total flowers per population, and estimated average flowers per stem. Average number of seeds per calyx, number of developed seeds, and viviparous seeds collected per population. Percent viviparous seed is a percent of total seed set (developed seed + viviparous seed) for each population. Percent calyces per stem with at least one incidence of vivipary across sampled individuals per population.

Population n = Indv	Population Size	Total Flowers (Avg/Stem) *	Avg Seeds per Calyx	Developed Seed	Viviparous Seed	% Viviparous Seed	% Calyces with Vivipary ** ±se
1 n = 49	800	570 (12)	0.3	78	80	51	32.7 ± 5.9 [a]
2 n = 48	102	332 (7)	0.1	15	3	17	2.5 ± 2.1 [b]
3 n = 19	170	177 (9)	0.3	36	18	33	21.5 ± 8.6 [ab]
4 n = 52	250	546 (11)	0.1	33	7	18	2.9 ± 2.0 [b]
5 n = 48	250	420 (9)	0.1	40	3	7	3.2 ± 2.2 [b]
6 n = 98	445	865 (9)	0.2	164	20	11	4.8 ± 1.3 [b]
7 n = 48	67	452 (9)	0.1	39	4	10	4.4 ± 2.9 [b]

* Flowers are estimated based on number of calyces. ** Letter denoting significant differences are based on the KW-Dunn's test results for medians.

Table 2. Summary of the number of sampled individuals and the average fruit set and seed set (including developed seed and viviparous seed) per stem for each population. Percent calyces per stem and percent of stems per population with at least one incidence of herbivory. The average percent germination for each population as well the range of germination across replicates within a population.

Population n = Indv	% Fruit Set * ±se	% Seed Set * ±se	% Calyces with Herbivore Damage * ±se	% Stems with Herbivore Damage	% Germination ** ±se (Range)
1 n = 49	14 ± 3 [a]	7 ± 2 [a]	7.8 ± 3.8 [a]	73.5	62.6 ± 9.2 [ab] (33.0–87.0)
2 n = 48	4 ± 1 [b]	1 ± 0 [b]	21.2 ± 3.7 [ab]	54.2	33.3 ± 17.6 [b] (0.0–60.0)
3 n = 19	18 ± 5 [ac]	8 ± 2 [ac]	5.5 ± 3.2 [b]	15.8	83.4 ± 7.4 [a] (67.0–100.0)
4 n = 52	3 ± 1 [bc]	1 ± 1 [bc]	13.7 ± 3.3 [b]	32.7	43.4 ± 8.5 [b] (17.0–67.0)
5 n = 48	7 ± 3 [bc]	3 ± 1 [bc]	36.5 ± 4.3 [a]	77.1	63.2 ± 9.7 [ab] (33.0–83.0)
6 n = 98	9 ± 2 [abc]	5 ± 1 [abc]	33.3 ± 3.2 [a]	68.4	72.0 ± 5.6 [ab] (50.0–80.0)
7 n = 48	5 ± 2 [bc]	2 ± 1 [bc]	9.9 ± 2.0 [b]	22.9	60.0 ± 6.6 [ab] (50.0–83.0)

* Letter denoting significant differences are based on the KW-Dunn's test results for medians. ** Letter denoting significant differences are based on the ANOVA-Tukey test results for means.

2.2. Germination Trials

The difference in the average germination among populations was significant ($F_{6,26} = 3.2$, $p < 0.05$), and overall, germination was high with five of seven populations exhibiting successful germination of 60% or more (Table 2). The range across replicates and populations was as low as zero and as high as 100% germination with the highest average at 83% (Table 2). The lowest germination across all populations was an average of 33%. Of the 137 transplants from germination trials, approximately 22% of seedlings survived after ~14 months. However, by the end of the study, all seedlings perished (i.e., ~20 months).

2.3. Correlations

Population size ranged from 67 to 800 with total number of flowers ranging from 177 to 865 across populations (Table 1). Reproductive output (i.e., fruit set, and seed set), germination, and herbivory were not correlated to population size (all r values < 0.676 and all p values > 0.096) or total number of flowers (all r values < -0.236 and all p values > 0.610).

3. Discussion

For most federally listed species, few will have long term demographic, reproductive, and/or ecological data available to inform conservation strategy. For many other rare species, decades may pass between surveys. This study aimed to document reproductive metrics for *Macbridea alba*, a rare Florida mint, after more than two decades since previously published work. Throughout this time, populations numbers have remained stable; however, the number of extant populations have become primarily restricted to public lands, specifically within Apalachicola National Forest [12]. The conversion of natural habitat to cattle pasture or improperly managed timberlands has enabled fragmentation, fire suppression, and woody encroachment: all factors that could impact long-term survival and recruitment of *M. alba* populations [12,13]. Across populations sampled in our study, fruit set and seed set were low despite high floral output. Germination success varied across populations but was high overall. The inspection of seeds confirmed and quantified the occurrence of vivipary and documented potential threats to *M. alba* recruitment and survival, including the presence of a natural seed predator prevalent across *M. alba* populations. These results show that populations can display significant variation in reproductive output, which has important implications for collection and ex situ propagation efforts. Importantly, our research both supports and adds to the existing body of work on *M. alba*'s reproductive ecology. By providing up-to-date data concerning *M. alba* reproductive biology, our results can help to prioritize recovery and safeguarding efforts for the species, as well as informing the species' current protected status. Beyond the immediate applications for *M. alba*, this approach is useful when seeking new data to inform conservation efforts, status assessments, and potential delisting of other rare and listed species.

3.1. Fruit Set, Seed Set, and Herbivory

The observed fruit and seed set were low across all surveyed *M. alba* populations. Although our study used different metrics from a previous work by Madsen (1999, [14]), we observed similar numbers in reproductive output for the species. Madsen referred to individual clumps and reported seeds per flower while we estimated seeds per flower by remnant calyces by stem per individual. In our study, we documented an average range of 0.1 to 0.3 seeds per calyx across populations compared to an average of 0.45 to 1.49 seeds per flower across populations in Madsen's study (1999, [14]). In addition, as in the case of Madsen (1999, [14]), higher floral production did not necessarily equate to an increase in fruit or seed set, and various other sources of variation at the population level likely determine ovule success. It is documented that *M. alba* population numbers and floral production commonly vary year to year depending on environmental conditions and burn history [24]. Studies that monitor floral and seed production over time will improve our understanding of patterns in variations of seed production.

Regardless of population size or number of flowers available, there were few developed fruits and seeds observed across populations, and an increase was not correlated with an increase in reproductive output. While there are many potential explanations, including temporal variability, the low reproductive output could in some part be explained by a low occurrence of pollinator visitations. Pitts-Singer et al. (2002, [15]) noted that although bumblebees and other bees were visiting *M. alba*, visitation rates were low. During site visits in 2019–2020, pollinators and their visits were rarely observed (Sara Johnson and Brenda Molano-Flores, personal observations). A recent review by Sheehan and Klepzig (2022, and citations therein, [25]) highlighted the resilience of the bee communities in the longleaf pine ecosystem and the benefits of habitat management for pollinators. With 73% of plants in the longleaf pine savanna relying on insects for pollination [26], additional observations are needed to better quantify pollinator visitation rates and reproductive output within the context of habitat management for *M. alba* and other rare plants. Depending on the species, different patterns could be observed throughout longleaf pine ecosystems (e.g., *Pinguicula ionantha*, [27]).

Another potential explanation for the low fruit and seed set is the prevalence of a newly documented seed herbivore for the species. *Endothenia hebesana* (Walker) is a polyphagous micro-lepidoptera species that feeds on the developing seeds of host plants [28]. This species has been documented to feed on other genera in the mint family, such as *Scutellaria*, *Veronica*, and *Physostegia*, but has not yet been documented on *M. alba* (James Hayden, Florida Museum of Natural History, personal communication). While this naturally occurring herbivore is unlikely the sole cause of low seed set in *M. alba* populations, herbivory was abundant and present in over 15% (ranging up to 77%) of stems across all sampled populations (Table 2). It is unknown whether *M. alba* individuals compensate for increased herbivory during the flowering season [29,30]. Continued pre-dispersal seed predation in perennials like *M. alba* may contribute to overall mortality, reduced recruitment, and limited population growth [31–33]. Resource limitation (i.e., lack of light or moisture) exacerbated by competition and encroachment may also contribute to the low observed seed set within *M. alba* populations, as encroachment is a major issue and important focus of habitat management in this region [12].

3.2. *Vivipary*

Vivipary has been documented previously in *M. alba* individuals with around 20% of collected seeds germinating in the calyx [16]. In our study, we found populations with higher or lower levels of vivipary than previously reported (Table 1). In addition, pre-germinated seed accounted for ~25% of all seeds collected across populations. It has been suggested that vivipary acts as a reproductive strategy allowing seedlings to overcome limiting growing conditions [34,35]. While increases in humidity and moisture [34], and references therein may lead to an increase in vivipary, in this study, we did not measure these environmental variables and as such, we cannot say whether these variables play the same role in vivipary for *M. alba*. However, *M. alba* calyces are positioned with an open cup shape at the top of the infructescence, creating a location for water to collect during rainy or humid weather. Based on our data, it is uncertain whether this vivipary is adaptive or incidental, and if the adaptive potential could be context dependent based on local conditions. Future research to document the consequence of pre-germinated seed will help provide clues to the success of vivipary for *M. alba*. For example, do seeds successfully fall to the ground and establish or die due to desiccation.

3.3. *Germination*

The ex situ germination success for *M. alba* was over 50% in five of seven populations. While germination results in Schulze et al. (2002, [16]) for *M. alba* focus on varying treatments of age, stratification, and incubation techniques, overall, final average percent germination across the study ranged from 67 to 85%. Mean percent germination was similar in this study at the maximum range (83%) when compared to Schulze et al. (2002, [16]). Two *M. alba* populations had germination ranging from 33–43%, on the lower range of germinability.

Additional work by Schulze et al. (2002, [16]) highlights the temporal nature of *M. alba* seed viability and the lack of persistence in the seed bank. *Macbridea alba* appears tolerant of seed burial only to a depth of less than five cm, as deep burial likely inhibits the emergence of cotyledons from the soil surface (i.e., germination may occur; however, seedlings do not emerge at the soil level and perish due to lack of light, which may be important for *M. alba* germination) [16]. Additionally, fire suppression, competition by invasive species, or woody encroachment could account for low observed seedlings in some populations [12]. These factors, combined with knowledge of *M. alba* seed ecology, suggest that seedling emergence and survivorship could be a limiting recruitment issue in wild populations, not seed germination. Seedlings are infrequently documented in the field, and it is possible that

M. alba seedlings require a select set of temporal and habitat conditions to germinate and survive in the wild. As noted by other studies (e.g., [36]), these combined limitations may present a narrow window of opportunity for successful sexual reproduction in this species.

3.4. Future Work

As with any listed rare plant species, additional work is needed to facilitate delisting. For example, due to the increased pressure from frequent stochastic weather events, isolation of natural populations, and shrinkage of the *M. alba* natural range outside of protected areas, collection and protection should strive to maintain and enhance genetic diversity where possible in both in situ and ex situ conservation efforts. Previous research implicates inbreeding depression as a potential risk to successful *M. alba* recruitment [17]; however, additional research is required to understand the current distribution of genetic diversity across populations and to reassess if inbreeding depression is still a concern. In addition, unknown implications of low outcrossing and limited genetic variability caused by the fragmentation of populations may leave existing *M. alba* populations vulnerable to continued habitat and climate change.

Macbridea alba also has been documented to spread frequently by rhizome via asexual reproduction, and clonal establishment may vary among years or sites. Further work is necessary to specify the primary reproductive strategy of *M. alba* and the frequency of sexual and asexual reproduction, as well as the environmental drivers related to the prevalence of reproductive strategy. For example, how does asexual reproduction change how we define relatedness within populations, and what is the role of vivipary as part of the reproductive strategy of this species? In addition, understanding the primary form of reproduction may help to explain the infrequency of seedlings encountered throughout populations [12,14]. It is possible that *M. alba*'s tendency towards sexual or asexual reproduction may fluctuate due to habitat condition and may vary at different times or seasons [37–39]. Developing a better understanding of these reproductive strategies in conjunction with current population genetics, habitat conditions, and frequency of fires will provide insights into the most productive strategy for maintaining diverse populations at in situ or ex situ locations. Having these data in combination with other datasets, such as long-term monitoring data, could facilitate the development of population viability analyses [40,41] or matrix projection models [42] to inform whether *M. alba* populations are stable, increasing, or declining overall. In addition, demographic models can help link abiotic and biotic factors in the environment to vital rates and overall fluctuations in abundance and reproductive effort [43].

4. Materials and Methods

4.1. Study Species

Macbridea alba Chapman (Lamiaceae, white birds-in-a-nest) is a federally threatened and state endangered perennial herbaceous mint restricted to Bay, Franklin, Gulf, and Liberty counties in the Florida panhandle region, or northwestern portion of the state of Florida in the United States of America [44]. The species was listed as threatened under the ESA in 1992 as threats of habitat degradation caused by poor management practices for timber and cattle were increasing [12]. This fire-adapted and disturbance-dependent species is monitored and managed by state and federal agencies throughout public lands where it persists. Populations are associated with grassy pine flatwoods of the longleaf pine (*Pinus palustris*) ecosystem, but individuals are commonly observed across a range of conditions from wet savannas and sand hills to disturbed roadsides [12]. Individuals typically produce one or more, often branched stems up to ~45 cm in height and are conspicuous when in bloom from May through July. Flowers are bisexual with bright white corollas arranged in a terminal inflorescence (Figure 1), and seeds mature from July to September. Seeds likely have low germination rates in the field and recruitment (i.e., and seedlings) has not been recorded in the wild [12,16].

Figure 1. Photos of plant and reproductive parts of *Macbridea alba* individuals with one or multiple stems per rosette. Photos from left to right: (**1**) basal rosettes in situ; (**2**) basal rosette and root system of ex situ grown individual; (**3**) multi-stemmed inflorescences of single individual (side-profile); (**4**) single inflorescence; (**5**) multi-stemmed inflorescences of multiple individuals (top-profile), (**6**) *Macbridea alba* seeds in petri dish for germination trial; (**7**) bagging single stem of multi-stemmed individual in the field; (**8**) multiple pre-germinated seeds within single calyx; (**9**) evidence of herbivory on calyces of *Macbridea alba* infructescence; (**10**) specimen of *Endothenia hebesana* (Walker) or the Verbena Bud Moth discovered in specimen bag during collection.

Macbridea alba reproduces via rhizome and by seed, producing up to four nutlets per flower. Vivipary occurs occasionally with seeds germinating within the calyx [16]. In addition, *M. alba* is self-compatible and pollinated by bumblebees [15]. Genetic research shows about 92% of genetic diversity is found within populations [17], and genetic diversity may be lower than other perennial Florida mint species [45]. Drought or extreme weather may reduce reproductive output or result in temporary dormancy until conditions improve [15,17]. Research suggests that *M. alba* may be a poor competitor with other plants, as it may require bare ground to germinate and could be restricted by its inability to tolerate shade [46]. Lastly, stored and buried seeds remain viable for up to six months; however, viability rapidly declines after one year. The absence of a persistent seed bank and lack of innate dormancy create a narrow temporal recruitment window for the species [16].

4.2. Study Area

Apalachicola National Forest (ANF) is home to over two-thirds of occurrence records for *M. alba*, as well as multiple long-term monitoring plots maintained by the Florida Natural Areas Inventory (FNAI) [47,48]. Fire suppression and habitat modification has fragmented the once extensive longleaf pine ecosystem of the coastal plain, but intact habitat persists within the protection of ANF and surrounding public lands [12]. Exacerbated by fire suppression and poor forest management practices on both private and public lands, encroachment by woody species has introduced competition, which challenges the survival of *M. alba* and associate herbaceous species within this fire prone region [12,24]. Furthermore, this area is managed by state and federal entities with mechanical and chemical removal of woody species, mowing, and frequent burning; however, burn frequency varies across compartments within ANF.

Preliminary surveys were conducted at a selection of previously reported records (n = 98) to estimate population size. Due to the quantity of sites surveyed during the study period, some population sizes were estimated, and some represent exact counts. Seven populations (at least one kilometer apart) were selected in ANF for seed collection based on their approximate population size and varying habitat and management conditions. Habitat condition ranged across populations in terms of the level of woody encroachment, the cover of vegetation at the canopy, understory, and ground levels, and the microtopography of the site from upland to wetland habitat. Fire compartment data for ANF [49] was utilized to determine the burn history (e.g., time in years since the last burn) for each population.

4.3. Seed Collection

A range of approximately 20 to 98 *M. alba* individuals were haphazardly selected for seed collection from each population based on estimated abundance. In July 2019, after flowering but before fruit development and seed dispersal, individuals were identified by tracing the stem to the base or basal rosette of each plant. One flowering stem per rosette was bagged with a mesh bag. Because a flowering stem often exhibits a branching inflorescence, all flower heads within that flowering stem were bagged. There is a low chance of shading from mesh bags to the infructescence of each plant. In September of the same year, stems were clipped below each bagged infructescence and were brought to the lab for dissection. The total infructescences collected (measured by infructescences per mesh bag) were counted and the total number of calyces (which also estimates flower production) per infructescence were removed and counted. Calyces that contained at least one seed were counted as a fruit. Calyces were dissected to expose seeds, which were then removed, counted, and pooled at the population level. Fully developed seeds were plump and a light tan, whereas any appearing soft or dark in color were considered dead or undeveloped and were discarded.

4.4. Reproductive and Population Metrics

For each individual (i.e., stem), the number of infructescences per stem, the total number of calyces (to estimate floral output per stem), and the number of fruits (i.e., total number of calyces containing at least one seed) were recorded. Fruit set was estimated by counting the number of fruits as a percent of the total number of calyces produced per stem:

$$\%\ fruit\ set = \#\ fruits / \#\ calyces\ per\ stem \times 100$$

In addition to developed seeds, viviparous seeds were encountered during seed extraction, and therefore, fully developed, and viviparous seeds were combined to calculate total set per stem. Each *M. alba* flower produces one ovary with a gynobasic style that may produce a fruit with up to four nutlets, as each of the four ovary lobes may produce a seed/nutlet. Potential seed set per stem was determined by multiplying the total number of calyces per stem by four. Therefore, seed set was defined as a percent of total set from potential seed set.

$$\textit{total set} = \textit{developed seed} + \textit{viviparous seed}$$

$$\%\ \textit{seed set} = \textit{total set/potential seed set} \times 100$$

The percentage of calyces with and without viviparous seed was documented, as well as the total percentage of viviparous seeds of total set per population. All viviparous seedlings gathered during collection were transplanted into 5 × 5 × 5 cm pots in a potting mixture consisting of 3-parts potting soil (peat and perlite), 3-parts sand, 2-parts perlite, 1-part lava rock, 1-part horticultural grit, and 1/2-part white pine (*Pinus strobus*) needles (collected from the researcher's neighborhood) and 1/2 part fine orchid bark (Dalton's Orchiata, Matamata, New Zealand), in ratio of volume. Transplants were raised in a temperature-controlled (see *Germination Trials*) greenhouse from September 2019 onward, and survivorship was documented.

Herbivore damage was documented in several *M. alba* individuals, and it was considered to be herbivore damage if there were holes or damage present on the calyces of the infructescence. The number of calyces with damage per stem was counted during dissection. The number of stems with at least one incidence of herbivory was also summarized for each population. Upon inspection of calyces, insect herbivores (i.e., adults and pupal cases) were collected.

4.5. Germination Trials

Greenhouse germination trials were conducted in the fall of 2019 for a period of two months at the University of Illinois at Urbana-Champaign Plant Sciences Lab Greenhouse. Germination trials were conducted within 2 months of collection, as work by Schulze et al. (2002, [16]) documented successful germination for seeds up to 6 months in age and a lack of dormancy for the species. For germination trials, the number of replicates and the number of seeds per replicate varied by population based on pooled seed collection totals. Seeds were placed in 100 mm by 15 mm plastic Petri dishes lined with one sheet of 90 mm diameter Whatman™ grade 1 filter paper before adding 2 mL of distilled water (dH20) to each. Petri dish edges were sealed with Parafilm™ to reduce evaporation and dishes were placed on a bench in a controlled greenhouse set to a 14 h light (7:15 a.m. to 9:15 p.m.) and 10 h dark period with day temperature set to 22/25 °C and a night temperature of 8/12 °C. Every day or every other day when seeds were checked for germination, position was randomized to avoid a "block effect". Germination was considered as the emergence of the radicle. Few germinants were observed by day six of the trial, and an additional two pieces of filter paper were added to help retain moisture and prevent seeds from drying out. At this time, Captan™ (an antifungal agent, 50% Wettable Powder, BONIDE Products LLC, Oriskany, NY, USA) was sprinkled over the filter paper and seeds to inhibit mold growth. Additional dH20 was added as needed and petri dishes were resealed with Parafilm™ each time. Sprouted seeds were immediately removed and transplanted into 5 × 5 × 5 cm pots in a soil mixture as outlined above. Transplants were raised in greenhouse conditions described above and survivorship was documented. Transplants were watered as needed (approximately every other day or so), and transplants were fertilized monthly with MSU orchid fertilizer (19-4-23, N-P-K, Michigan State University (MSU) formula, East Lansing, MI, USA). Pesticides were used at least twice during this time by greenhouse staff to control greenhouse pests, such as mealybug and thrips.

4.6. Statistical Analysis

To examine differences among populations in fruit set, seed set, amount of viviparous seed, and the number of calyces with herbivore damage, Kruskal–Wallis tests followed by Dunn post hoc tests were conducted. A one-way ANOVA followed by a Tukey post hoc test was used to test differences among populations regarding the overall germination percentage. To standardize the presented data, the average proportions ± SEs per stem for each population are reported, and significance was determined at alpha = 0.05. To determine if reproductive output, germination, and herbivory are correlated to population size and total number of flowers, Pearson Product Moment and Spearman Rank Order correlations were conducted. All analyses were performed using R Version 3.6.3 [50]. The following packages were used: *agricolae* [51], *car* [52], *dplyr* [53], and *rcompanion* [54].

5. Conclusions

This study provides up-to-date data for an endemic mint, *Macbridea alba*, that has remained stagnant on the Endangered Species Act (ESA) list since its listing. Research conducted over 20 years ago provided important baseline data for the species; however, conditions have changed significantly in the species' native range throughout that time. For rare plant species, particularly ESA protected species, updated data can provide useful information for evaluating the effectiveness of current conservation and management plans by documenting potential changes in biology, reproductive output, and recruitment. The information in this study contributes to the outlined recovery needs for the species, particularly the recovery goal of improving in situ and ex situ propagation and reintroduction efforts. With updated data, we hope that conservation practitioners can prioritize recovery goals to protect populations where they persist, and if possible, delist species.

Author Contributions: Conceptualization: S.A.J., J.C. and B.M.-F.; Methodology: S.A.J., J.C., B.M.-F. and D.N.Z.; Formal Analysis: S.A.J., B.M.-F. and D.N.Z.; Investigation: S.A.J., B.M.-F., J.C. and D.N.Z.; Data Curation: S.A.J., J.C. and B.M.-F.; Writing—Original Draft Preparation: S.A.J.; Writing—Review & Editing: S.A.J., J.C., B.M.-F. and D.N.Z.; Funding Acquisition: B.M.-F. and J.C. All authors have read and agreed to the published version of the manuscript.

Funding: This work was supported by U.S. Fish and Wildlife Service Section 6 under the Florida Forest Service, FDACS contract number 025436. Any opinions, findings, conclusions, or recommendations are those of the authors and do not necessarily reflect the views of the U.S. Fish and Wildlife Service.

Data Availability Statement: Due to the nature of species listing status, raw and/or locality data can only be made available through the permission of the U.S. Fish and Wildlife Service. For other questions, please contact the corresponding author.

Acknowledgments: Thanks to Ann Johnson, Amy Jenkins, Jenna Annis, and the staff at Florida Natural Areas Inventory for providing data critical to this project, and Michael Jenkins (Florida Forest Service), Allix North, Sandra Chafin, and Dylan Shoemaker (St. Joseph Bay State Buffer Preserve), and Tate's Hell State Forest for access to public lands and logistical support. Thanks to Eric Janssen and Matthew Candeias (University of Illinois: Urbana-Champaign), and Vivian Negrón-Ortiz (U.S. Fish and Wildlife Service) for field and logistic support. Thank you to James Hayden (Florida Museum of Natural History) for identification of insect specimens. These specimens were retained and deposited into the Florida State Collection of Arthropods, housed in the McGuire Center for Lepidoptera and Biodiversity.

Conflicts of Interest: The authors declare no conflict of interest.

References

1. Estill, J.C.; Cruzan, M.B. Phytogeography of rare plant species endemic to the Southeastern United States. *Castanea* **1999**, *66*, 3–23. Available online: https://www.jstor.org/stable/4033879 (accessed on 1 April 2021).
2. Chichorro, F.; Juslén, A.; Cardoso, P. A review of the relation between species traits and extinction risk. *Biol. Conserv.* **2019**, *237*, 220–229. [CrossRef]
3. Rabinowitz, D. Seven forms of rarity. In *The Biological Aspects of Rare Plant Conservation*; Synge, H., Ed.; John Wiley and Sons: New York, NY, USA, 1981; pp. 205–215.
4. Bevill, R.L.; Louda, S.M. Comparisons of related rare and common species in the study of plant rarity. *Conserv. Biol.* **2001**, *13*, 493–498. [CrossRef]
5. Le Roux, J.J.; Hui, C.; Castillo, M.L.; Iriondo, J.M.; Keet, J.H.; Khapugin, A.A.; Médail, F.; Rejmánek, M.; Theron, G.; Yanelli, F.A.; et al. Recent anthropogenic plant extinctions differ in biodiversity hotspots and coldspots. *Curr. Biol.* **2019**, *29*, 2912–2918. [CrossRef]

6. Lughadha, E.N.; Bachman, S.P.; Leão, T.C.C.; Forest, F.; Halley, J.M.; Moat, J.; Acedo, C.; Bacon, K.L.; Brewer, R.F.A.; Gâteblé, G.; et al. Extinction risk and threats to plants and fungi. *Plants People Planet* **2020**, *2*, 389–408. [CrossRef]
7. Schemske, D.W.; Husband, B.C.; Ruckelshaus, M.H.; Goodwille, C.; Parker, I.M.; Bishop, J.G. Evaluating approaches to the conservation of rare and endangered plants. *Ecology* **1994**, *75*, 584–606. [CrossRef]
8. Raven, P.H. *Ex Situ Plant Conservation: Supporting Species Survival in the Wild*; Island Press: Washington, DC, USA, 2004; Volume 3.
9. Corlett, R.T. Safeguarding our future by protecting biodiversity. *Plant Divers.* **2020**, *42*, 221–228. [CrossRef] [PubMed]
10. Monks, L.; Coates, D.; Bell, T.; Bowles, M.L. Determining success criteria for reintroductions of threatened long-lived plants. In *Plant Reintroduction in a Changing Climate; The Science and Practice of Ecological Restoration*; Maschinski, J., Haskins, K.E., Raven, P.H., Eds.; Island Press: Washington, DC, USA, 2012. [CrossRef]
11. Blaustein, R.J. Biodiversity hotspot: The Florida panhandle. *BioScience* **2008**, *58*, 784–790. [CrossRef]
12. Negrón-Ortiz, V. *Macbridea alba (White Birds-in-a-Nest) 5-Year Review: Summary and Evaluation*; USFWS Technical Report: Panama City, FL, USA, 2020. Available online: https://ecos.fws.gov/docs/tess/species_nonpublish/3070.pdf (accessed on 15 January 2021).
13. Walker, J.L.; White, D. Morphological and flower production changes with time since burning in *Macbridea alba*. *Suppl. Bull. Ecol. Soc. Am.* **1994**, *75*, 240.
14. Madsen, D.L. Seed Production and Germination Studies of *Macbridea alba*. M.S. Thesis, Clemson University, Clemson, SC, USA, 1999.
15. Pitts-Singer, T.L.; Hanula, J.L.; Walker, J.L. Insect pollinators of three rare plants in a Florida longleaf pine forest. *Fla. Entomol.* **2002**, *85*, 308–316. [CrossRef]
16. Schulze, D.; Walker, J.; Spira, T. Germination and seed bank studies of *Macbridea alba* (Lamiaceae), a federally threatened plant. *Castanea* **2002**, *67*, 280–289. Available online: https://www.jstor.org/stable/4034350 (accessed on 18 August 2021).
17. Godt, M.J.; Walker, J.; Hamrick, J.L. Allozyme diversity in *Macbridea alba* (Lamiaceae), an endemic Florida mint. *J. Hered.* **2004**, *95*, 244–249. [CrossRef]
18. United States Fish & Wildlife Service (USFWS). *Recovery Plan for Four Plants of the Lower 423 Apalachicola Region, Florida. Euphorbia telephioides (Telephus spurge), Macbridea alba (White Birds-in-a-Nest), Pinguicula ionantha (Godfrey's butterwort), Scutellaria floridana (Florida 425 Skullcap)*; Southeast Region: Atlanta, GA, USA, 1994.
19. Elmendorf, S.C.; Henry, G.H.R.; Hollister, R.D.; Fosaa, A.M.; Gould, W.; Hermanutz, L.; Hofgaard, A.; Jónsdóttir, I.S.; Jorgenson, J.C.; Lévesque, E.; et al. Experiment, monitoring, and gradient methods used to infer climate change effects on plant communities yield consistent patterns. *Proc. Natl. Acad. Sci. USA* **2015**, *112*, 448–452. [CrossRef] [PubMed]
20. Matthies, D.; Bräuer, I.; Maibom, W.; Tscharntke, T. Population size and the risk of local extinction: Empirical evidence from rare plants. *Oikos* **2004**, *105*, 481–488. [CrossRef]
21. Carr, M.H.; Zwick, P.D. *Florida 2070: Mapping Florida's Future–Alternative Patterns of Development in 2070*; Technical Report, FDACS & 1000 Friends of Florida; Geoplan Center at the University of Florida: Gainesville, FL, USA, 2016. Available online: http://1000friendsofflorida.org/florida2070/wp-content/uploads/2016/09/florida2070technicalreportfinal.pdf (accessed on 18 August 2021).
22. Rhodes, C.J. Pollinator decline–An ecological calamity in the making? *Sci. Prog.* **2018**, *101*, 121–160. [CrossRef]
23. Zaya, D.N.; Leicht-Young, S.A.; Pavlovic, N.B.; Ashley, M.V. Heterospecific pollination by an invasive congener threatens the native American bittersweet, *Celastrus scandens*. *PLoS ONE* **2021**, *16*, e0248635. [CrossRef]
24. Anderson, C.T.; Dietz, S.; Jenkins, A.; Drake, J. *The Effects of Fire Season, Frequency, and Forest Structure on the Flowering Abundance of Macbridea alba*; Florida Natural Areas Inventory: Tallahassee, FL, USA, 2020.
25. Sheehan, T.N.; Klepzig, K.D. Arthropods and fire within the biologically diverse longleaf pine ecosystem. *Ann. Entomol. Soc. Am.* **2022**, *115*, 69–94. [CrossRef]
26. Folkerts, G.W.; Deyrup, M.A.; Sisson, D.C. Arthropods associated with xeric longleaf pine habitats in the southeastern United States: A brief overview. In Proceedings of the Tall Timbers Fire Ecology Conference, Tallahassee, FL, USA, 3–6 November 1993; Tall Timbers Research Inc.: Tallahassee, FL, USA, 1993; Volume 18, pp. 159–192.
27. Molano-Flores, B.; Primer, S.; Annis, J.; Feist, M.A.; Coons, J.; Digges, R. Reproductive ecology of three rare North American *Pinguicula* species. *Plant Species Biol.* **2018**, *33*, 129–139. [CrossRef]
28. Mulvaney, C.R.; Molano-Flores, B.; Whitman, D.W. Is insect herbivory contributing to the threatened status of *Agalinis auriculata* (Orobanchaceae) in Illinois? *J. Torrey Bot. Soc.* **2006**, *133*, 560–565. [CrossRef]
29. Kettenring, K.M.; Weekley, C.W.; Menges, E.S. Herbivory delays flowering and reduces fecundity of *Liatris ohlingerae* (Asteraceae), an endangered, endemic plant of the Florida scrub. *J. Torrey Bot. Soc.* **2009**, *136*, 350–362. [CrossRef]
30. Menges, E.S.; Waller, D.M.; Gawler, S.C. Seed set and seed predation in *Pedicularis furbishiae*, a rare endemic of the St. John River, Maine. *Am. J. Bot.* **1986**, *73*, 1168–1177. [CrossRef]
31. Leimu, R.; Lehtilä, K. Effects of two types of herbivores on the population dynamics of a perennial herb. *Basic Appl. Ecol.* **2006**, *7*, 224–235. [CrossRef]
32. Ancheta, J.; Heard, S. Impacts of insect herbivores on rare plant populations. *Biol. Conserv.* **2011**, *144*, 2395–2402. [CrossRef]
33. Leja, M.; Chi, K.; Molano-Flores, B. Presence and intensity of pre-dispersal seed predation in a rare plant in response to habitat quality and population metrics. *Nat. Areas J.* **2015**, *35*, 542–549. [CrossRef]
34. Lee, J.; Harmer, R. Vivipary, a reproductive strategy in response to environmental stress? *Oikos* **1980**, *35*, 254–265. [CrossRef]
35. Elmqvist, T.; Cox, P. The evolution of vivipary in flowering plants. *Oikos* **1996**, *77*, 3–9. [CrossRef]

36. Eriksson, O.; Fröborg, H. "Windows of opportunity" for recruitment in long-lived clonal plants: Experimental studies of seedling establishment in *Vaccinium* shrubs. *Can. J. Bot.* **1996**, *74*, 1369–1374. [CrossRef]
37. Watkinson, A.; Powell, J. Seedling recruitment and the maintenance of clonal diversity in plant populations–a computer simulation of *Ranunculus repens*. *J. Ecol.* **1993**, *81*, 707–717. [CrossRef]
38. Barrett, S.C.H. Clonality and plant sexual reproduction. *Proc. Natl. Acad. Sci. USA* **2015**, *112*, 8859–8866. [CrossRef]
39. Palagi, J.M.; Ashley, M.V. Deer florivory is associated with changes in clonal structure of the woodland plant Bluebead Lily. *Int. J. Plant Sci.* **2019**, *180*, 357–365. [CrossRef]
40. Menges, E. Population viability analyses in plants: Challenges and opportunities. *Trends Ecol. Evol.* **2000**, *15*, 51–56. [CrossRef] [PubMed]
41. Schwartz, M.W. Assessing population viability in long-lived plants. In *Population Viability in Plants*; Brigham, C.A., Schwartz, M.W., Eds.; Springer: Berlin/Heidelberg, Germany, 2003; Volume 165, pp. 239–266. [CrossRef]
42. Merow, C.; Dahlgren, J.P.; Metcalf, C.J.E.; Childs, D.Z.; Evans, M.E.; Jongejans, E.; Record, S.; Rees, M.; Salguero-Gómez, R.; McMahon, S.M. Advancing population ecology with integral projection models: A practical guide. *Methods Ecol. Evol.* **2014**, *5*, 99–110. [CrossRef]
43. Dahlgren, J.P.; Ehrlén, J. Linking environmental variation to population dynamics of a forest herb. *J. Ecol.* **2009**, *97*, 666–674. Available online: http://www.jstor.org/stable/20528897 (accessed on 12 December 2022). [CrossRef]
44. Atlas of Florida Plants. Available online: http://florida.plantatlas.usf.edu (accessed on 15 May 2021).
45. McDonald, D.B.; Hamrick, J.L. Genetic variation in some plants of Florida scrub. *Am. J. Bot.* **1996**, *83*, 21–27. [CrossRef]
46. Walker, J. Rare vascular plant taxa associated with the longleaf pine ecosystems: Patterns in taxonomy and ecology. In Proceedings of the Tall Timbers Fire Ecology Conference, Tallahassee, FL, USA, 3–6 November 1993; Tall Timbers Research Inc.: Tallahassee, FL, USA, 1993; Volume 18, pp. 105–126.
47. Florida Natural Areas Inventory (FNAI). (Tallahassee, FL, USA). Macbridea alba element of occurrence spatial data. Private communication, 2018.
48. Florida Natural Areas Inventory (FNAI). (Tallahassee, FL, USA). *Florida Conservation Lands*. 2019. Available online: https://www.fnai.org/gisdata.cfm (accessed on 1 August 2019).
49. United States Department of Agriculture (USDA); Michael, J. (Florida Forest Service, Tallahassee, FL, USA). Fire compartment data for Apalachicola National Forest. Private communication, 2020.
50. R Core Team. *R: A Language and Environment for Statistical Computing*; R Foundation for Statistical Computing: Vienna, Austria, 2021. Available online: https://www.R-project.org/ (accessed on 30 January 2021).
51. de Mendiburu, F. *Agricolae: Statistical Procedures for Agricultural Research, R Package Version 1.3-3*; Universidad Nacional Agraria: La Molina, Peru, 2020. Available online: https://CRAN.R-project.org/package=agricolae (accessed on 30 January 2023).
52. Fox, J.; Weisberg, S. *An {R} Companion to Applied Regression*, 3rd ed.; Sage: Thousand Oaks, CA, USA, 2019. Available online: https://socialsciences.mcmaster.ca/jfox/Books/Companion/ (accessed on 30 January 2021).
53. Wickham, H.; François, R.; Henry, L.; Müller, K. dplyr: A Grammar of Data Manipulation. In *R Package Version 1.0.3*; Posit PBC: Boston, MA, USA, 2021. Available online: https://CRAN.R-project.org/package=dplyr (accessed on 30 January 2021).
54. Mangiafico, S. R package version 2.3.27. In *Rcompanion: Functions to Support Extension Education Program Evaluation*; The Comprehensive R Archive Network, Institute for Statistics and Mathematics of WU (Wirtschaftsuniversität Wien): Vienna, Austria. Available online: https://CRAN.R-project.org/package=rcompanion (accessed on 30 January 2021).

plants

MDPI

Article

Defining Populations and Predicting Future Suitable Niche Space in the Geographically Disjunct, Narrowly Endemic Leafy Prairie-Clover (*Dalea foliosa*; Fabaceae)

Ashley B. Morris [1,2,*], Clayton J. Visger [3], Skyler J. Fox [1,4], Cassandra Scalf [2], Sunny Fleming [5] and Geoff Call [6]

1 Department of Biology, Furman University, Greenville, SC 29613, USA; skylerjfox18@gmail.com
2 Independent Researcher, San Antonio, TX 78247, USA; scalf.cassandra@gmail.com
3 Department of Biological Sciences, California State University, Sacramento, CA 95819, USA; clayton.visger@csus.edu
4 Department of Biology, Georgia Southern University, Statesboro, GA 30458, USA
5 Environmental Systems Research Institute, Inc. (ESRI), Redlands, CA 92373, USA; sfleming@esri.com
6 Tennessee Ecological Services Field Office, U.S. Fish and Wildlife Service, Cookeville, TN 38501, USA; geoff_call@fws.gov
* Correspondence: ashley.morris@furman.edu

Abstract: Conservation actions for rare species are often based on estimates of population size and number, which are challenging to capture in natural systems. Instead, many definitions of populations rely on arbitrarily defined distances between occurrences, which is not necessarily biologically meaningful despite having utility from a conservation management perspective. Here, we introduce a case study using the narrowly endemic and highly geographically disjunct leafy prairie-clover (*Dalea foliosa*), for which we use nuclear microsatellite loci to assess the current delimitations of populations and management units across its entire known range. We model future potential suitable niche space for the species to assess how currently defined populations could fare under predicted changes in climate over the next 50 years. Our results indicate that genetic variation within the species is extremely limited, particularly so in the distal portions of its range (Illinois and Alabama). Within the core of its range (Tennessee), genetic structure is not consistent with populations as currently defined. Our models indicate that predicted suitable niche space may only marginally overlap with the geology associated with this species (limestone glades and dolomite prairies) by 2070. Additional studies are needed to evaluate the extent to which populations are ecologically adapted to local environments and what role this could play in future translocation efforts.

Keywords: conservation genetics; *Dalea foliosa*; dolomite prairies; limestone glades; microsatellites; population boundaries; suitable niche

Citation: Morris, A.B.; Visger, C.J.; Fox, S.J.; Scalf, C.; Fleming, S.; Call, G. Defining Populations and Predicting Future Suitable Niche Space in the Geographically Disjunct, Narrowly Endemic Leafy Prairie-Clover (*Dalea foliosa*; Fabaceae). *Plants* **2024**, *13*, 495. https://doi.org/10.3390/plants13040495

Academic Editors: Brenda Molano-Flores and James Cohen

Received: 21 December 2023
Revised: 30 January 2024
Accepted: 6 February 2024
Published: 9 February 2024

1. Introduction

Global biodiversity is threatened at an unprecedented rate by complex interactions between anthropogenic activities and climate change, but increased investment in effective conservation actions has the potential to combat the anticipated number of future species extinctions [1,2]. In the US, analyses have shown that individual species' recovery progress over time is positively correlated with conservation funding but that the current federal budget is insufficient to address the costs associated with recovery actions across all listed species [3–5]. These financial limitations require those responsible for managing species recovery at both federal and state levels to make difficult decisions with respect to prioritizing which actions will both be cost-effective and have the greatest impact on the likelihood of species recovery. Perhaps the most challenging of these decisions are defining populations and characterizing the resiliency of those populations as a strategy for prioritizing recovery actions, neither of which is a trivial task [6–11]. Given these challenges and the overwhelming workload facing state and federal conservation decision makers, developing effective species conservation strategies often requires the involvement of government and non-government partners working together to develop lines of research inquiry that prioritize explicit recovery actions and implement active adaptive management strategies [12–14]. Here, we present a case study focused on the federally endangered plant *Dalea foliosa* (Gray) Barneby (leafy prairie-clover; Fabaceae), in which we (academic researchers and state and federal decision makers) used genetic approaches to assess the biological relevance of distance-delimited population boundaries, and we used climate models and current species distribution data to predict shifts in potential future available habitat based on current ecological requirements. We then discuss the importance of defining populations in biologically meaningful ways when instituting recovery actions.

The US Endangered Species Act of 1973 (H.R. 5961, 117th Congress) indicates that recovery plans should be developed and implemented for any species listed under the Act, and such plans should include "objective, measurable criteria" that, if met, would support a determination to remove the species from the list. These plans should also identify site-specific management actions needed for the species' conservation and survival, as well as cost and time estimates to achieve recovery goals. While the term population is used at least 20 times in the Act, it is never defined. The United States Fish and Wildlife Service (USFWS), which is the entity charged with the implementation of the Act, has more recently developed a Species Status Assessment (SSA) Framework [11,15], which is intended to provide a scientific basis for policy application under the Act regarding species classification and recovery planning and implementation. The authors of the SSA Framework explicitly addressed the importance of and the challenges associated with defining populations, specifically highlighting methods such as genetic analysis and arbitrarily defined distances between groups as examples used by some researchers and practitioners. In the absence of other data, many state agencies charged with monitoring rare species use the arbitrarily defined distance via the concept of Element Occurrence (EO), as defined by NatureServe [16], which typically uses a default 1 km separation between units as a proxy for population delimitation. However, the authors noted that this is an inappropriate measure for many plant species, and when life history or ecological data are available to support modifications in separation distance, such changes should be considered. The issues raised here illustrate how traditional guidance on defining populations or attempts to standardize the definition of populations across plant species are not straightforward and may not be the most appropriate ways to describe biologically meaningful boundaries when attempting to allocate resources for recovery action prioritization.

Recovery actions often include the transplanting of individuals or seeds from one location to another to improve population resiliency by increasing the population size or genetic diversity or increasing redundancy by establishing new populations in perceived suitable habitat [17,18]; therefore, linking the spatial distribution of population genetic diversity on the landscape to the ecological factors regulating those populations is key to predicting potential future changes in species distributions [19,20]. Global climate change is predicted to have significant impacts on species distributions within the next 50 years, with some effects already being observed [21]. Plants are obviously incapable of migrating in the sense that animals are to escape changing environments. However, plants are capable of responding to environmental changes through either the natural selection of existing adaptive genetic variation or phenotypically plastic responses to environmental change [22]. Either of these types of responses is complex and difficult to predict without extensive experimental manipulation, but a first step in understanding how plant populations may respond to climate change is to model future changes in potentially suitable climate niches [23,24]. Such predictive models, when posited in the caveats of their limitations and coupled with population genetic studies, can be powerful tools for conservation decision makers working to prioritize recovery actions.

In this study, we present a case study of *Dalea foliosa* (leafy prairie-clover; Fabaceae), a federally endangered legume associated with the limestone glades and barrens of north Alabama and Middle Tennessee and the dolomite prairies of northern Illinois, using genetic data and future climate modeling to inform current recovery action decision making. We used nine nuclear microsatellite loci to assess genetic variation across 617 individuals from 29 different sampling locations spanning the known distribution of the species (Table 1 and Figure 1). Our analyses were performed with the goal of determining the extent to which current EO or other population boundary determinations used by managing agencies mirror the patterns observed from genetic data. Additionally, we modeled the climatic niche space of current *D. foliosa* populations, projected those models onto the forecasted climate of 2070, and overlaid these projections on top of limestone and dolomite geography to better understand where suitable climate and suitable geology intersect. In combination, we discuss the implications of the results of these two approaches for current recovery actions for this species.

Table 1. *Dalea foliosa* localities sampled for the present study.

Site Name		Site Code	Month Sampled	Collector(s)
Dalea foliosa sites sampled in Illinois				
Keepataw		DF01	July 2014	Pollack
Material Services		DF08	Aug 2014	Pollack
Dellwood		DF03	July 2014	Pollack
Midewin		DF02	Aug 2014	Pollack
Dalea foliosa sites sampled in Tennessee				
Hamilton Creek		DF24	Aug 2015	Fleming
Lebanon		DF33	2022	Elam
Cedar Forest Rd W	Cedars of Lebanon Complex	DF29	June 2021	Elam, Fox, Morris
Vesta		DF31	June 2021	Elam, Fox, Morris
Rowland Barrens E		DF20	Sept 2015	Bishop, Fleming, Williams
Rowland Barrens W		DF19	Sept 2015	Bishop, Fleming, Williams
Richmond Shop Rd		DF21	Sept 2015	Fleming
Cedar Forest Rd S		DF28	June 2021	Elam, Fox, Morris
Cedars Powerline		DF32	June 2021	Elam, Fox, Morris
Hidden Springs		DF30	June 2021	Elam, Fox, Morris
Lane Farm		DF13	Aug 2015	Bishop, Williams
HWY 452 (Lane Farm)		DF27	Sept 2015	
Holly Grove Rd		DF23	Sept 2015	Bishop, Fleming
Hall Farm		DF25	Aug 2015	Bishop, Crabtree
Flat Rock		DF12	2015	
Burnt Hill Rd		DF10	Sept 2015	Bishop, Williams, Call
TVA Powerline (Chapel Hill)		DF26	Sept 2015	Bishop, Fleming
Berlin Glade		DF34	Aug 2022	Elam, Call, Cogburn
Blue Springs	Duck River Complex	DF22	Sept 2015	Bishop, Fleming
Columbia Glade W		DF11	Sept 2015	Fleming
Columbia Glade E		DF18	Sept 2015	Bishop, Williams
Dalea foliosa sites sampled in Alabama				
Franklin Co., AL		DF14	July 2015	Barger
Lawrence Co., AL TVA ROW		DF15	July 2015	Barger
Lawrence Co., AL Paved roadside		DF16	July 2015	Barger
Lawrence Co., AL Dirt roadside		DF17	July 2015	Barger

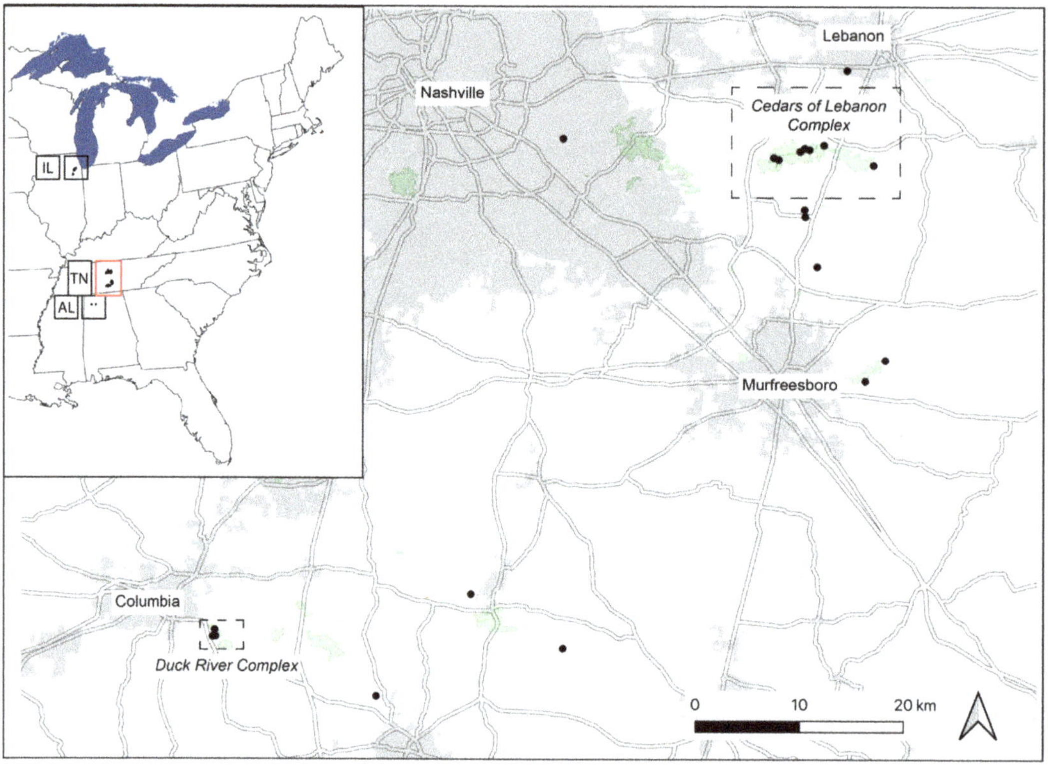

Figure 1. Localities of *Dalea foliosa* genotyped for nine nuclear microsatellite loci in the present study. The inset shows the known species range, with black dots indicating sampling localities; disjunct regions are defined using standard state codes: AL = Alabama; TN = Tennessee; IL = Illinois. The red box in the inset represents the area that is shown in the larger map. Again, black dots indicate sampling localities, and two areas of focus for current recovery actions in Tennessee (Cedars of Lebanon Complex and Duck River Complex) are delineated by dotted lines. Details for each sampled locality are provided in Table 1.

2. Results

2.1. Summary Statistics

Allele frequencies by locus and sampling location are provided in Supplemental Table S1. All four Illinois populations were monomorphic for loci *Dfol082*, *Dfol092*, *Dfol241*, *Dfol016*, *Dfol171*, and *Dfol133*. Only one Illinois population was variable at locus *Dfol023* (population DF03) and at locus *Dfol003* (population DF02). Locus *Dfol005* was variable for three of the four Illinois populations, with population DF03 being the only one monomorphic at this locus. Only one Tennessee population (DF32 in the Cedars of Lebanon Complex) was monomorphic at all loci. All Alabama populations were monomorphic for loci *Dfol023*, *Dfol241*, and *Dfol171*. Three of the four Alabama populations were monomorphic for loci *Dfol005* and *Dfol133*, with the one exception in each case being population DF14. No private alleles were detected in Illinois; one private allele was detected in Tennessee (population DF24, *Dfol005*, $f = 0.021$); and four private alleles were detected in Alabama (population DF15, *Dfol092*, $f = 0.042$; DF16, one each at *Dfol082* and *Dfol016*, $f = 0.042$ and 0.083, respectively; DF17, *Dfol003*, $f = 0.042$). Significant deviation from HWE was detected for all loci except *Dfol171*. The number of populations that deviated from HWE varied by locus, ranging from *Dfol133* with 2 populations to *Dfol005* with 10 populations (Supplemental Table S2).

Summary statistics are provided in Table 2 Percent polymorphism averaged across loci varied from 0.00% (DF32) to 100% (DF23). Regionally, Illinois populations exhibited the lowest levels of polymorphism, with three of the four populations having only 11.11% polymorphic loci. In comparison, the Alabama populations ranged between 11.11% (DF15) and 44.44% (DF16 and DF17). Tennessee populations exhibited both the lowest (0.00%) and highest (100%) values, with much variation in between (Table 2). Observed heterozygosity was consistently low across Illinois populations (H_o = 0.000 in DF03 to 0.023 in DF02), with Alabama populations exhibiting slightly higher values (H_o = 0.009 in DF15 to 0.148 in DF14). Tennessee populations exhibited higher observed heterozygosity values than either of the other two regions, with considerable variation by population (Table 2). Consistently positive values of F across all three regions (but see negative values for DF01, DF25, and DF15) are indicative of inbreeding within populations.

Table 2. Summary statistics for nine microsatellite loci over 617 individuals across 29 sites sampled for *Dalea foliosa*.

Site Name	N	$\%P$	N_a	N_e	H_o	H_e	F	
Illinois sampling locations								
DF01	16	11.11	1.111	1.007	0.007	0.007	−0.032	
DF08	24	11.11	1.222	1.026	0.005	0.021	0.781	
DF03	24	11.11	1.111	1.020	0.000	0.017	1.000	
DF02	24	22.22	1.222	1.091	0.023	0.065	0.632	
Tennessee sampling locations								
DF24	23	66.67	2.444	1.708	0.219	0.330	0.400	
DF33	20	44.44	1.556	1.314	0.106	0.144	0.252	
DF29	21	11.11	1.111	1.077	0.000	0.045	1.000	Cedars of Lebanon Complex
DF31	21	44.44	1.444	1.283	0.101	0.158	0.285	
DF20	23	44.44	1.444	1.057	0.033	0.047	0.127	
DF19	24	55.56	1.556	1.234	0.046	0.138	0.744	
DF21	24	33.33	1.444	1.109	0.056	0.077	0.455	
DF28	23	22.22	1.333	1.269	0.108	0.119	0.080	
DF32	12	0.00	1.000	1.000	0.000	0.000		
DF30	23	33.33	1.333	1.052	0.037	0.040	0.019	
DF13	22	88.89	2.333	1.389	0.147	0.250	0.417	
DF27	23	66.67	2.000	1.316	0.044	0.201	0.764	
DF23	23	100.00	2.333	1.640	0.235	0.363	0.412	
DF25	23	44.44	1.667	1.229	0.130	0.132	−0.021	
DF12	23	11.11	1.111	1.107	0.010	0.055	0.823	
DF10	21	55.56	1.667	1.396	0.119	0.214	0.359	
DF26	19	66.67	2.000	1.412	0.163	0.243	0.373	
DF34	21	44.44	1.444	1.310	0.069	0.175	0.580	
DF22	23	44.44	1.556	1.386	0.169	0.206	0.185	Duck River Complex
DF11	22	22.22	1.222	1.205	0.087	0.107	0.188	
DF18	23	55.56	1.778	1.344	0.110	0.182	0.424	
Alabama sampling locations								
DF14	11	33.33	1.667	1.546	0.148	0.206	0.286	
DF15	11	11.11	1.111	1.010	0.009	0.009	−0.043	
DF16	11	44.44	1.556	1.184	0.058	0.114	0.267	
DF17	12	44.44	1.556	1.092	0.028	0.076	0.635	

N = number of individuals sampled and genotyped; $\%P$ = percentage of polymorphic loci; N_a = mean number of alleles per locus; N_e = number of effective alleles per locus, calculated as $1/(\Sigma p_i^2)$; H_o = observed heterozygosity, calculated as number of heterozygotes/N; H_e = expected heterozygosity, calculated as $1 - \Sigma p_i^2$; F = fixation index, calculated as $H_e - H_o/H_e$.

The lowest pairwise population F_{ST} values were between populations within the Cedars of Lebanon Complex in Tennessee (F_{ST} = 0.008 for DF30 and DF20) and between populations within Illinois (F_{ST} = 0.017 for DF01 and DF08). The highest pairwise population F_{ST} values were consistently between Alabama population DF15 and populations from Tennessee (DF32, F_{ST} = 0.981) and Illinois (DF01, F_{ST} = 0.966). All pairwise population F_{ST} values are provided in Supplemental Table S3. With respect to PCoA results based on the analysis of individuals, 20.64% of the observed variation was explained by the first axis, 34.72% by the first and second axes combined, and 45.66% by all three axes combined. The PCoA results for axes one and two are presented in Figure 2 These results indicate a considerable overlap among populations within the Cedars of Lebanon Complex in Tennessee; there is also a considerable overlap among populations within the Duck River Complex. Illinois populations fall within the space of Tennessee populations, primarily in the area of the Cedars of Lebanon Complex, while Alabama populations exhibit more differentiation but still overlap with Tennessee populations, primarily in the Duck River Complex. Similar but slightly different patterns were observed for PCoA based on populations (see Supplemental Figure S1). There is greater differentiation between Alabama and Tennessee, although Illinois and Tennessee still show some overlap. The results of the Mantel test support isolation by distance with a p-value of 0.010.

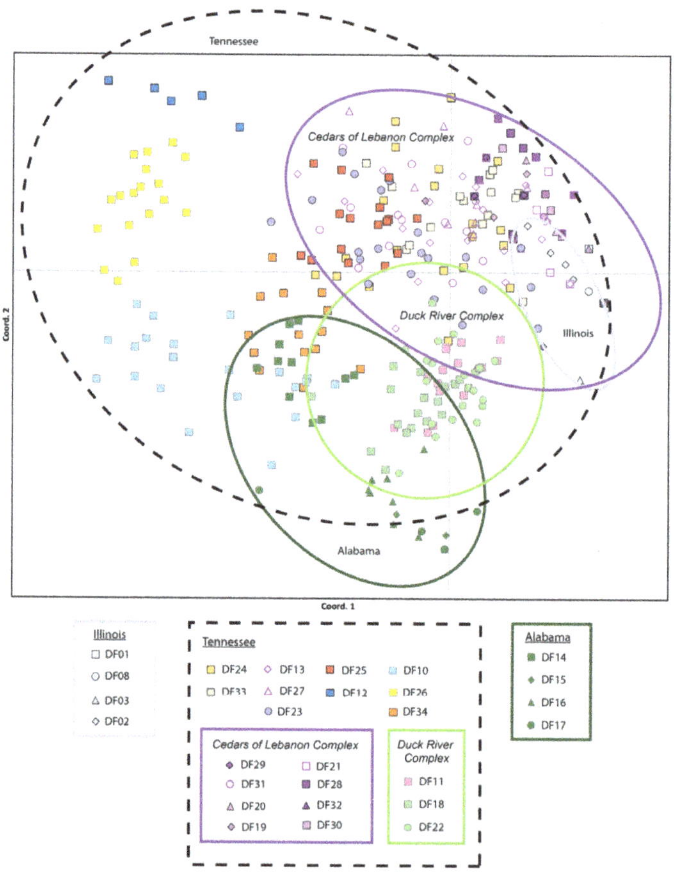

Figure 2. Principal Coordinate Analysis (PCoA) of the 29 populations of *Dalea foliosa* sampled across the species range and genotyped for nine nuclear microsatellite loci. The percentage of variation explained by the first and second axes combined was 34.72%. Color coding is consistent with the STRUCTURE clusters presented in Figure 3, where Illinois sites are coded in pale blue, Tennessee sites are encircled with a black dotted line and further grouped by management unit (Cedars of Lebanon Complex is encircled in purple, Duck River Complex in light green), and Alabama sites are coded in dark green.

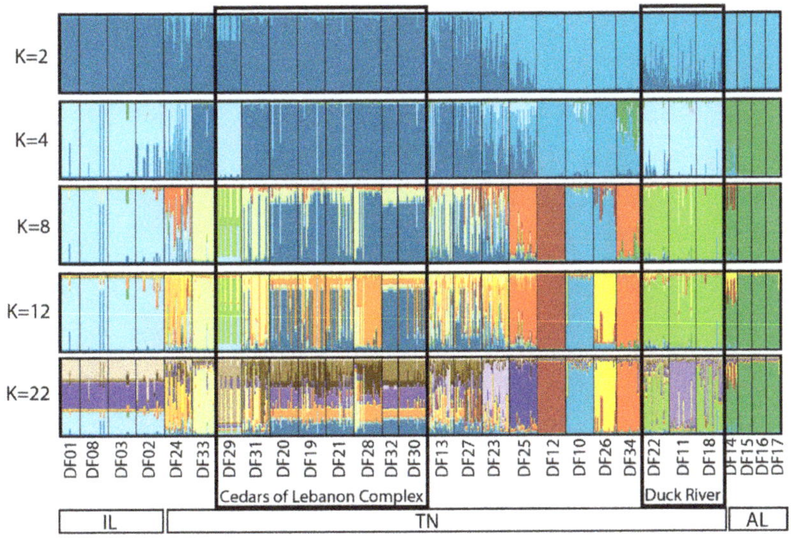

Figure 3. STRUCTURE analysis of the 29 sampled sites of *Dalea foliosa* for the nine microsatellite loci included in the present study. The best approximation of K using Evanno et al. [25] was $K = 2$; the best approximation of K using Pritchard et al. [26] was $K = 22$. Values of K between 2 and 22 presented here are those that illustrate the greatest change between values within that range. Sampling locations from left to right are consistent with the order of sites from north to south, as listed in Tables 1 and 3.

2.2. Individual and Population Assignment

The estimated best K based on Evanno et al. [25] was determined to be $K = 2$, whereas the best K based on Pritchard et al. [26] was $K = 22$. Due to the similarity among values of K between 2 and 22, we chose to present the output for selected values of K (2, 4, 8, 12, and 22) that reflect the range of structure observed (Figure 3). Based on this output, the three geographic regions form distinct clusters (each designated with a distinct color), with no real structure observed among populations within Illinois or among populations within Alabama. Within Tennessee, similarity among populations within the Cedars of Lebanon Complex can be seen, and similarity among populations within the Duck River Complex can be seen. Other populations that are more geographically distinct appear to form distinct clusters. To further illustrate what we view to be the most biologically meaningful representation of these data, a more detailed map view of sampling locations as they relate to the STRUCTURE output for $K = 12$ is shown in Figure 4, where clusters can be clearly separated by the geographic distance between populations.

2.3. Current and Future Climatic Conditions

Climatic models generated for Tennessee and Illinois *D. foliosa* had area under the curve (AUC) scores of 0.997 and 0.948, respectively. The currently predicted suitable climate for both sets of *D. foliosa* populations appears to exceed (and greatly exceed in the case of Illinois) their realized distribution (Figure 5, two left panels), even when considering the intersection of suitable climate with suitable geology. The future predicated suitable climate is substantially shifted and decreased in both disjunct populations, with very little overlap projected to co-occur with suitable geology (Figure 5, two right panels).

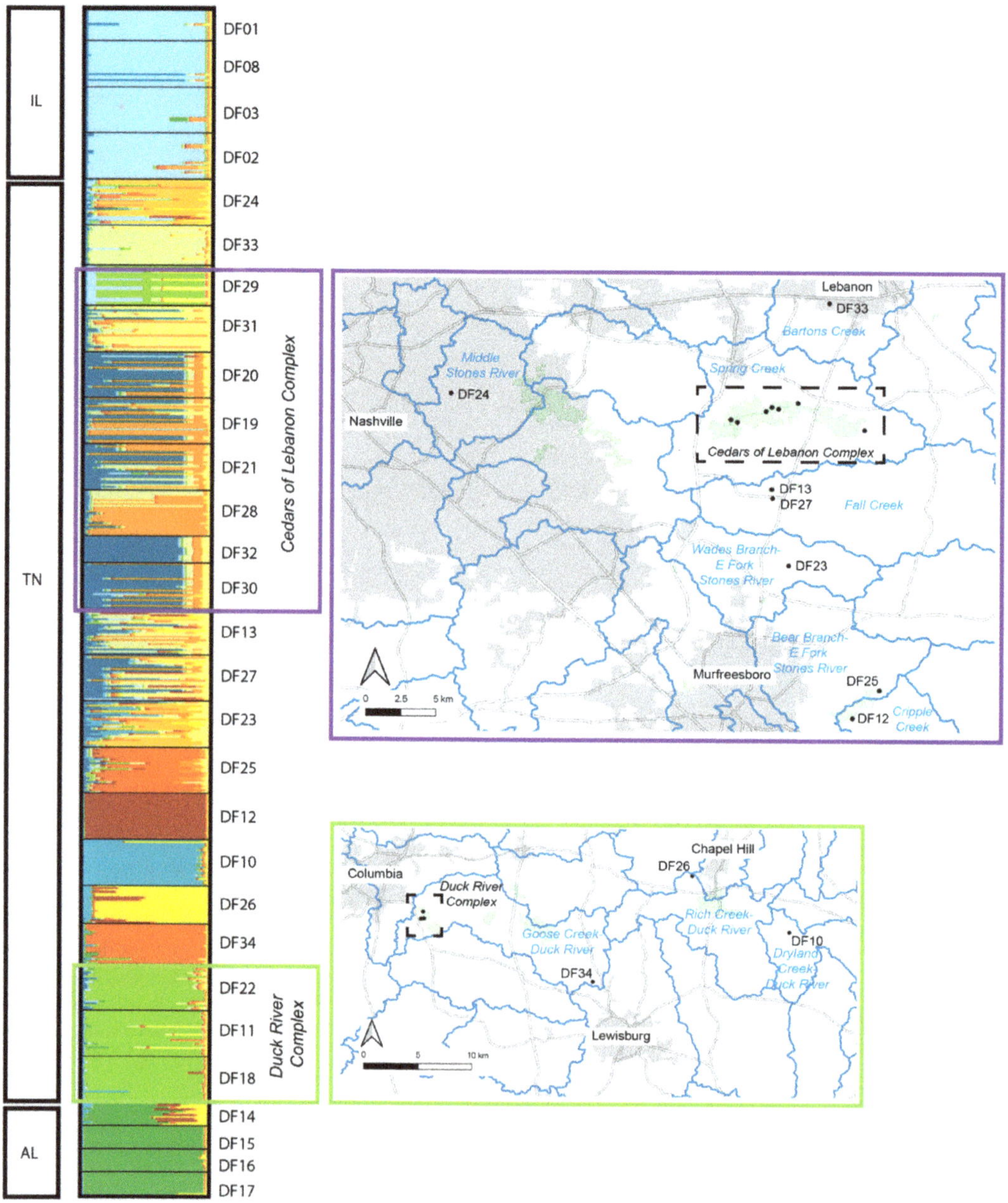

Figure 4. STRUCTURE analysis of *Dalea foliosa* localities sampled in Tennessee for the present study. The two maps presented here provide details for each of the two managed complexes (Cedars of Lebanon Complex and Duck River Complex) shown in Figure 1. The STRUCTURE output here is for $K = 12$, with each value of K represented as a different color. Within the maps, HUC-12 watersheds are delineated by blue lines.

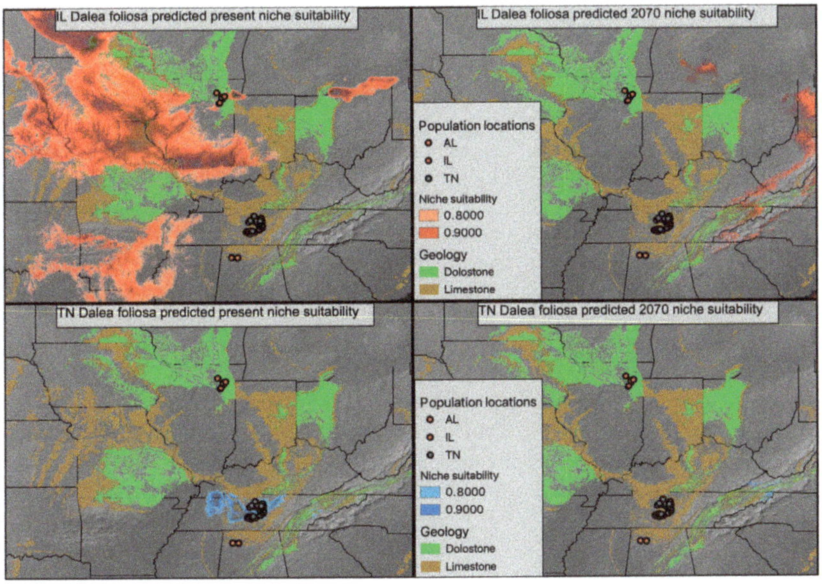

Figure 5. Present niche suitability for *D. foliosa* predicted using Illinois populations (**top left panel**) and Tennessee populations (**bottom left panel**) in red and blue, respectively. The climate models of the Illinois and Tennessee populations were then projected onto the forecasted climate of 2070 in red (**top right panel**) and blue (**bottom right panel**), respectively. A geologic base layer was used to represent the occurrence of dolostone and limestone, the preferred geologies of *D. foliosa*.

3. Discussion

In the US, federally designated endangered and threatened species are subject to management considerations as defined by recovery plans, and the primary emphasis for recovery typically relates to alleviating threats to maintain or increase population resiliency and redundancy and conserve the representation of adaptive variation across a species range [11], yet the definition of population is often arbitrarily defined as a result of limited available biological information. The data we present here indicate that current population designations do not reflect the ways in which sampled sites of *D. foliosa* are connected and/or isolated from one another via gene flow, which can have considerable impacts on the allocation of limited resources for recovery actions.

In Illinois, as previously shown in Morris et al. [27], genetic variation is extremely limited both within and among populations. Based on the data presented here, the breadth of diversity observed in Illinois is fully captured within Tennessee. Tennessee is clearly both the geographic and genetic center of diversity for the species, harboring both the largest numbers of occurrences of individuals and the greatest genetic diversity within and among occurrences. Alabama, much like Illinois, exhibits relatively low levels of genetic variation but also carries private alleles relative to the other two regions. This information, combined with the results of current and future climate modeling, is used below to make explicit recommendations for defining population boundaries for recovery actions.

3.1. Defining Populations in *Dalea foliosa*

As noted, genetic variation within *D. foliosa* is exceedingly low. Edwards et al. [28] previously noted that both *D. foliosa* and a more widespread congener, *D. purpurea*, had lower levels of genetic diversity than were observed in other members of Fabaceae. McMahon and Hufford [29] also noted minimal sequence divergence among *Dalea* species in a phylogenetic analysis based on the nuclear ribosomal internal transcribed spacer (nrDNA ITS), possibly indicating that extant species are evolutionarily young. With this in mind, even subtle genetic variation within and among populations of *D. foliosa* relative to the overall level of diversity within the species may prove important for future translocation efforts.

Occurrences sampled within Illinois exhibit some of the lowest levels of diversity both within and among sites sampled across the species range, with allelic diversity so low that outcrossing between sites within Illinois is unlikely to improve genetic diversity across that portion of the range [27]. This essentially means that the Illinois sites sampled here are inferred to be functioning as a single population, with all extant occurrences found within a 16 km radius. The neighboring sites (mapped as distinct EOs) sampled here are between approximately 5 km and 20 km apart. The lack of genetic variation observed here is likely the result of a genetic bottleneck caused by Pleistocene glacial expansion and retreat, as suggested by Edwards et al. [28]. This idea of Illinois sites as a Pleistocene relict is consistent with the observation that all alleles detected in Illinois were also detected in Tennessee, suggesting that the Illinois sites emerged from a Tennessee center of diversity at some point in the evolutionary past. It is somewhat surprising that, given the hypothesized amount of time that would have passed, genetic drift or mutation has not resulted in differentiation between Illinois and Tennessee. Again, if *D. foliosa* is in fact an evolutionarily young species (which could be further tested using a time-calibrated phylogenetic analysis of *Dalea* and related genera), that could perhaps explain the patterns observed here (i.e., overall low genetic variation, resulting in even lower genetic variation in Pleistocene refugial populations). Future studies using higher-resolution genetic datasets should provide additional insight into the dynamics of these genetically depauperate sites. Based on the data presented here, without additional genetic input from other sources (i.e., Tennessee or Alabama), genetic diversity within Illinois is not expected to increase over time through natural processes, which could ultimately increase the likelihood of local extinction in the region.

In Tennessee, there are at least 20 EOs, as defined by NatureServe protocols [16], included in the present genetic dataset. Based on the combined analyses presented here, only a subset of these represent distinct genetic clusters, or populations. For example, the majority of EOs within the Cedars of Lebanon Complex appear to be functioning as a single population, with either continuous modern-day gene flow among them or a past signature of connectivity that no longer exists. According to pairwise population F_{ST} values (Supplemental Table S3), the sampled sites within the Cedars of Lebanon Complex are as similar to each other as the sampled sites within the Illinois complex. Additionally, sites within the Cedars of Lebanon Complex are consistently more similar to each other than are other Tennessee sites within close geographic proximity to each other (e.g., Duck River Complex). The sites within the Cedars of Lebanon Complex are in relatively close proximity to each other, with distances between most neighboring sites ranging between 0.5 km and 2 km, with the greatest distance between neighboring sites being approximately 5 km. Some of the highest levels of genetic diversity observed within Tennessee are found in sites just 7 km (DF13 and DF27) and 14 km (DF23) south of the Cedars of Lebanon Complex (Table 2, Figures 3 and 4). A portion of this area south of Cedars (DF13 and DF27) is known as Lane Farm, which interestingly shows some genetic similarity to Illinois sites (Supplemental Table S3). Additionally, Lane Farm shows genetic similarity to other sites across Tennessee (DF11, DF25, and DF33), which could suggest that Lane Farm represents a historic genetic source population for the species. However, we acknowledge that additional genetic resolution could change our interpretation of these results, and more data are needed to further support this idea.

As with the Cedars of Lebanon Complex, the Duck River Complex appears to form a tight genetic cluster, while the more geographically isolated sites in the same region appear to reflect distinct clusters, which could be a consequence of the geographic distance between locations there (Figures 3 and 4; Supplemental Table S3). Distances between sampled sites within the Duck River Complex are quite limited (~350–800 m), with sites in the region being separated by 11 to 70 km, in contrast to distances between 550 m and 5 km between neighboring sites in the Cedars of Lebanon Complex.

In Alabama, all four sites sampled are genetically very similar to each other and exhibit very little diversity overall. Three of the four sites (DF15–DF17) are separated by approximately 0.5 km, while the fourth site (DF14) is approximately 25 km away. Based on population pairwise F_{ST} results, there are some similarities between Alabama (DF17) and the Duck River Complex in Tennessee (DF22), as well as to those sites in and around the Lane Farm area (DF13 and DF23). Again, this could suggest that the area around Lane Farm represents a historically central source population for the broader distribution of the species. Additionally, the connectivity between the Duck River Complex and Alabama has been suggested in other glade species (see discussion below), indicating the possibility that the Duck River Complex is another important source population for the species.

Overall, based on the data presented here, Illinois is functionally a single population, as is Alabama. Both Illinois and Alabama suffer from extremely low levels of genetic diversity that warrant immediate concern. The patterns within Tennessee are more complex, with each currently recognized management complex (Cedars of Lebanon and Duck River) functioning as a single population and all other sites being sufficiently geographically isolated to largely warrant their treatment as individual populations. There is no specific distance that appears to be appropriate for defining the separation of populations. The evolutionary histories of both Illinois and Alabama mean that sites separated by as much as 20–25 km still exhibit a great deal of genetic similarity. In contrast, within Tennessee, we observed differentiation among sites within 7 km of each other. Our interpretation of the data presented here is that there is a need for connectivity corridors between geographically distant sites within Tennessee to facilitate gene flow among them and to avoid the potential for the localized extinction of isolated populations. What remains unclear is the role of pollinators in gene flow among sites, what those pollinators are, how far they can travel, and how efficient they are at successful pollen transfer. Preliminary work in Tennessee and Alabama suggests that *D. foliosa* is pollinated by a host of bee species, and these species likely forage within a mile of their nest sites (pers. comm., Bashira Chowdhury). Previous work by Molano-Flores (unpublished data) indicates that, at least in Illinois, *D. foliosa* is self-compatible, and there were no significant differences in fruit set between hand-pollinated and open-pollinated flowers. It is unclear how widespread selfing is in this species and what fitness impacts may result in subsequent generations. A greater understanding of reproductive ecology in this system would provide more insight into what constitutes evolutionarily meaningful distances between populations to facilitate gene flow.

3.2. Comparison with Other Calcareous Glade Endemics

To our knowledge, the only other species with a similar geographic distribution to *D. foliosa* for which genetic data have been published are *Astragalus tenneseensis* A. Gray ex Chapman (Fabaceae) [28] and *Leavenworthia stylosa* A. Gray (Brassicaceae) [30]. *Astragalus tennesseensis* is a perennial herb associated with limestone glades in north Alabama, Middle Tennessee, and dolomite glades in northern Illinois. Based on the allozyme work of Edwards et al. [28], the Illinois populations of *A. tennesseensis* were the least genetically variable, just as was observed in our study for *D. foliosa*. Additionally, the Alabama populations of *A. tennesseensis* appeared most similar to a Tennessee population they called Blue Spring, which occurs within the region we refer to as the Duck River Complex. This is notable because we observed a similar pattern in *D. foliosa*, with pairwise population F_{ST} values between Alabama and Tennessee being most similar between our Blue Springs site (DF22) and Lane Farm (DF13; closer to the Cedars of Lebanon Complex) sites and an Alabama site (DF17).

Leavenworthia stylosa is an annual herbaceous species endemic to the limestone glades of the Central Basin of Tennessee. Dixon et al. [30] used nuclear microsatellite loci to test the assumptions of the abundant-center hypothesis (ACH, also known as the central-marginal hypothesis) in *L. stylosa*. The ACH has received a great deal of attention in recent years [31–33] due to the number of species comparisons that do not support it, as well as the various complicating factors that result in a study not meeting the assumptions of the hypothesis. Therefore, we chose not to formally test this hypothesis in our work. However, the results of Dixon et al. [30] contain several key findings that are relevant to the results we present here. First, the center of the species range for *L. stylosa* is within the Stones River watershed, which is also true for *D. foliosa*. The authors found evidence for individuals of *L. stylosa* from the same watershed sharing ancestry from the same genetic cluster, which is consistent with a hypothesis of water-based dispersal of seeds in that species. According to Baskin and Baskin [34,35], *D. foliosa* in Tennessee fruits into mid-October, at which point the shoots die and hold onto the seed through the winter. The authors described the "freshly matured seed coats" of *D. foliosa* as being impermeable to water and found that very few seeds of a cohort would germinate within the first year. In the species recovery plan [36], potential dispersal agents for *D. foliosa* were listed as wind, gravity, birds, and small mammals, with no mention of dispersal by water. If this is true, then there would be no reason that the watershed should drive genetic structure in *D. foliosa*. However, the species does tend to occur along ephemeral washes that experience localized flooding in winter to early spring, which could serve as a mechanism for dispersal, although there are no empirical studies that have documented seed dispersal mechanisms in this species. To some extent, *D. foliosa* does appear to cluster by watershed (Figure 4), although we suggest that a more viable explanation for what we observe in our data is simply a consequence of the structure by geographic distance. Locations within close geographic proximity to each other tend to exhibit shared ancestry with respect to genetic clusters. This seems likely to be a consequence of limited gene flow between more distant sites due to either pollination limitation or through limited dispersal, rather than being driven by differences in watershed. Finally, Dixon et al. [30] detected higher admixture among populations of *L. stylosa* within the Stones River watershed, concluding that this pattern was indicative of a larger influx of migrants (i.e., pollen or seeds) in that portion of the species range. We also observed greater admixture within the Stones River watershed for *D. foliosa*, specifically within the Cedars of Lebanon Complex (Figures 3 and 4). It is notable that the Cedars of Lebanon Complex (~9000 acres of protected land) is considered the largest undeveloped contiguous area of Central Basin limestone glades and barrens in the world. Additionally, this complex harbors the largest number and highest density of occurrences of *D. foliosa*. Given these facts and the potential for spatial connectivity among extant sites, it should be no surprise that we observe the greatest genetic admixture in this portion of the species range.

3.3. Predicted Future Suitable Climatic Conditions for *Dalea foliosa*

As a narrow endemic, ecological models of *Dalea foliosa* must come with a substantial caveat. Specifically, *D. foliosa* likely has very specific microhabitat requirements, which could include shallow topsoil and flooding frequency [35–37], that are not reflected in broad global models. Therefore, the models we discuss here are better interpreted as predicted suitable climatic conditions and their intersection with the limestone and dolomite geology that defines the glades this species inhabits. Based on both the present and predicted future climatic models, it appears that by 2070, following a moderate climate change scenario, limestone/dolomite geology will only intersect with a modest region of climate similar to the current Illinois *D. foliosa* experience [38,39]. Conversely, in 2070, very little area is predicted to have climatic conditions similar to the current Tennessee *D. foliosa* population conditions. Are there substantive physiological differences between Tennessee and Illinois *D. foliosa* that led to them inhabiting climatically divergent regions? Or are there little to no physiological differences between the two regions and the climatic difference is indicative of a broader climatic tolerance that reflects the pre-disjunction range? Future work involving either reciprocal transplants and/or common garden experiments will be critical for understanding the actual impact climate change will impose on the realized niche space of *D. foliosa*. If Illinois populations of *D. foliosa* are empirically shown to differ in climatic preference from Tennessee or Alabama populations and are better suited to survive the forecasted climatic changes, the limited genetic diversity of these populations may become an area where management officials take a more active role in cultivating genetic diversity via transplants. Furthermore, active management to prevent the encroachment of woody vegetation within extant and potentially suitable *D. foliosa* sites will be needed in order to conserve existing genetic variation and promote gene flow among sites.

4. Methods

4.1. Study Species

Dalea foliosa is a short-lived perennial associated with the highly fragmented limestone glades and barrens of northwest Alabama and Middle Tennessee and the equally imperiled dolomite prairies of northern Illinois. The species was federally listed as endangered in 1991, and at that time, there were 29 "known populations" in three states: Alabama (2), Illinois (3), and Tennessee (24) [40]; recovery criteria indicated that the species could be "considered recovered and eligible for delisting when at least 3 high-viability populations in each Illinois and Alabama and 12 high-viability populations in Tennessee are protected and managed". The primary threat to the species is competition from woody encroachment under restricted fire regimes, followed by habitat destruction and anthropogenic development throughout its geographic range. Furthermore, Tennessee is considered the center of its geographic distribution and the "reservoir of genetic diversity", leading to a recommendation to conserve as many Tennessee populations as possible. All sites in the species range included here are mapped as EOs within Biotics, a national data management platform from NatureServe (pers. comm., Cathy Pollack [USFWS–Illinois], Caitlin Elam [Tennessee Department of Environment and Conservation (TDEC) Division of Natural Areas (DNA)], and Wayne Barger [Alabama Department of Conservation and Natural Resources (ALDCNR)]). The earliest possible recovery date was estimated to be 2005 [40]; as of the most recent 5-Year Review [41], none of the recovery criteria had been met.

Previous genetic work by Edwards et al. [28] surveyed allozyme diversity (nine enzyme systems) among 240 individuals from 10 populations (3 in Illinois, 6 in Tennessee, and 1 in Alabama) across the range of the species. They identified lower-than-expected levels of isozyme diversity, with the Tennessee populations exhibiting the highest levels of variation. Furthermore, the authors concluded that the observed genetic patterns were most likely a result of past evolutionary history related to glacial expansion and retreat rather than to modern-day population dynamics or the genetic makeup of the soil seed bank. More recently, members of our team surveyed nuclear microsatellite diversity at six loci for 226 individuals from 11 populations (9 from Illinois and 2 from Tennessee) to assess genetic diversity within and among managed Illinois sites to evaluate the impacts of augmentation and introduction efforts [27]. As observed by Edwards et al. [28], our results indicated that there are extremely low levels of genetic diversity among Illinois populations, with Tennessee populations exhibiting much higher levels of diversity. Furthermore, we observed less genetic variation within and among the nine Illinois sites surveyed than within one of the Tennessee sites surveyed. We concluded that the Illinois soil seed bank likely has insufficient genetic variation to rescue the current populations but that additional understanding of ecophysiology is needed before we can consider the translocation of material from Tennessee as a possible recovery action.

4.2. Field Sampling

The leaf material for this study was primarily sampled between 2014 and 2022, with the majority of sites being sampled in 2014 and 2015. In the present study, we included a total of 29 sites from across the species range, including 4 in Illinois, 21 in Tennessee, and 4 in Alabama (Table 1). All sites are thought to be naturally occurring, with the exception of one site in Illinois (Keepataw), which is naturally occurring but was augmented with seedlings sourced from other Illinois sites (see [27]). All Illinois sites included here are within an approximately 16 km radius and are mapped as separate EOs; Tennessee sites occur within an approximately 45 km radius and are each mapped as separate EOs; and in Alabama, three of the four sites are within 800 m of each other, with the fourth site being approximately 25 km away, and each of the four is mapped as a separate EO. The southernmost Illinois site is approximately 590 km from the northernmost Tennessee site, while the southernmost Tennessee site is approximately 130 km from the northernmost Alabama site (Figure 1). In Tennessee, the Tennessee Division of Natural Areas (DNA) within the Tennessee Department of Environment and Conservation (TDEC) defines two distinct management units: the Cedars of Lebanon Complex and the Duck River Complex. The Cedars Complex includes eight sites from the current study, while the Duck River Complex includes three sites from the current study, all of which are indicated in Table 1. Note that these management unit designations were not designated as such to imply any biological relationship; instead, they represent EOs in close geographic proximity to each other that are a contiguous state-owned management unit (pers. comm., Caitlin Elam). The remaining sites are treated as separate management units. The GPS coordinates of sites across the range are not provided here for the protection of the species. All material was collected by staff working with USFWS, which is responsible for the monitoring of the species. A leaf or several leaflets were collected from each of a maximum of 30 individuals at each site and immediately stored in silica-gel desiccant for DNA preservation.

4.3. Microsatellite Development and Genotyping

Total genomic DNA was extracted from leaf material using either the Qiagen DNeasy Plant Mini Kit (Qiagen, Valencia, CA, USA) or the Qiagen DNeasy Plant Pro Kit following the manufacturer's protocols with minor modifications; recommendations for difficult species with high concentrations of polyphenolic compounds were followed. Seven nuclear microsatellite loci previously developed for *D. foliosa* [27], as well as three previously unpublished loci, were selected for the present study (Table 3). Each locus was amplified individually following the protocols outlined in Morris et al. [42]. We used the three-primer approach of Schuelke [43] to fluorescently label PCR products. For each locus, a 17-base tail (5′-GTA AAACGACGGCCAGT-3′) was added to the 5′ end of each forward primer, and a 7-base "pigtail" (5′-GTTTCTT-3′) was added to the 5′ end of each reverse primer. A third primer was designed to match the 17-base tail and was fluorescently labeled with either FAM, VIC, or NED fluorophores. The final concentrations for each reaction included 1X Platinum Taq buffer (Life Technologies, Foster City, CA, USA), 2 mM $MgCl_2$, 0.5 μM forward primer with the 5′-M13 tail, 0.15 μM fluorescently labeled M13 primer, 0.2 μM pig-tailed reverse primer, 0.2 mM dNTPs, 0.5 U Platinum Taq (Life Technologies), and 1 μL of DNA. PCR reactions were performed in 96-well plates, with approximately 1/3 of each plate containing positive controls to facilitate consistency in the sizing of alleles across genotyping runs. Two to four loci were poolplexed for genotyping using the LIZ-500 size standard on an ABI 3130 (Life Technologies, Foster City, CA, USA) either at Cornell University Institute of Biotechnology or in the Department of Biology at Middle Tennessee State University. Alleles were sized using GeneMarker MTP software v. 2.6.0 (Softgenetics LLC, State College, PA, USA).

4.4. Summary Statistics

As previously noted, approximately 1/3 of samples were replicated across genotyping runs to monitor consistency in genotyping calls. Through this process, we identified inconsistencies in run patterns for one locus (*Dfol020*) to the extent that we chose not to retain this locus in the dataset. All subsequent analyses were based on a reduced, nine-locus dataset (Table 3). Genetic diversity metrics were calculated using GenAlEx 6.5 [44,45]. Summary statistics were calculated for each population averaged over all loci as well as by individual locus. Loci were tested for Hardy Weinberg Equilibrium, and summary statistics included percent polymorphic loci ($\%P$), the mean number of alleles (N_a), the mean number of effective alleles (N_e), the mean number of private alleles (P_A), observed heterozygosity (H_o), unbiased expected heterozygosity (uH_e), and the fixation index (F). F values near zero indicate random mating; positive values (up to a value of 1.00) indicate inbreeding or undetected null alleles; and negative values indicate heterozygote excess as a result of either negative assortative mating or heterotic selection [44]. Pairwise population F_{ST} values were calculated, and Principal Coordinate Analysis (PCoA) was performed on genetic distance matrices for both individuals and populations, as calculated in GenAlEx following the methods of Peakall et al. [46] and Smouse and Peakall [47], using the covariance-standardized method to further characterize structure within and among populations. A Mantel test for isolation by distance was performed across all populations using a pairwise geographic distance matrix and the output from the population pairwise F_{ST} analysis described above, using 99 permutations.

Table 3. Characterization of nuclear microsatellite loci developed for the federally endangered *Dalea foliosa* (leafy prairie-clover; Fabaceae).

Locus	Primer Sequences (5′-3′)	Repeat Motif	Allele Size (bp)
Dfol003	F: GACATGGGTGGGTATGATTGAAG R: CGCGTGATGAGACCCTTATAAAG	$(AG)_8$	260
Dfol005	F: ATGAAGGAAGATAATACCCGGCC R: CTTGCTGCTTTCGAATCATTCAC	$(AG)_{13}$	164
Dfol016	F: CACACAAACAGGAAGAGAGATGG R: AACTAATGATTCCACCAGCCAAC	$(AG)_{10}$	203
Dfol020 [1]	F: TCAGCGTCTTTGATCATCTGTTC R: TTTCAGGGTGTTTGACAAGGATC	$(AT)_{15}$	255
Dfol023	F: ACCGATGATAGAAGAAAGCAAGG R: TGCTTTCATAGTCTTCAACGTCC	$(AAC)_7$	194
Dfol082 [2]	F: CACACAAACAGGAAGAGAGATGG R: AACTAATGATTCCACCAGCCAAC	$(AG)_8$	201
Dfol092	F: TTTCGCATCGTAACCTGAAGAAG R: GTCTCTGTGTCCTTCATTCCTTG	$(AGG)_7$	213
Dfol133 [2]	F: CACACCGTGGAATCTTACTGTG R: ACCCTCCTTTCCACAAACAATAAG	$(AC)_8$	190
Dfol171 [2]	F: TTCTTCACCTGCGTTGATTATGG R: ATCCAGCAAAGTCTATGAAGCTG	$(AG)_9$	251
Dfol241	F: TGTGACACAAGTTGAACAAGATC R: AGAAATCGCTGTTTCCTTCCAAC	$(AAG)_6$	173

[1] Locus *Dfol020* was removed from the analysis due to inconsistent performance. See text for details. [2] Previously unpublished loci; all other loci previously described in Morris et al. [27].

4.5. Individual and Population Assignment

Individuals were assigned to genetic clusters using STRUCTURE 2.3.4 [26,48] running 150,000 Markov chain Monte Carlo (MCMC) replicates following a burn-in period of 50,000, admixture assumed, and values of K ranging from 1 to 29, with 40 replicates per K value, all using STRUCTURE ver. 2.3.4 [26,49] as implemented in ParallelStructure [50] on the CIPRES Science Gateway [51]. The output was then submitted to CLUMPAK [52] to estimate the best value of K using both Evanno et al. [25] and Pritchard et al. [26]. According to Funk et al. [53], the Pritchard et al. method was more consistent at predicting known horse breeds from a large dataset than was the Evanno method, leading us to compare the results of both methods here. The visual output for K was generated using the CLUMPAK main pipeline with default settings for CLUMPP [54] and DISTRUCT [55].

4.6. Current and Future Climatic Conditions

Occurrence data for Tennessee and Illinois *D. foliosa* populations (n = 82 and n = 9, respectively) were acquired from the state and federal agencies responsible for the annual monitoring of the species. Climatic niche modeling was not conducted on Alabama *D. foliosa* due to the low number of populations and their close proximity to one another. Current and future (2071-2100 ipsl-cm6a-lr ssp370) climatic layers based on those used in Morris et al. [27] were obtained from CHELSA at a 30-arc-second resolution (Bio 10—mean air temp of the warmest quarter; Bio 11–mean air temp of the coldest quarter; Bio 16—mean monthly precipitation of the wettest quarter; Bio 17—mean monthly precipitation of the driest quarter; gst—mean temperature of all growing season days based on TREELIM; and gsp—precipitation sum accumulated on all days during the growing season based on TREELIM) [56,57]. A polygon mask including only the Eastern United States where *D. foliosa* occurs was created using QGIS ver. 3.30.3 [58] and used to trim GIS layers in R ver. 4.3.2 [59].

The logistic output of MaxEnt ver. 3.4.4 [60,61] was used to generate climatic niche models. In addition to default settings, a 15% random test percentage and a maximum of 5000 iterations were used to generate the current model, which was then projected on the future climate dataset. The resulting current and future distributions were visualized in QGIS [58] with a base map plus a GIS layer illustrating the distribution of limestone and dolomite obtained from the State Geologic Map Compilation [62].

5. Conclusions

Dalea foliosa has an unexpectedly low level of genetic variation for a species with such a broad geographic distribution. Additionally, the unique association of this species with specific geological substrates, coupled with the observed modern-day geographic disjunction, raises challenging questions for conservation biologists and land managers: Are the observed patterns innate to the pre-human evolutionary history of the species? Or are they the result of human impacts within the last several hundred years? The current state of the species is most likely a combination of these two scenarios, and recovery actions should consider these histories moving forward. As discussed above, there is no clear geographic distance that can be used to delineate populations of this species. While it is likely valuable to those who are responsible for annual demographic monitoring to define EOs, these should not be considered biologically meaningful units in *D. foliosa*. We note that while we did not discuss the population viability index (PVI) here, this tool is being used as a way to assess recovery in this species [39]. There are many ways in which population viability can be evaluated through models [9,63–65], such that a careful re-evaluation of the strategy used to assess viability in *D. foliosa* may be warranted. In particular, the incorporation of genetics into population viability analysis (PVA) models could be a valuable exercise. Given the potential for individual plants to live at least eight years, the genetic monitoring of populations on a 10-year cycle using higher-resolution genetic datasets will provide an assessment of the level of genetic turnover within populations, which would inform long-term strategies regarding the potential for recovery of the species leveraging the soil seed bank (if possible).

It is unclear whether the low levels of neutral genetic variation observed in the present study will have a significant impact on long-term fitness within *D. foliosa* populations. Furthermore, our future predictions of suitable climatic conditions for the species rely on an assumption that the ecophysiological range of the species is fairly stable, with limited plasticity under future predicted changes in climate. To assess concerns related to fitness and physiological plasticity, future studies should focus on adaptive genetic variation in both common garden experiments and controlled experimental settings. Such studies will provide additional insight into the factors most likely to drive or inhibit germination and reproductive output under modified environmental stresses. Recovery in *D. foliosa* will depend on a coordinated, collaborative effort among partners to better understand the ecological determinants of evolutionary success in this species.

Supplementary Materials: The following supporting information can be downloaded at https://www.mdpi.com/article/10.3390/plants13040495/s1, Table S1: Microsatellite allele frequencies by locus, population, and geographic region. Table S2: Summary of Chi-Square tests for Hardy Weinberg Equilibrium. Table S3: Pairwise population FST values. Figure S1: Principal Coordinates Analysis (PCoA) of the 29 populations of *Dalea foliosa* sampled across the species range and genotyped for nine nuclear microsatellite loci.

Author Contributions: Conceptualization, A.B.M., C.J.V. and G.C.; Methodology, A.B.M., C.J.V., S.J.F. and C.S.; Formal Analysis, A.B.M., C.J.V. and S.J.F.; Investigation, A.B.M., C.J.V. and S.J.F.; Resources,

A.B.M., S.J.F., S.F. and G.C.; Data Curation, A.B.M., C.J.V., S.J.F. and C.S.; Writing—Original Draft Preparation, A.B.M. and C.J.V.; Writing—Review and Editing, A.B.M., C.J.V., S.J.F., C.S., S.F. and G.C.; Visualization, A.B.M. and C.J.V.; Supervision, A.B.M.; Project Administration, A.B.M.; Funding Acquisition, A.B.M. All authors have read and agreed to the published version of the manuscript.

Funding: This research was funded by U.S. Fish and Wildlife Service grant numbers F13AC00084 to A.B.M. and F21AC00647 to the Tennessee Department of Environment and Conservation (TDEC) Division of Natural Areas (DNA) with a subcontract to A.B.M. Additional support was provided by the MTSU Undergraduate Research Experience and Creative Activity (URECA) Committee to C.S.

Data Availability Statement: Data are available from the corresponding author upon request. The data are not publicly available due to sensitive locality information.

Acknowledgments: The authors thank Caitlin Elam (TDEC DNA), Wayne Barger (ALDCNR), and Cathy Pollack (Chicago Field Office USFWS) for guidance throughout this project and for facilitating field work and collections in their respective areas. The authors would also like to thank the following individuals for field collections: Andrea Bishop, Stephanie Williams, Todd Crabtree, and Robby Cogburn, all currently or formerly with TDEC DNA; Randall Boisvert (Material Service Corporation, now Heidelberg Materials, in Illinois), Michelle Pearion (Midewin National Tallgrass Prairie), and Staff of Forest Preserve District of Will County, Illinois. We thank the following individuals for their support in the lab: Stephanie Cooper, Kevin Trostel, Austin Burleyson, Logan Whiles, and Shelby Watkins, all former students of MTSU. We thank Steve Bogdanowicz (Cornell University) for continued support on microsatellite development and troubleshooting. The findings and conclusions in this article are those of the authors and do not necessarily represent the views of the U.S. Fish and Wildlife Service.

Conflicts of Interest: Author Sunny Fleming was employed by the company ESRI. The remaining authors declare that the research was conducted in the absence of any commercial or financial relationships that could be construed as a potential conflict of interest.

References

1. Caro, T.; Rowe, Z.; Berger, J.; Wholey, P.; Dobson, A. An Inconvenient Misconception: Climate Change Is Not the Principal Driver of Biodiversity Loss. *Conserv. Lett.* **2022**, *15*, e12868. [CrossRef]
2. Isbell, F.; Balvanera, P.; Mori, A.S.; He, J.S. Expert Perspectives on Global Biodiversity Loss and Its Drivers and Impacts on People. *Front. Ecol. Environ.* **2023**, *21*, 94–103. [CrossRef]
3. Gerber, L.R. Conservation Triage or Injurious Neglect in Endangered Species Recovery. *Proc. Natl. Acad. Sci. USA* **2016**, *113*, 3563–3566. [CrossRef] [PubMed]
4. Buxton, R.T.; Avery-Gomm, S.; Lin, H.-Y.; Smith, P.A.; Cooke, S.J.; Bennett, J.R. Half of Resources in Threatened Species Conservation Plans Are Allocated to Research and Monitoring. *Nat. Commun.* **2020**, *11*, 4668. [CrossRef] [PubMed]
5. Martin, T.G.; Kehoe, L.; Mantyka-Pringle, C.; Chades, I.; Wilson, S.; Bloom, R.G.; Davis, S.K.; Fisher, R.; Keith, J.; Mehl, K.; et al. Prioritizing Recovery Funding to Maximize Conservation of Endangered Species. *Conserv. Lett.* **2018**, *11*, e12604. [CrossRef]
6. Moritz, C. Defining "Evolutionarily Significant Units" for Conservation. *Trends Ecol. Evol.* **1994**, *9*, 373–375. [CrossRef] [PubMed]
7. Crandall, K.A.; Bininda-Emonds, O.R.; Mace, G.M.; Wayne, R.K. Considering Evolutionary Processes in Conservation Biology. *Trends Ecol. Evol.* **2000**, *15*, 290–295. [CrossRef]
8. Kardos, M.; Armstrong, E.E.; Fitzpatrick, S.W.; Hauser, S.; Hedrick, P.W.; Miller, J.M.; Tallmon, D.A.; Funk, W.C. The Crucial Role of Genome-Wide Genetic Variation in Conservation. *Proc. Natl. Acad. Sci. USA* **2021**, *118*, e2104642118. [CrossRef]
9. Reed, J.M.; Mills, L.S.; Dunning, J.B., Jr.; Menges, E.S.; McKelvey, K.S.; Frye, R.; Beissinger, S.R.; Anstett, M.-C.; Miller, P. Emerging Issues in Population Viability Analysis. *Conserv. Biol.* **2002**, *16*, 7–19. [CrossRef]
10. Menges, E.S. The Application of Minimum Viable Population Theory to Plants. *Genet. Conserv. Rare Plants* **1991**, *45*, 158–164
11. Smith, D.R.; Allan, N.L.; McGowan, C.P.; Szymanski, J.A.; Oetker, S.R.; Bell, H.M. Development of a Species Status Assessment Process for Decisions under the U.S. Endangered Species Act. *J. Fish Wildl. Manag.* **2018**, *9*, 302–320. [CrossRef]
12. Moritz, C. Conservation Units and Translocations: Strategies for Conserving Evolutionary Processes. *Hereditas* **2004**, *130*, 217–228. [CrossRef]
13. McCarthy, M.A.; Possingham, H.P. Active Adaptive Management for Conservation. *Conserv. Biol.* **2007**, *21*, 956–963. [CrossRef] [PubMed]
14. Kadykalo, A.N.; Buxton, R.T.; Morrison, P.; Anderson, C.M.; Bickerton, H.; Francis, C.M.; Smith, A.C.; Fahrig, L. Bridging Research and Practice in Conservation. *Conserv. Biol.* **2021**, *35*, 1725–1737. [CrossRef]
15. USFWS. *USFWS Species Status Assessment Framework: An Integrated Analytical Framework for Conservation*; Version 3.4; U.S. USFWS: Atlanta, GA, USA, 2016.
16. NatureServe. *Habitat-Based Plant Element Occurrence Delimitation Guidance*; Version 1.0; NatureServe: Arlington, VA, USA, 2020.

17. Van Rossum, F.; Le Pajolec, S.; Raspé, O.; Godé, C. Assessing Population Genetic Status for Designing Plant Translocations. *Front. Conserv. Sci.* **2022**, *3*, 829332. [CrossRef]
18. Novak, B.J.; Phelan, R.; Weber, M.U.S. Conservation Translocations: Over a Century of Intended Consequences. *Conserv. Sci. Pract.* **2021**, *3*, e394. [CrossRef]
19. Draper, D.; Marques, I.; Iriondo, J.M. Species Distribution Models with Field Validation, a Key Approach for Successful Selection of Receptor Sites in Conservation Translocations. *Glob. Ecol. Conserv.* **2019**, *19*, e00653. [CrossRef]
20. Maschinski, J.; Falk, D.A.; Wright, S.J.; Possley, J.; Roncal, J.; Wendelberger, K.S. Optimal Locations for Plant Reintroductions in a Changing World. In *Plant Reintroduction in a Changing Climate: Promises and Perils*; Maschinski, J., Haskins, K.E., Raven, P.H., Eds.; Island Press/Center for Resource Economics: Washington, DC, USA, 2012; pp. 109–129, ISBN 9781610911832.
21. Parmesan, C.; Hanley, M.E. Plants and Climate Change: Complexities and Surprises. *Ann. Bot.* **2015**, *116*, 849–864. [CrossRef]
22. Anderson, J.; Song, B.-H. Plant Adaptation to Climate Change—Where Are We? *J. Syst. Evol.* **2020**, *58*, 533–545. [CrossRef] [PubMed]
23. Hällfors, M.H.; Liao, J.; Dzurisin, J.; Grundel, R.; Hyvärinen, M.; Towle, K.; Wu, G.C.; Hellmann, J.J. Addressing Potential Local Adaptation in Species Distribution Models: Implications for Conservation under Climate Change. *Ecol. Appl.* **2016**, *26*, 1154–1169. [CrossRef] [PubMed]
24. Schwartz, M.W. Using Niche Models with Climate Projections to Inform Conservation Management Decisions. *Biol. Conserv.* **2012**, *155*, 149–156. [CrossRef]
25. Evanno, G.; Regnaut, S.; Goudet, J. Detecting the Number of Clusters of Individuals Using the Software STRUCTURE: A Simulation Study. *Mol. Ecol.* **2005**, *14*, 2611–2620. [CrossRef] [PubMed]
26. Pritchard, J.K.; Stephens, M.; Donnelly, P. Inference of Population Structure Using Multilocus Genotype Data. *Genetics* **2000**, *155*, 945–959. [CrossRef] [PubMed]
27. Morris, A.B.; Visger, C.; Watkins, S.; Pollack, C. Extremely Low Levels of Genetic Variation and Predicted Shifts in Suitable Niche Space for a Geographically Disjunct, Federally Endangered Legume, Leafy Prairie-Clover (*Dalea foliosa*). *Castanea* **2023**, *88*, 91–110. [CrossRef]
28. Edwards, A.L.; Wiltshire, B.; Nickrent, D.L. Genetic Diversity in *Astragalus tennesseensis* and the Federal Endangered *Dalea foliosa* (Fabaceae). *J. Torrey Bot. Soc.* **2004**, *131*, 279–291. [CrossRef]
29. McMahon, M.; Hufford, L. Phylogeny of Amorpheae (Fabaceae: Papilionoideae). *Am. J. Bot.* **2004**, *91*, 1219–1230. [CrossRef]
30. Dixon, A.L.; Herlihy, C.R.; Busch, J.W. Demographic and Population-Genetic Tests Provide Mixed Support for the Abundant Centre Hypothesis in the Endemic Plant *Leavenworthia stylosa*. *Mol. Ecol.* **2013**, *22*, 1777–1791. [CrossRef] [PubMed]
31. Abeli, T.; Gentili, R.; Mondoni, A.; Orsenigo, S.; Rossi, G. Effects of Marginality on Plant Population Performance. *J. Biogeogr.* **2014**, *41*, 239–249. [CrossRef]
32. Kennedy, J.P.; Preziosi, R.F.; Rowntree, J.K.; Feller, I.C. Is the Central-Marginal Hypothesis a General Rule? Evidence from Three Distributions of an Expanding Mangrove Species, *Avicennia germinans* (L.) L. *Mol. Ecol.* **2020**, *29*, 704–719. [CrossRef]
33. Sporbert, M.; Keil, P.; Seidler, G.; Bruelheide, H.; Jandt, U.; Aćić, S.; Biurrun, I.; Campos, J.A.; Čarni, A.; Chytrý, M.; et al. Testing Macroecological Abundance Patterns: The Relationship between Local Abundance and Range Size, Range Position and Climatic Suitability among European Vascular Plants. *J. Biogeogr.* **2020**, *47*, 2210–2222. [CrossRef]
34. Baskin, C.C.; Baskin, J.M. Germination Ecophysiology of Herbaceous Plant Species in a Temperate Region. *Am. J. Bot.* **1988**, *75*, 286–305. [CrossRef]
35. Baskin, J.M. Greenhouse and Laboratory Studies on the Ecological Life Cycle of *Dalea foliosa* (Fabaceae), a Federal Endangered Species. *Nat. Areas J.* **1998**, *18*, 54–62.
36. US Fish and Wildlife Service. *Recovery Plan for the Leafy Prairie-Clover (Dalea foliosa)*; US Fish and Wildlife Service: Atlanta, GA, USA, 1996; pp. 1–74.
37. Thompson, J.N.; Walck, J.L.; Hidayati, S.N. Microhabitat Requirements of the Federally Endangered *Dalea foliosa*, with Recommendations on Establishment of New Populations. *Castanea* **2006**, *71*, 94–104. [CrossRef]
38. Molano-Flores, B.; Zaya, D.N.; Baty, J.; Spyreas, G. An Assessment of the Vulnerability of Illinois' Rarest Plant Species to Climate Change. *Castanea* **2019**, *84*, 115–127. [CrossRef]
39. Molano-Flores, B.; Bell, T.J. Projected Population Dynamics for a Federally Endangered Plant under Different Climate Change Emission Scenarios. *Biol. Conserv.* **2012**, *145*, 130–138. [CrossRef]
40. USFWS. *Recovery Plan for Dalea foliosa (Leafy Prairie-Clover) (Gray) Barneby*; U.S. Fish and Wildlife Service: Atlanta, GA, USA, 1996.
41. USFWS. *Leafy Prairie-Clover (Dalea foliosa) 5-Year Review: Summary and Evaluation*; U.S. Fish and Wildlife Service, Tennessee Ecological Services Field Office: Cookeville, TN, USA, 2022.
42. Morris, A.B.; Scalf, C.; Burleyson, A.; Johnson, L.T.; Trostel, K. Development and Characterization of Microsatellite Primers in the Federally Endangered *Astragalus bibullatus* (Fabaceae). *Appl. Plant Sci.* **2016**, *4*, 1500126. [CrossRef]
43. Schuelke, M. An Economic Method for the Fluorescent Labeling of PCR Fragments. *Nat. Biotechnol.* **2000**, *18*, 233–234. [CrossRef]
44. Peakall, R.; Smouse, P.E. Genalex 6: Genetic Analysis in Excel. Population Genetic Software for Teaching and Research. *Mol. Ecol. Notes* **2006**, *6*, 288–295. [CrossRef]
45. Peakall, R.; Smouse, P.E. GenAlEx 6.5: Genetic Analysis in Excel. Population Genetic Software for Teaching and Research—An Update. *Bioinformatics* **2012**, *28*, 2537–2539. [CrossRef]

46. Peakall, R.; Smouse, P.E.; Huff, D.R. Evolutionary Implications of Allozyme and RAPD Variation in Diploid Populations of Dioecious Buffalograss Buchloë Dactyloides. *Mol. Ecol.* **1995**, *4*, 135–148. [CrossRef]
47. Smouse, P.E.; Peakall, R. Spatial Autocorrelation Analysis of Individual Multiallele and Multilocus Genetic Structure. *Heredity* **1999**, *82 Pt 5*, 561–573. [CrossRef]
48. Falush, D.; Stephens, M.; Pritchard, J.K. Inference of Population Structure Using Multilocus Genotype Data: Linked Loci and Correlated Allele Frequencies. *Genetics* **2003**, *164*, 1567–1587. [CrossRef]
49. Hubisz, M.J.; Falush, D.; Stephens, M.; Pritchard, J.K. Inferring Weak Population Structure with the Assistance of Sample Group Information. *Mol. Ecol. Resour.* **2009**, *9*, 1322–1332. [CrossRef]
50. Besnier, F.; Glover, K.A. ParallelStructure: A R Package to Distribute Parallel Runs of the Population Genetics Program STRUCTURE on Multi-Core Computers. *PLoS ONE* **2013**, *8*, e70651. [CrossRef]
51. Miller, M.A.; Pfeiffer, W.; Schwartz, T. Creating the CIPRES Science Gateway for Inference of Large Phylogenetic Trees. In Proceedings of the 2010 Gateway Computing Environments Workshop (GCE), New Orleans, LA, USA, 14 November 2010; pp. 1–8.
52. Kopelman, N.M.; Mayzel, J.; Jakobsson, M.; Rosenberg, N.A.; Mayrose, I. Clumpak: A Program for Identifying Clustering Modes and Packaging Population Structure Inferences across K. *Mol. Ecol. Resour.* **2015**, *15*, 1179–1191. [CrossRef]
53. Funk, S.M.; Guedaoura, S.; Juras, R.; Raziq, A.; Landolsi, F.; Luís, C.; Martínez, A.M.; Musa Mayaki, A.; Mujica, F.; do Mar Oom, M.; et al. Major Inconsistencies of Inferred Population Genetic Structure Estimated in a Large Set of Domestic Horse Breeds Using Microsatellites. *Ecol. Evol.* **2020**, *10*, 4261–4279. [CrossRef] [PubMed]
54. Jakobsson, M.; Rosenberg, N.A. CLUMPP: A Cluster Matching and Permutation Program for Dealing with Label Switching and Multimodality in Analysis of Population Structure. *Bioinformatics* **2007**, *23*, 1801–1806. [CrossRef] [PubMed]
55. Rosenberg, N.A. Distruct: A Program for the Graphical Display of Population Structure. *Mol. Ecol. Notes* **2004**, *4*, 137–138. [CrossRef]
56. Paulsen, J.; Körner, C. A Climate-Based Model to Predict Potential Treeline Position around the Globe. *Alp. Bot.* **2014**, *124*, 1–12. [CrossRef]
57. Karger, D.N.; Conrad, O.; Böhner, J.; Kawohl, T.; Kreft, H.; Soria-Auza, R.W.; Zimmermann, N.E.; Linder, H.P.; Kessler, M. Climatologies at High Resolution for the Earth's Land Surface Areas. *Sci Data* **2017**, *4*, 170122. [CrossRef]
58. QGIS Development Team. QGIS Geographic Information System. Open Source Geospatial Foundation Project. Available online: http://qgis.org (accessed on 19 December 2023).
59. R Core Team. *R: A Language and Environment for Statistical Computing*; R Core Team: Vienna, Austria, 2021. Available online: https://www.R-project.org (accessed on 19 December 2023).
60. Phillips, S.J.; Anderson, R.P.; Schapire, R.E. Maximum Entropy Modeling of Species Geographic Distributions. *Ecol. Model.* **2006**, *190*, 231–259. [CrossRef]
61. Phillips, S.J.; Dudík, M.; Schapire, R.E. A Maximum Entropy Approach to Species Distribution Modeling. In Proceedings of the Twenty-First International Conference on Machine Learning, Banff, AB, Canada, 4–8 July 2004; Association for Computing Machinery: New York, NY, USA, 2004; p. 83.
62. Horton, J.D.; San Juan, C.A.; Stoeser, D.B. *The State Geologic Map Compilation (SGMC) Geodatabase of the Conterminous United States*; Data Series; Version 1.0: Originally Posted on 30 June 2017; Version 1.1: August 2017; U.S. Geological Survey: Reston, VA, USA, 2017.
63. Brook, B.W.; O'Grady, J.J.; Chapman, A.P.; Burgman, M.A.; Akcakaya, H.R.; Frankham, R. Predictive Accuracy of Population Viability Analysis in Conservation Biology. *Nature* **2000**, *404*, 385–387. [CrossRef] [PubMed]
64. Coulson, T.; Mace, G.M.; Hudson, E.; Possingham, H. The Use and Abuse of Population Viability Analysis. *Trends Ecol. Evol.* **2001**, *16*, 219–221. [CrossRef] [PubMed]
65. Akçakaya, H.R.; Sjögren-Gulve, P. Population Viability Analyses in Conservation Planning: An Overview. *Ecol. Bull.* **2000**, *48*, 9–21.

plants

MDPI

Review

Pityopsis ruthii: An Updated Review of Conservation Efforts for an Endangered Plant

Phillip A. Wadl [1,*], Adam J. Dattilo [2], Geoff Call [3], Denita Hadziabdic [4] and Robert N. Trigiano [4,*]

1 U.S. Vegetable Laboratory, United States Department of Agriculture, Agricultural Research Service, Charleston, SC 29414, USA
2 Tennessee Valley Authority, Knoxville, TN 37902, USA; ajdattilo@tva.gov
3 United States Fish and Wildlife Service, Cookeville, TN 38501, USA; geoff_call@fws.gov
4 Department of Entomology and Plant Pathology, The University of Tennessee, 2505 EJ Chapman Drive, Knoxville, TN 37996, USA; dhadziab@utk.edu
* Correspondence: phillip.wadl@usda.gov (P.A.W.); rtrigian@utk.edu (R.N.T.); Tel.: +1-8434025388 (P.A.W.); +1-8659744744 (R.N.T.)

Abstract: *Pityopsis ruthii* (Small) Small, Ruth's golden aster, is an endangered Asteraceae species that grows in the riparian zone along small sections of two rivers in the Southern Appalachian Mountains of the United States of America (USA). Since 1985, the species has been listed under the Endangered Species Act by the United States Fish and Wildlife Service (USFWS). The mission of the USFWS is to conserve, protect, and enhance fish, wildlife, and plants and their habitats for the continued benefit of the American people. The agency provides national leadership in the recovery and conservation of imperiled plant species by working with the scientific community to protect important habitats, increase species' populations, and identify and reduce threats to species survival with the goal of removal from federal protection. Over the past 35 years, research efforts have focused on studies designed to delineate the range and size of populations, determine habitat requirements, reproductive and propagation potential, and understand the demographic, ecological, and genetic factors that may increase vulnerability to extinction for *P. ruthii*. Cooperative partnerships have driven the completion of actions called for in the strategy to recover *P. ruthii*, and in this review, we highlight these efforts within the context of species conservation.

Keywords: Ruth's golden aster; Asteraceae; plant conservation; endangered species; ripraian

Citation: Wadl, P.A.; Dattilo, A.J.; Call, G.; Hadziabdic, D.; Trigiano, R.N. *Pityopsis ruthii*: An Updated Review of Conservation Efforts for an Endangered Plant. *Plants* **2023**, *12*, 2693. https://doi.org/10.3390/plants12142693

Academic Editors: Brenda Molano-Flores and James Cohen

Received: 1 June 2023
Revised: 6 July 2023
Accepted: 17 July 2023
Published: 19 July 2023

1. Introduction

Anthropogenic activities [1], including those that induce climate change, have had profound negative effects on populations and distributions of plant species across the globe [2–4], and in 2017, the extinction of thousands of species worldwide was predicted [5]. Declines or extinctions of species due to climate changes could be exacerbated by the loss or limitations of appropriate habitats [6], competition among species occupying similar niches [7], competition for pollinators [8], and low seed dispersal and success rates, among others [9]. However, the increase in maximum annual temperatures appears to be a principal driving force associated with many local extinctions along with, to a lesser extent, changes in precipitation variables [3].

In the USA, over 900 plant species have been listed under the Endangered Species Act (ESA) and the U.S. Fish and Wildlife Service (USFWS) oversees the protection of these species. The mission of the USFWS is to conserve, protect, and enhance fish, wildlife, and plants and their habitats for the continued benefit of the American people. The agency provides national leadership focused on the recovery and conservation of imperiled plant species. This is accomplished through coordination with the scientific community to protect important habitats, increase species' population sizes, and identify and reduce threats to species survival with the goal of removal from federal protection. All federally listed

Figure 2. The endangered *Pityopsis ruthii* (Ruth's golden aster) growing along the riparian corridor on the Hiwassee River in Southeastern Tennessee, United States of America. (**A**) An individual plant exhibiting erect to ascending habit with lanceolate leaves that are silvery pubescent and overlap in tight spirals. Inflorescences are either solitary or in a cyme (1.5–2 cm diameter heads), composed of 8–15 ray florets, with bright yellow petals and numerous disk florets; (**B**) the florets produce ~4 mm long pubescent achenes each with a 4–5 mm long pappus; (**C**) the herbaceous perennial is characterized by habitat specialization, as distribution is restricted to discrete locations on exposed phyllite boulders within and along the banks of the Hiwassee and Ocoee rivers.

4. Genetic Diversity Studies

It has been hypothesized that inbreeding depression or limitation of genetically compatible individuals impacts sexual reproduction both spatially and temporally in *P. ruthii* [24,25]. Knowledge of genetic diversity, population structure, and gene flow is critical for the delineation of what constitutes a viable population and in completing Recovery Tasks 6.3 and 7.4 (determine whether Hiwassee and Ocoee River populations, respectively, are self-sustaining).

The first effort to determine what constitutes a self-sustaining population using population genetic analyses was conducted in 1994 [26]. Analyses of plants from discrete locations on the Hiwassee and Ocoee rivers with two polymorphic allozyme markers found that 84% of the genetic differentiation was within occurrence (sampling location), 15% of differentiation was between occurrences, and 1% was between river systems [26]. Further investigation of chloroplast variation revealed the presence of three haplotypes within plants from the discrete locations: two associated with the upper and lower regions of the Hiwassee River and one associated with the Ocoee River [27]. To further study the genetic diversity and population genetics of *P. ruthii*, Wadl et al. [28] developed 12 nuclear and 5 chloroplast microsatellite markers. Preliminary analyses of individuals from two discrete Hiwassee River locations and one discrete Ocoee River location with these markers agreed with the findings of Sloan [26]; 90% of genetic differentiation was within occurrence and the individuals on the Hiwassee and Ocoee River should be defined as different populations.

By 2011, an exhaustive census and delineation of all known occurrences (discrete locations) of *P. ruthii* was completed [11]. The Hiwassee River population consisted of ~12,000 individuals across 57 discrete locations and the Ocoee River population consisted of ~1000 individuals across 9 discrete locations [11]. Subsequently, the population structure and genetic diversity were assessed for 814 individuals from 33 discrete locations using

7 chloroplast and 12 nuclear microsatellite markers [13]. Higher levels of gene flow and lower levels of population differentiation were discovered for the Hiwassee River when compared to the Ocoee River [13]. The authors recommended the management of the species using a framework of two-to-four subpopulations along the Hiwassee River and two-to-three subpopulations along the Ocoee River. This should guide the selection of seeds or individuals used for augmentation, reintroduction, and/or translocation should these actions be needed for the species' conservation.

More recently, the genetic diversity of half-to full- siblings of *P. ruthii* from one Hiwassee River population and three Ocoee River populations (n = 170) and four populations of *P. graminifolia* (Michx.) Nutt. var. *latifolia* (Fernald) Semple & F.D.Bowers (n = 148) were characterized using nine microsatellite markers [29]. *P. graminifolia* var. *latifolia* is a herbaceous perennial that is widely distributed throughout the southeastern USA that grows in close proximity to *P. ruthii*. Genetic diversity estimates were the highest for the *P. graminifolia* var. *latifolia* populations as assessed by the number of alleles per locus and Nei's diversity index. For the *P. ruthii* populations, these same diversity estimates were the highest in the Hiwassee River population compared to the three Ocoee River populations. Discriminant analysis of principal components using the multiallelic genetic data for the *P. ruthii* populations clearly differentiated the Hiwassee River populations from the Ocoee River populations. This supports the finding of the previous population genetics studies that the Hiwassee and Ocoee rivers should be managed as separate populations. Furthermore, Boyd et al. [29] provided evidence for two genetic subpopulations on the Ocoee River and these results support the recommendation of Hatmaker et al. [13] suggesting that the Ocoee River should be managed as two-to-three viable breeding subpopulations.

5. Propagation Methods

The ability to generate plant materials of *P. ruthii* through asexual and sexual reproduction is critical for ex situ and in situ conservation of the species [30,31] and ultimately supports augmentation, reintroduction, and/or translocation efforts should they be needed to maintain viable, self-sustaining subpopulations on the Hiwassee and Ocoee rivers. There was minimal knowledge of the sexual reproduction of *P. ruthii* prior to the publication of the recovery plan. Effective propagation methods are critical for completing Recovery Tasks 3.2 (determine life history, seed germination, and seedling establishment requirements), 6.4 and 7.5 (establish on unoccupied suitable habitat), and 6.5 and 7.6 (establish cultivated populations of plants from the Hiwassee and Ocoee Rivers and provide for long-term seed storage).

Sexual reproduction is required for the establishment of new plants within existing populations and the findings of White [32] indicated that a reproductive barrier was not present in the species as viable seed production was similar in wild-grown vs. greenhouse-grown plants obtained from the Hiwassee River. Although viable seeds are produced, White [32] observed no seedlings in the field and concluded that *P. ruthii* has difficulty establishing on the phyllite boulders. Farmer [33] investigated seed propagation of the species for nursery production as a means toward the expansion of natural populations. Seed heads were collected from plants in late September and dried at ambient laboratory conditions for 24 h. Visual examination indicated that ~5% of the seed were viable. Germination experiments investigating the effect of light and temperature were conducted, and the results indicated that *P. ruthii* seeds had no chilling or special light requirements. Rather, seeds germinate more completely and rapidly at low-to-moderate temperatures (7–24 °C) than at 24–29 °C. Seed germination in the field has been observed between November and January [34], but the seedlings often fail to persist [24,25,32,34,35]. Multiple studies observed pollinators within *P. ruthii* populations but provide no information for identification of the species responsible [14,23,34]. Variable seed set was observed between Hiwassee and Ocoee River populations, with the Ocoee River populations exhibiting higher rates of seed set and seed viability as assessed by germination tests [24]. Boyd et al. [29] collected viable seeds from one Hiwassee River population and three Ocoee River populations and reported highly successful germination from all populations, although germination rates were not provided in their report.

The wild collected seed germinated and grown in a common garden setting produces a dense mat and produce viable seed, but viability declined to 75% within 6 months in seeds that were stored in glass containers at 3 °C [33]. Wadl et al. [25] tested the viability of seeds from long-term storage and germination ranged from 0 to 38% and they attributed the low germination rate to the effects of long-term storage or variability inherent in the collected seeds. There is no way to determine this because there are no records for baseline germination rates or treatment of seeds prior to long-term storage. Regardless, it appears that seed viability is highly variable between river systems and the year of collection.

Production of low numbers of viable seeds is a barrier to the establishment of new plants. Information on the impact of pollinators on seed propagation was absent until recently. To better understand the highly variable sexual reproductive capacity of *P. ruthii*, Moore et al. [36] assessed which insect species may be contributing roles as potential pollinators at in situ and ex situ locations. Species of the Halictidae were common at the ex situ locations and infrequent at the in situ locations. Honeybees (*Apis mellifera* L.) and *Bombus impatiens* Cresson were commonly observed at the in situ locations. The most abundant floral visitor was *Toxomerus geminatus* Say but very little pollen was carried and no pollen was carried by lepidopteran species. Seeds were collected from three Hiwassee River populations and low germination rates were observed for viable seeds. Seeds were considered viable as determined by staining with 2,3,5-triphenyl tetrazolium chloride and germination was defined as successful if the seed produced a green cotyledon. Evidence of inbreeding depression was found as seed viability and germination were higher in controlled crosses made between geographically separated, but genetically similar populations compared to crosses of individuals within a population [36].

Asexual reproduction through stem regeneration of the subaerial root–rhizome crown has been reported and was speculated as the primary method of reproduction for *P. ruthii* [32,37]. A methodology for asexual propagation of the species has been developed recently [25,38]. The in vitro regeneration of plants from flower receptacles and leaf tissue cultured on Murashige and Skoog [39] tissue culture medium supplemented with growth regulators was the first report of asexual propagation for the species [38]. A method for rapid in vitro multiplication of new plants derived from lateral shoots has been developed [25]. Furthermore, Wadl et al. [25] optimized surface sterilization methods to demonstrate the feasibility of in vitro seed germination and developed simple and robust methods for rapid vegetative propagation of terminal stem cuttings. These methods provide a foundation for providing a disease and insect-free source of propagules for use in recovery efforts or germplasm conservation. The long-term limitation of reliance on asexual propagation solely for population development is that clonal propagation of specific individuals alone, without propagation of many individuals, cannot sustain a robust and genetically variable population. However, mass asexual reproduction of many individuals coupled with the development of successful breeding and seed germination schemes can establish a diverse population.

6. Conclusions: Recovery Implementation and Future Directions

The life history strategy and strict habitat specialization of *P. ruthii* suggest that the species will always be geographically restricted, but since the species was formally listed as endangered by the USFWS in 1985 [10], a suite of monitoring and research efforts have focused on its conservation. The most impactful of these conservation actions was linked to tasks outlined in the recovery plan for the species [12]. As the primary agency responsible for the recovery of species listed under the ESA, the USFWS uses the best available science to delineate specific tasks that collectively outline a conservation strategy that, if implemented, will help ensure the species is no longer at risk of extinction. The strategy for recovering *P. ruthii* consists broadly of actions intended to improve understanding of the species' distribution, ecology, and life history; identify positive and negative influences on the viability of populations; develop approaches for managing habitat and populations; and monitor the status of populations in the Ocoee and Hiwassee rivers.

Cooperative partnerships have driven the completion of actions called for in the strategy to recover *P. ruthii*, with contributions from the Tennessee Valley Authority (TVA), USDA Forest Service, Tennessee State Parks, USFWS, academia, botanical gardens, and private industry. The first five-year review of *P. ruthii* [40] summarized the results of several projects that were completed during the 1990s and early 2000s to implement tasks described in the species' recovery plan. Beginning around 2010, researchers began a more recent phase of recovery projects to better understand the ecology of *P. ruthii* and address ongoing and potential future threats to the species [10,13,25,28,29,38,41–44]. These efforts focused on multiple topics including completion of on-the-ground delineation of populations along the Hiwassee River, evaluating propagation and reintroduction techniques, permanent monitoring, pests and pathogens, pollination ecology, and population genetics. While results from these efforts have advanced species recovery in new ways, areas remain where future work is needed. The most pertinent unresolved issues surrounding recovery, including the need for improving understanding of how plant releases in the Hiwassee River location affect occupied habitat, and developing strategies to manage flows beneficially for *P. ruthii*, are discussed below.

The TVA operates dams upstream of *P. ruthii* populations on both the Hiwassee and Ocoee Rivers. Since the dam closure, the hydrology of these systems has been altered to support the two primary goals of flood control and the generation of electricity. This has resulted in fundamental changes in habitat for a species that occupies a narrow ecological niche—shallow soils within cracks in rock outcrops that occur in a frequently disturbed zone between the river channel and the surrounding forest. Plants require sunny conditions to establish and reproduce, but cannot tolerate frequent, prolonged inundation with water. While it has long been understood that frequent fluctuations in river flow historically served to scour flooded areas of trees that could shade- out *P. ruthii* [12,15,32,37,40,45], the role of cyclical drought in maintaining open habitat for the species had not been highlighted until Moore et al. [11].

This new understanding that river flows were not the sole mechanism for maintaining all habitats supporting substantial numbers *P. ruthii* plants was encouraging. It suggested that at least some sites were more resistant to the threat of encroaching woody vegetation resulting from altered hydrology, presumably lowering the risk of extirpation for the Hiwassee River population. However, this does not explain how the river interacts with plants at sites where flow appears to be the primary ecological mechanism maintaining habitat for *P. ruthii*.

Determining how flows of varying magnitude, duration, and frequency on both the Hiwassee and Ocoee rivers interact with plants in occupied habitats may be the single most important data gap in understanding the ecology of *P ruthii* and making future conservation decisions. Understanding river flows and how they drive *P. ruthii* population dynamics features prominently in the recovery plan and intersects several specific tasks [12]. River flows of an appropriate magnitude, duration, and frequency are needed for maintaining suitable habitat by eliminating competing vegetation, dispersing seed, and depositing soil in bedrock crevices where *P. ruthii* grows.

Long-term monitoring data provide insight into how future studies could be conducted to determine optimum and operationally achievable flow regimes that would benefit *P. ruthii*. TVA began an annual census on the Ocoee River in 1987, just a few years after entering into an agreement with the state of Tennessee to provide recreational releases of water for up to 116 days per year along the portion of the river supporting *P. ruthii*. Since that time, the total population along the Ocoee River has increased demonstrably from a low of 523 individuals in 1993 to 1388 individuals in 2022; the population has been greater than 1000 plants every year since 2008 [11,46]. Plants have not expanded their distribution along the Ocoee River since 1987 because habitat is inherently limited, but there are more plants in areas where the species has occurred since monitoring began.

The collection of comprehensive population census data on the Hiwassee River began in 2011. While many of the sites on the Hiwassee River appear relatively stable, others have been declining for many years [11]. Sites kept open by drought cycles appear more stable than sites that rely on high-flow water events. Unlike the Ocoee River, the Hiwassee River does not have regular recreational flows and only receives a 25 cubic feet per second (cfs) base flow from Apalachia Dam. The overall population along the Hiwassee River does not appear to be at risk of extirpation, but some sites could be lost over time under current flow management guidelines.

From November 2015 until March 2016, TVA performed scheduled maintenance on the Apalachia powerhouse and switchyard, which is situated along the Hiwassee River. This work necessitated the closure of the powerhouse and diversion of all flows through the "dry" section of the Hiwassee River where *P. ruthii* occurs. Typically, flow in this section of the river consists only of base releases of 25 cfs from Apalachia Dam plus tributary inputs, which are driven by local rainfall. Therefore, under normal operating conditions, higher flows in this section of the river are intermittent and short-term in duration. During the maintenance work, flows were elevated for 132 days. Several high rainfall events also occurred during this time, which necessitated higher-than-planned releases from Appalachia Dam. Daily averages varied, but releases ranging from 2500 to 5000 cfs were common; the highest and lowest releases during that time period were 8386 and 541 cfs, respectively. At some sites, individual *P. ruthii* plants were inundated for nearly all of this period, whereas other sites where *P. ruthii* occurs well above the river channel were not inundated. Census counts conducted in the fall of 2016 indicated a decrease in plants on the Hiwassee River, whereas counts on the Ocoee River remained stable. This suggested that population declines observed along the Hiwassee River were related to the elevated flows present during the Apalachia powerhouse maintenance period rather than the simultaneously occurring drought conditions that affected both watersheds during 2016.

Releasing periodic higher flows has been cited as an action that could reduce woody plant encroachment that degrades *P. ruthii* habitat along the Hiwassee River, but observations of occupied habitat along the river during and after the maintenance on Apalachia powerhouse suggested that managed high releases alone are not likely to be effective at reversing this threat. The magnitude and duration of this flow event was unparalleled since Apalachia Dam was completed in 1943. These historically high flows did not appreciably change the extent of woody vegetation along portions of the Hiwassee River supporting *P. ruthii*. Monitoring data do suggest periodic elevated flows might stimulate population increases at currently occupied sites, presumably through increased recruitment and establishment of seedlings, but it is unlikely that releases from Apalachia Dam could reduce established woody plant encroachment on a meaningful scale. Factors that limit the effectiveness of this approach include both operational and ecological constraints. The presence of downstream homes and businesses in the Hiwassee River watershed places an upper limit on releases that can be sustained, even for short periods, without increasing the risk of harm to life and property due to flooding. This operational constraint is compounded by the alteration, over time, of the riparian vegetation community along the river. The fact that mature woody vegetation has become well established in many parts of the narrow riparian zone where phyllite outcrops would otherwise be available for the recruitment and establishment of *P. ruthii* presents an ecological constraint. Alteration of flows since the late 1940s has resulted in the establishment of vegetation that is resistant to low-frequency, moderately high flows that are operationally possible due to downstream flooding concerns.

Recreational flows have likely helped to maintain the habitat for *P. ruthii* and increase the population size along the Ocoee River, although there is no way to prove causality from the correlation between increased periodic flow events and the overall population increase. A methodology used on the Youghiogheny River in Pennsylvania to look at the effects of river flow on another species in the Asteraceae, *Marshallia pulchra* W.M.

Knapp, D.B. Poind & Weakley [47] could be used to better understand the relationship between flows and inundation of *P. ruthii*'s habitat on the Ocoee River and inform flow management for the Hiwassee River. Examining the frequency and duration of inundation events along the Ocoee River, which has stable or increasing population numbers, and then using that information to assess flow regimes at a mix of stable and declining sites on the Hiwassee River may offer important guidance on how possible future changes to flows from Apalachia Dam may affect the species. It is possible that flow management, combined with periodic drought, could provide habitat conditions sufficient and conducive to maintaining demographic structure and genetic variation within the subpopulations now recognized in the Hiwassee River, such that they are self-sustaining. However, additional management to reduce woody vegetation encroachment could be needed in some sites to restore open conditions where flow alterations over the past eight decades have allowed woody riparian vegetation to become established to an extent that will be irreversible through flow alteration alone. In addition, given the susceptibility *P. ruthii* to extended periods of inundation, any proposed changes to the 25 cfs base flow in the Hiwassee River would need to be carefully considered to ensure the increase would not negatively impact the species, particularly clusters of plants that occur at lower elevation near the current active channel.

Because of the very restricted range and evident lack of competitiveness in suitable habitats, it is obvious that any change in the home habitat, either minor or major, of *P. ruthiii* would probably negatively affect the survival of this plant. Climate changes as manifested in temperature changes, and especially the amount of rainfall could possibly influence the recovery and stabilization of the existing populations. Increased or decreased rainfall and the resultant fluctuations of water flow on the two home river systems of *P. ruthii* could potentially alter habitat characteristics and seed distribution and limit the total area available to support and establish populations of Ruth's Golden Aster. The number of plants and genetic diversity of *P. ruthii* subpopulations must be monitored, cataloged frequently, and correlated with local climatic variables over time to assess changes in population structure and survival of this unique species.

Furthermore, the reasons why *P. ruthii* is rare are still unknown. If the answers to these questions about rarity were known, arguably, we would be in a much better position to fully recover the species. However, the factors thought to be responsible are (1) association with a specific, not widely available geologic substrate; (2) dependence on a now-gone hydrologic regime to (a) provide sufficient influx of sediment to support plant establishment in the crevice habitats the species occupies and (b) regulate populations of other herbaceous taxa competing for resources in those crevices or woody/vine taxa that cast excessive shade and limit growth/reproductive output; and (3) variable production of filled seeds that are capable of germination and successfully establishing new plants in a taxing environment that likely limits the number of individuals able to transition from seedling to later life history stages, presumably due to inbreeding as a result of low numbers of compatible mates. Boyd et al. [29] attempted to shed some light on this question of rarity through ecophysiological and genetic comparisons with the more common *P. graminifolia* var. *latifolia*.

The purpose of this review is to highlight the relevant research efforts that have delineated the range and size of populations, determined habitat requirements, documented reproductive and propagation potential, and comprehend the demographic, ecological, and genetic factors that may increase vulnerability to extinction for *P. ruthii*. Cooperative partnerships have led to the successful completion of specific actions called for in the species recovery plan for *P. ruthii*. These partnerships have been invaluable in filling knowledge gaps and providing a foundation for the USFWS to guide conservation and management decisions for the species. A major lesson learned is that consistent effort/attention over time to recovery efforts for a given species is difficult to maintain, but crucial for maintaining momentum. Maintaining regular communication among recovery partners is key, and the recovery work for *P. ruthii* reflects what has been successful for other species as well: when the partners are working together with effective communication, progress happens more quickly. Whether led by the USFWS or by other engaged partners, effective communication is needed to make recovery happen.

Author Contributions: R.N.T. conceptualized the manuscript. All authors contributed equally to the writing. All authors have read and agreed to the published version of the manuscript.

Funding: This work was supported by a Non-Assistance Cooperative Agreement between the University of Tennessee and USDA, ARS (NACA 58-6062-6).

Data Availability Statement: Not applicable.

Acknowledgments: Mention of trade names or commercial products in this article is solely for the purpose of providing specific information and does not imply recommendation or endorsement by The Tennessee Valley Authority, The University of Tennessee, the USDA, or the U.S. Fish and Wildlife Service. The USDA is an equal opportunity employer. The findings and conclusions in the article are those of the authors and do not necessarily represent the views of the U.S. Fish and Wildlife Service.

Conflicts of Interest: The authors declare no conflict of interest.

References

1. Ellis, E.C.; Goldewijk, K.K.; Siebert, S.; Lightman, D.; Ramankutty, N. Anthropogenic transformation of the biomes, 1700–2000. *Glob. Ecol. Biogeogr.* **2010**, *19*, 589–606. [CrossRef]
2. Garza, G.; Rivera, A.; Barrera, C.S.V.; Martinez-Avalos, J.G.; Dale, J.; Arroyo, T.P.F. Potential effects of climate change on the geographical distribution of the endangered plant species *Manihot walkerae*. *Forests* **2020**, *11*, 689. [CrossRef]
3. Roman-Palacios, C.; Wiens, J.J. Recent response to climate change reveal the drivers of species extinction and survival. *Proc. Natl. Acad. Sci. USA* **2020**, *117*, 4211–4217. [CrossRef]
4. Paul, C.; Brandl, S.; Friedrich, S.; Falk, W.; Hartl, F.; Knoke, T. Climate change and mixed forest: How do altered survival probabilities impact economically desirable species proportions of Norway spruce and European beech? *Ann. Forest Sci.* **2019**, *76*, 14. [CrossRef]
5. Lovejoy, T.E. Extinction tsunami can be avoided. *Proc. Natl. Acad. Sci. USA* **2017**, *114*, 8440–8441. [CrossRef]
6. Ellis, S.; Seal, U.S. *Conservation Assessment and Management Plan (CAMP) Process*; IUCN/SSC Conservation Breeding Specialist Group: Apple Valley, MN, USA, 1996.
7. Rodriquez-Robles, U.; Arredondo, J.T.; Huber-Sannwald, E.; Yepez, E.A.; Ramos-Leal, J.A. Coupled plant traits adapted to wetting/drying cycles of substrates co-define niche multidimensionality. *Plant Cell Environ.* **2020**, *43*, 2394–2408. [CrossRef]
8. Phillips, R.D.; Peakall, R.; van der Niet, T.; Johnson, S.D. Niche perspectives on plant-pollinator interactions. *Trends Plant Sci.* **2020**, *25*, 779–793. [CrossRef]
9. Urban, M.C.; Tewksbury, J.J.; Sheldon, K.S. On a collision course: Competition and dispersal differences create no-analogue communities and cause extinctions during climate change. *Proc. R. Soc. B* **2012**, *279*, 2072–2080. [CrossRef]
10. U.S. Fish and Wildlife Service (USFWS). Endangered and threatened wildlife and plants; determination of *Pityopsis ruthii* (Ruth's golden aster) to be an endangered species. *Fed. Regist.* **1985**, *50*, 29341–29344.
11. Moore, P.A.; Wadl, P.A.; Skinner, J.A.; Trigiano, R.N.; Bernard, E.C.; Klingeman, W.E.; Dattilo, A.J. Current knowledge, threats, and future efforts to sustain populations of *Pityopsis ruthii* (Asteraceae), an endangered southern Appalachian species. *J. Torrey Bot. Soc.* **2016**, *143*, 117–134. [CrossRef]
12. U.S. Fish and Wildlife Service (USFWS). Ruth's Golden Aster Recovery Plan. 1992. Available online: http://ecos.fws.gov/docs/recovery_plan/920611.pdf (accessed on 15 May 2023).
13. Hatmaker, E.A.; Staton, M.E.; Dattilo, A.J.; Hadziabdic, D.; Rinehart, T.A.; Schilling, E.E.; Trigiano, R.N.; Wadl, P.A. Population structure and genetic diversity within the endangered species *Pityopsis ruthii* (Asteraceae). *Front. Plant Sci.* **2018**, *9*, 943. [CrossRef]
14. Small, J.K. Studies in the botany of the Southeastern United States—XII. *Bull. Torrey Bot. Club* **1897**, *24*, 487–496. [CrossRef]
15. Bowers, F.A. The existence of *Heterotheca ruthii* (Compositae). *Castanea* **1972**, *37*, 130–132.
16. Shinners, L.H. The north Texas species of *Heterotheca* including *Chrysopsis* (Compositae). *Field Lab.* **1951**, *19*, 66–71.
17. Semple, J.C. Cytotaxonomy of *Chrysopsis* and *Heterotheca* (Compositae-Asteraceae): A new interpretation of phylogeny. *Can. J. Bot.* **1977**, *55*, 2503–2513. [CrossRef]
18. Semple, J.C.; Blok, V.C.; Heiman, P. Morphological, anatomical, habit and habitat differences among the golden aster genera *Chrysopsis*, *Heterotheca*, and *Pityopsis* (Compositae-Asteraceae). *Can. J. Bot.* **1980**, *58*, 17–163. [CrossRef]
19. Semple, J.C.; Bowers, F.D. A revision of the goldenaster genus *Pityopsis* Nutt. (Compositae:Asteraceae). *Univ. Waterloo Biol. Ser.* **1985**, *28*, 1–34.
20. Gowe, A.K.; Brewer, J.S. The evolution of fire-dependent flowering in goldenasters (*Pityopsis* spp.). *J. Torrey Bot. Soc.* **2005**, *132*, 384–400. [CrossRef]
21. Semple, J.C. Chrysopsis; Pityopsis; Heterotheca. In *Flora of North America Vol. 20, Asteraceae, Part 2. Astereae and Senecioneae*, 10th ed.; Oxford University Press: Oxford, UK, 2006; pp. 213–221, 222–226, 230–256.

22. Teoh, V.H. Phylogeny, Hybridization and the Evolution of Fire-Stimulated Flowering within the Grass-Leaved Goldenasters (*Pityopsis*, Asteraceae). Master's Thesis, University of Mississippi, Oxford, MS, USA, 2008.
23. Hatmaker, E.A.; Wadl, P.A.; Rinehart, T.A.; Carrol, J.; Lane, T.S.; Trigiano, R.N.; Staton, M.E.; Schilling, E.E. Complete chloroplast genome comparisons for *Pityopsis* (Asteraceae). *PLoS ONE* **2020**, *15*, e0241391. [CrossRef]
24. Cruzan, M. *Ecological Genetics of Pityopsis ruthii: Final Research Report I, Reproductive Ecology*; Tennessee Department of Environment and Conservation (TDEC)—Division of Natural Areas: Nashville, TN, USA, 2001.
25. Wadl, P.A.; Rinehart, T.A.; Dattilo, A.J.; Pistrang, M.; Vito, L.M.; Milstead, R.; Trigiano, R.N. Propagation for the conservation of *Pityopsis ruthii*, an endangered species from the southeastern United States. *HortScience* **2014**, *49*, 194–200. [CrossRef]
26. Sloan, S.A. Allozyme Variation within and between Populations of Ruth's Golden Aster, *Pityopsis ruthii* (Small) Small. Master's Thesis, The University of Tennessee, Knoxville, TN, USA, 1994.
27. Cruzan, M.; Estill, J.C. *Ecological Genetics of Pityopsis ruthii: Final Research Report II, Phylogeography*; Tennessee Department of Environment and Conservation (TDEC)—Division of Natural Areas: Nashville, TN, USA, 2001.
28. Wadl, P.A.; Dattilo, A.J.; Scheffler, B.; Trigiano, R.N. Development of microsatellite loci for the endangered species *Pityopsis ruthii* (Asteraceae). *Amer. J. Bot.* **2011**, *98*, 342–345. [CrossRef]
29. Boyd, J.N.; Odell, J.; Cruse-Sanders, J.; Rogers, W.; Anderson, J.T.; Baskauf, C.; Brzyski, J. Phenotypic plasticity and genetic diversity elucidate rarity and vulnerability of an endangered riparian plant. *Ecosphere* **2022**, *13*, e3996. [CrossRef]
30. Huang, Y.M.; Chen, C.-W.; Chiou, W.-L.; Chang, H.-M.; Lin, H.-Y. Ex situ conservation of threatened ferns and lycophytes in Taiwan, aspect of reproductive biology. *Int. J. Plant Reprod. Biol.* **2019**, *11*, 121–127.
31. Tang, R.; Li, Y.; Xu, T.; Schinner, J.; Sun, W.; Chen, G. In-situ and ex situ pollination biology of the four threatened plant species and the significance for conservation. *Biodivers. Conserv.* **2020**, *29*, 381–391. [CrossRef]
32. White, A.J. An Autoecological Study of the Endangered Species, *Heterotheca ruthii* (Small) Harms. Master's Thesis, The University of Tennessee, Knoxville, TN, USA, 1977.
33. Farmer, R.E. Seed propagation of *Heterotheca ruthii*. *Castanea* **1977**, *42*, 146–149.
34. Clebsch, E.; Sloan, A. *Final Report, Contract between The University of Tennessee and Tennessee Dept. Environ. Conserv. for the Study of Various Aspects of the Ecology and Life History of the Endangered Species Ruth's Golden-Aster (Pityopsis ruthii)*; Tennessee Department of Environment and Conservation (TDEC)—Division of Natural Areas: Nashville, TN, USA, 1993.
35. Cruzan, M.; Beaty, P. *Population Biology of Ruth's Golden Aster (Pityopsis ruthii): Final Report*; ID-96-05937-6-00; Tennessee Department of Environment and Conservation (TDEC)—Division of Natural Areas: Nashville, TN, USA, 1998.
36. Moore, P.A.; Klingeman, W.E.; Wadl, P.A.; Trigiano, R.N.; Skinner, J.A. Seed production and floral visitors to *Pityopsis ruthii* (Asteraceae: Asterales), and endangered aster native to the Southern Appalachians. *J. Kansas Ent. Soc.* **2020**, *93*, 327–348. [CrossRef]
37. Wofford, B.E.; Smith, D.K. *Status Report on Heterotheca ruthii (Small) Harms to the U.S. Department of the Interior, Fish and Wildlife Serv., Region 4*; U.S. Department of the Interior: Washington, DC, USA, 1980.
38. Wadl, P.A.; Dattilo, A.J.; Vito, L.M.; Trigiano, R.N. Shoot organogenesis and plant regeneration in *Pityopsis ruthii*. *Plant Cell Tiss. Org. Cult.* **2011**, *106*, 513–516. [CrossRef]
39. Murashige, T.; Skoog, F. A revised medium for rapid growth and bioassays with tobacco tissue cultures. *Physiol. Plant* **1962**, *15*, 473–497. [CrossRef]
40. U.S. Fish and Wildlife Service. Ruth's Golden Aster (*Pityopsis ruthii*) 5-Year Review: Summary and Evaluation. 2012. Available online: https://ecos.fws.gov/docs/tess/species_nonpublish/1908.pdf (accessed on 15 May 2023).
41. Wadl, P.A.; Saxton, A.M.; Call, G.; Dattilo, A.J. Restoration of the endangered Ruth's golden aster (*Pityopsis ruthii*). *Southeast. Nat.* **2018**, *17*, 19–31. [CrossRef]
42. Trigiano, R.N.; Dattilo, A.J.; Wadl, P.A. First report of powdery mildew on Ruth's golden aster (*Pityopsis ruthii*) caused by *Golovinomyces cichoracearum* (*Erysiphe cichoracearum*). *Plant Dis.* **2011**, *95*, 879. [CrossRef]
43. Trigiano, R.N.; Rinehart, T.A.; Dee, M.M.; Wadl, P.A.; Poplawski, L.; Ownley, B.H. First report of aerial blight of Ruth's golden aster (*Pityopsis ruthii*) caused by *Rhizoctonia solani* in the United States. *Plant Dis.* **2014**, *98*, 855. [CrossRef]
44. Edwards, T.P.; Trigiano, R.N.; Wadl, P.A.; Ownley, B.H.; Windham, A.S.; Hadziabdic, D. First report of *Alternaria alternata* causing leaf spot on Ruth's golden aster (*Pityopsis ruthii*) in Tennessee. *Plant Dis.* **2016**, *101*, 383. [CrossRef]
45. Thomson, D.; Schwartz, M.W. Using population count data to assess the effects of changing river flow on an endangered riparian plant. *Conserv. Biol.* **2006**, *20*, 1132–1142. [CrossRef]
46. Baxter, K.R.; Collins, J.L.; Cox, P.B. *Results of Large-Flowered Skullcap (Scutellaria montana), Ruth's Golden Aster (Pityopsis ruthii), and Green Pitcher Plant (Sarracenia oreophila) Surveys during 2004–2005*; Tennessee Department of Environment and Conservation (TDEC)—Division of Natural Areas: Nashville, TN, USA, 2005.
47. Tracey, C.; Zimmerman, E.A. *Investigation into Hydrodynamics and Plant Communities of the Younhiogheny River at Ohiopyle, Pennsylvania*; Pennsylvania Natural Hertiage Program: Harrisburg, PA, USA, 2021.

plants

MDPI

Article

Reproduction Modes and Conservation Implications in Three Polyploid *Sorbus* Stenoendemics in Eastern Slovakia (Central Europe)

Vladislav Kolarčik [1,*], Mária Mirková [1] and Vlastimil Mikoláš [2]

[1] Institute of Biology and Ecology, Faculty of Science, Pavol Jozef Šafárik University, Mánesova 23, SK-041 54 Košice, Slovakia
[2] Independent Researcher, Hanojská 4, SK-040 13 Košice, Slovakia
* Correspondence: vladislav.kolarcik@upjs.sk; Tel.: +421-552-342-318

Abstract: The remarkable species diversity of the genus *Sorbus* is a result of polyploidization and frequent hybridization between interacting species of different cytotypes. Moreover, hybridization is possible between several parental taxa. Gametophytic apomixis, which is common among polyploid *Sorbus* taxa, indicates the role of clonal reproduction in the evolutionary stabilization of hybridogeneous genotypes. The precise determination of the origin of seeds and their quantitative evaluation may elucidate inter-cytotype interactions, the potential role of mixed-cytotype populations in evolutionary success, and the long-term survival of some hybrid species. We investigated the reproduction modes of selected species of *Sorbus* in mixed-cytotype populations in eastern Slovakia, Central Europe. We determined the pollen quality, seed production rate, and the ploidy level of mature trees, as well as the origin of the embryo and endosperm in seeds of the stenoendemics *S. amici-petri*, *S. dolomiticola*, and *S. hornadensis*. The tetraploids *S. amici-petri* and *S. hornadensis* are characterized by regular and highly stainable pollen grains and reproduce predominantly via pseudogamous apomixis. In contrast, triploid *S. dolomiticola* usually has oval, heterogenous, and weakly stainable pollen grains, suggesting male meiotic irregularities. Although seeds originate via pseudogamous apomixis in *S. dolomiticola* as well, the ploidy level of sperm cells participating in the fertilization of central cells is usually determined by co-occurring species of different cytotypes. This suggests that maintaining mating partners is necessary for the long-term survival of a triploid species. We documented rare B_{III} hybrids and the residual sexuality in tetraploids. The distribution of seeds of meiotic and apomeiotic origins in *S. amici-petri* shows bimodal characteristics; however, genotypes with predominantly sexual seed types are rare. Reproduction modes documented in polyploid stenoendemics of *Sorbus* and inferred microevolutionary intercytotype relationships highlight the mixed-cytotype populations as the source of biodiversity in apomictic plant complexes. We suggest that conservation efforts should focus on maintaining the species and cytotypic diversity of *Sorbus* populations, especially when it comes to the conservation of triploid species.

Keywords: apomixis; Central Europe; endemism; fertilization; flow cytometry; hybridization; parthenogenesis; pollen stainability; polyploidy; pseudogamy

Citation: Kolarčik, V.; Mirková, M.; Mikoláš, V. Reproduction Modes and Conservation Implications in Three Polyploid *Sorbus* Stenoendemics in Eastern Slovakia (Central Europe). *Plants* **2023**, *12*, 373. https://doi.org/10.3390/plants12020373

Academic Editors: Brenda Molano-Flores and James Cohen

Received: 30 November 2022
Revised: 7 January 2023
Accepted: 10 January 2023
Published: 13 January 2023

1. Introduction

Reproduction is a trait of organisms that involves the development of male and female gametes and the origin of new individuals [1,2]. Additionally, it plays an important role in the diversification of plant groups and their subsequent evolution [3]. Plants can reproduce sexually, asexually, or hemisexually [4,5], and can even utilize combinations of these strategies; for instance, some offspring of an individual may originate sexually and others asexually [6]. Sexual reproduction produces new individuals with genetic features inherited from both parents. Asexual reproduction gives rise to clonal populations, a group of individuals that are genetically identical to their maternal or paternal individual [7–9].

Clones may originate vegetatively or through seeds (agamospermy) [10,11]. Hemisexuality is rather rare and involves the inheritance of part of the genetic material sexually and the other part clonally [5,12]. Plant populations or lineages may exhibit various modes of reproduction, which are reflected in their genetic, phytochemical, and morphological diversity [13–15]. Therefore, determining the reproduction mode may be crucial for understanding population and species microevolutionary dynamics. This is particularly important in the case of endemic, vulnerable, and threatened plants, and the knowledge of reproduction modes may even help to improve conservation planning and management.

The genus *Sorbus* L. represents a complex, taxonomically difficult group of plants [16]. Species differentiation in *Sorbus* is driven by polyploidization and hybridization, coupled with a common cytotype-specific reproductive segregation (diploids are sexual and polyploids are predominantly asexual) [16–23]. Reproductive systems in *Sorbus* are well understood. Several studies have shown that asexual reproduction in *Sorbus* is characterized by the presence of gametophytic apomixis, the parthenogenetic development of embryos from unreduced egg cells, and nutritive tissue endosperm produced via pseudogamy (fertilization of unreduced central cells) [18,20,24–26]. The frequent sympatry of diploids and polyploids facilitates the origin of hybridogeneous genotypes, which may become fixed through the development of apomictic traits [27] (the emergence of apomeiosis/apospory, parthenogenesis, pseudogamy). Recently, detailed biosystematic investigations of the genus revealed several new apomictic species in Bosnia and Herzegovina, the Czech Republic, Hungary, and Sweden [19,28–32], indicating the prevalence of high evolutionary dynamics in this genus. Further investigations in new regions may reveal similar diploid-polyploid (sexual-apomictic) patterns and document new species.

Common co-occurrences of the diploid sexual species, *S. aria* (L.) Crantz, *S. aucuparia* L., and *S. torminalis* (L.) Crantz [20,22], as well as of other polyploids, presumably apomictic taxa, such as *S. danubialis* (Jáv.) Prodan (s.l.) and *S. thaiszii* (Soó) Kárpáti (s.l.) [20,33], may contribute to ongoing hybridizations and the emergence of new species in the Stredné Pohornádie valley in eastern Slovakia. In 1996–2015, V. Mikoláš described three species, *S. dolomiticola* Mikoláš, *S. amici-petri* Mikoláš, and *S. hornadensis* Mikoláš, which are narrow endemics of the region [34–36], and all of them are considered endangered [37]. The first species described from the region, *S. dolomiticola*, is a triploid, stenotopical species (able to tolerate only a restricted range of habitats or ecological conditions) growing in forest-steppe habitats on xerothermous steep slopes that is presumably a hybrid of *S. torminalis* and *S. danubialis* s.l. *Sorbus dolomiticola* is characterized by pale gray/white tomentose lower side of broadly ovate to nearly rhombic leaves and orange-red globose fruits that are 10.5–12.5 × 10.5–12.5 mm in size (Figure 1). *Sorbus amici-petri* is a tetraploid species that is supposedly a hybrid of *S. torminalis* and *S. thaiszii* s.l. It is similar to *S. dolomiticola* and is characterized by leaves that are broadly ovate to rhombic, but differs slightly in size, and has larger red (orange) globose fruits (11.0–13.0 × 13.0–14.0 mm). Such leaf and fruit features may be associated with a higher ploidy level in *S. amici-petri*. The species *S. hornadensis* is tetraploid and morphologically distinct from *S. dolomiticola* and *S. amici-petri*, and has supposedly evolved via hybridization between *S. thaiszii* s.l. and *S. aucuparia*. The species has broadly ovate, shallowly lobed leaves (usually five–seven pairs of lobes) and dark red, shortly cylindric fruits that are 11.0–13.5 × 10–12.5 mm in size. All three species and their putative parental taxa grow sympatrically in a few sites, such as in S-W oriented slopes near Kysak, Obišovce, and Trebejov villages in Stredné Pohornádie valley. These species probably reproduce asexually, and would likely be aposporous pseudogamous apomicts; however, this remains unverified to date.

In the present study, we document the modes of reproduction of three endemic species, *S. dolomiticola*, *S. amici-petri*, and *S. hornadensis*. We aimed to (i) assess reproduction success, i.e., regular seed formation; (ii) determine the frequency of presumed apomeiosis and, if present, rare meiosis; (iii) reconstruct the endosperm origin and infer whether intercytotype interaction is necessary for regular seed formation; and (iv) test for differences in reproduction modes between taxa in this study, e.g., variation in the frequency of meiosis, pseudogamy, and the fertilization of egg cells (possibly including the origin of B_{III} hybrids). We discuss the microevolutionary consequences of these reproduction modes and their potential threats to endemic species. Previous studies have shown that the pseudogamic origin of functional endosperms in the seeds of self-incompatible species may have important conservation implications [17,38].

Figure 1. Three stenoendemic (taxon having very restricted distribution range) *Sorbus* species growing in eastern Slovakia (Central Europe) investigated in the present study. Leaves of short, sterile brachyblast and fruits (left inset) enabled precise identification of the species in the field. Example of flow cytometry (FCM) histogram (right inset) documenting variation in DNA content (relative fluorescence on x-axis) among different cytotypes: diploid *S. torminalis*, triploid *S. dolomiticola*, and tetraploid *S. amici-petri*. In FCM analyses, *Bellis perennis* L. was used as an internal reference standard. Note, *S. hornadensis* is also tetraploid.

2. Results

2.1. Flow Cytometric Determination of Ploidy Level of Mature Trees

The flow cytometry ploidy level measurements fully correspond to the morphology of each species. All of the trees of *S. amici-petri* were tetraploid (15 trees), those of *S. dolomiticola* were triploid (15 trees), and those of *S. hornadensis* were tetraploid (12 trees). Additionally, ploidy levels of supposedly diploid *S. aria* (4 trees) and *S. torminalis* (3 trees) were confirmed.

2.2. Pollen Stainability and Pollen Size

On average, pollen stainability was above 90% in tetraploids, 95.6% in *S. amici-petri*, and 91.9% in *S. hornadensis*. On average, triploid species *S. dolomiticola* showed a 60.6% decrease in pollen stainability (Figure 2). Stained pollen grains were mostly triangular; however, oval pollen grains were quite frequent, and a rare quadrangular pollen shape was also recorded in both tetraploids (Figure 2). In triploid *S. dolomiticola*, stained pollen grains were often oval, and triangular pollen grains were rare. Unstained pollen grains were usually irregular in shape and often shrunken. A size analysis of stained pollen grains showed that oval pollen grains of triploids were highly variable in size compared to those of tetraploids (Figure 2).

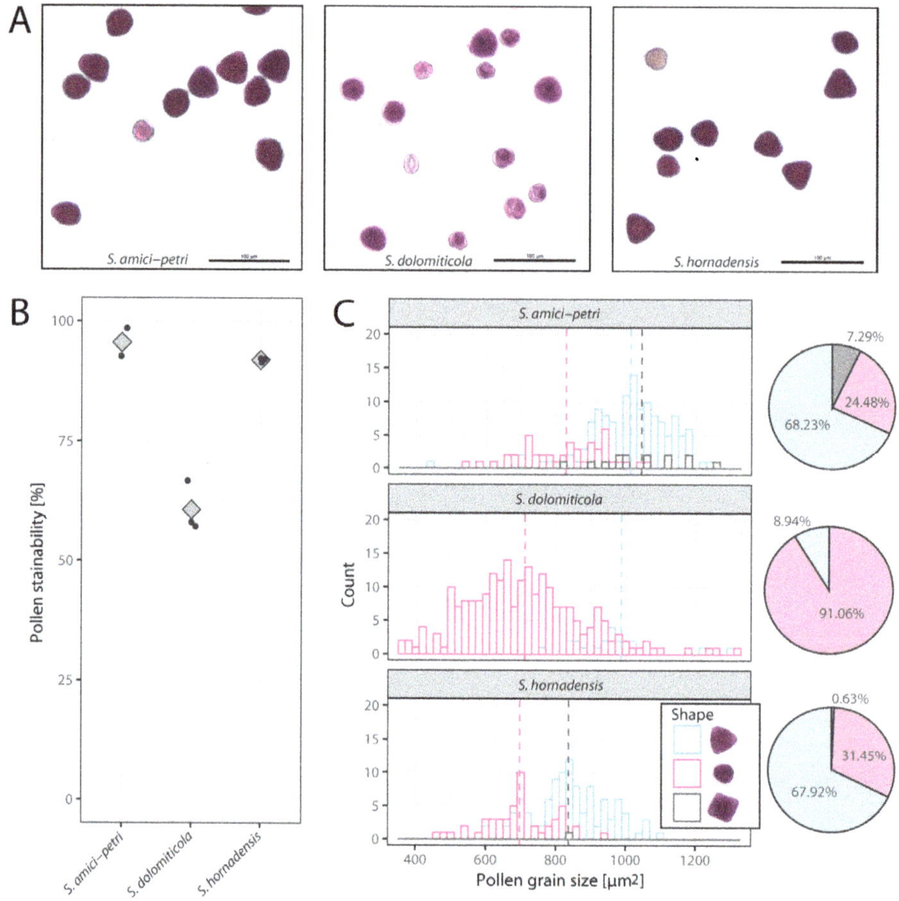

Figure 2. Pollen stainability and pollen size in polyploids *Sorbus amici-petri*, *S. dolomiticola*, and *S. hornadensis*. (**A**) Representative microscopic views of stained pollen grains of three species. (**B**) Variation in pollen stainability among species. (**C**) Pollen grain size and proportion of pollen shape types. In B, dots and diamonds represent pollen stainability per sample and their averages, respectively. Dashed lines are mean values in (**C**).

2.3. Seed Production Rate

Although all of the investigated individuals produced seeds successfully, the number of well-developed seeds varied among species (ANOVA, $F(2, 16) = 7.209$, $p < 0.01$). While the average seed production rate for triploid *S. dolomiticola* was 1.14 seeds per fruit, this parameter decreased in both tetraploids, with *S. amici-petri* and *S. hornadensis* showing average values of 0.36 and 0.29, respectively (Figure 3). Many seeds were infected by insect larvae, and this frequency was higher in tetraploids when compared to triploids, which may be the reason for the observed pattern.

2.4. Flow Cytometric Seed Screen

The results of the FCSS analyses showed that 8 out of 316 seeds (2.53%) had only one peak (for embryo) on FCSS histograms, which did not allow for the determination of reproduction modes in these cases. Three seeds showed more than two expected (embryo and endosperm) peaks on FCSS histograms, which suggests the origin of possible "twins"–two developed embryo sacs within a single ovule. These were not necessarily uninterpretable, but we decided to exclude them from further analyses. We retained 305 analyzed seeds for further analyses. Examples of FCSS histograms for the most frequent seed types (see below) are shown in Figure 4.

Calculations of embryo ploidy levels from DNA content data enable the identification of discrete euploid categories for 303 seeds (99.34%), and 2x, 3x, 4x, 5x, and 6x embryos were found. Only two seeds were assigned to the aneuploid category of ~3.5x (Figure 5). Similarly, the ploidy levels calculated for endosperms were discrete (Figure 6), although more aneuploid categories were recorded. We identified 3x and 4x endosperms in diploids. In triploids, 8x, 10x, and 11x endosperms were most frequent, and ~6.5x, 7x, ~7.5x, ~8.5x, 9x, ~9.5x, and ~10.5x were rare. The most complex situation was documented for tetraploids. While most of the observed endosperms were 6x, 10x, 12x, ~15.5x, and 16x, we also found a variety of other ploidy levels, namely 5x, ~7.5x, ~9.5x, 11x, ~11.5x, ~12.5x, 13x, ~13.5x, 14x, ~14.5x, ~18.5x, and ~19.5x. A methodological error must be considered here: the frequency of some aneuploid categories may be slightly overestimated in the present study, as is seen, for instance, in the case of ~11.5x or ~15.5x. Calculated values are very close to neighboring euploid categories, which may stem from errors in the ploidy level identification, which increases in the cases of >12x endosperms [18,39].

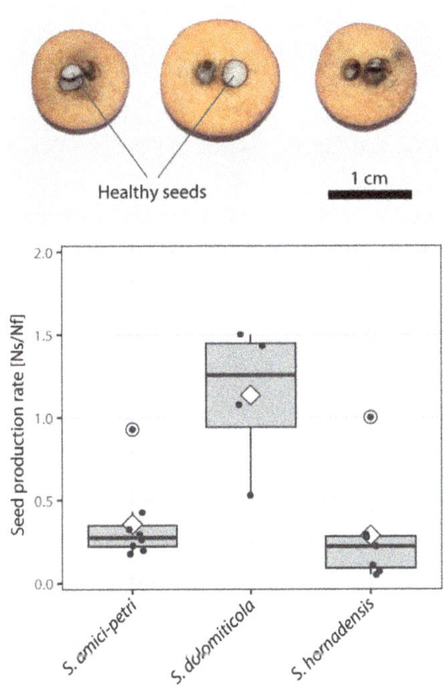

Figure 3. Seed production rate in polyploids *Sorbus amici-petri*, *S. dolomiticola*, and *S. hornadensis*. Examples of three halved fruits (*S. amici-petri*, basal halves) with well-developed seeds (upper subfigure). Variation in seed production rate (lower subfigure); dots represent seed production rate of individual trees; the lower and upper hinges of boxplot correspond to the first and third quartiles (delimit inter-quartile range; IQR), whiskers extend to $\pm 1.5 \times$ IQR, and empty circles represent outliers; horizontal lines and diamonds represent median and mean values, respectively. Ns–number of well-developed seeds, Nf–number of examined fruits.

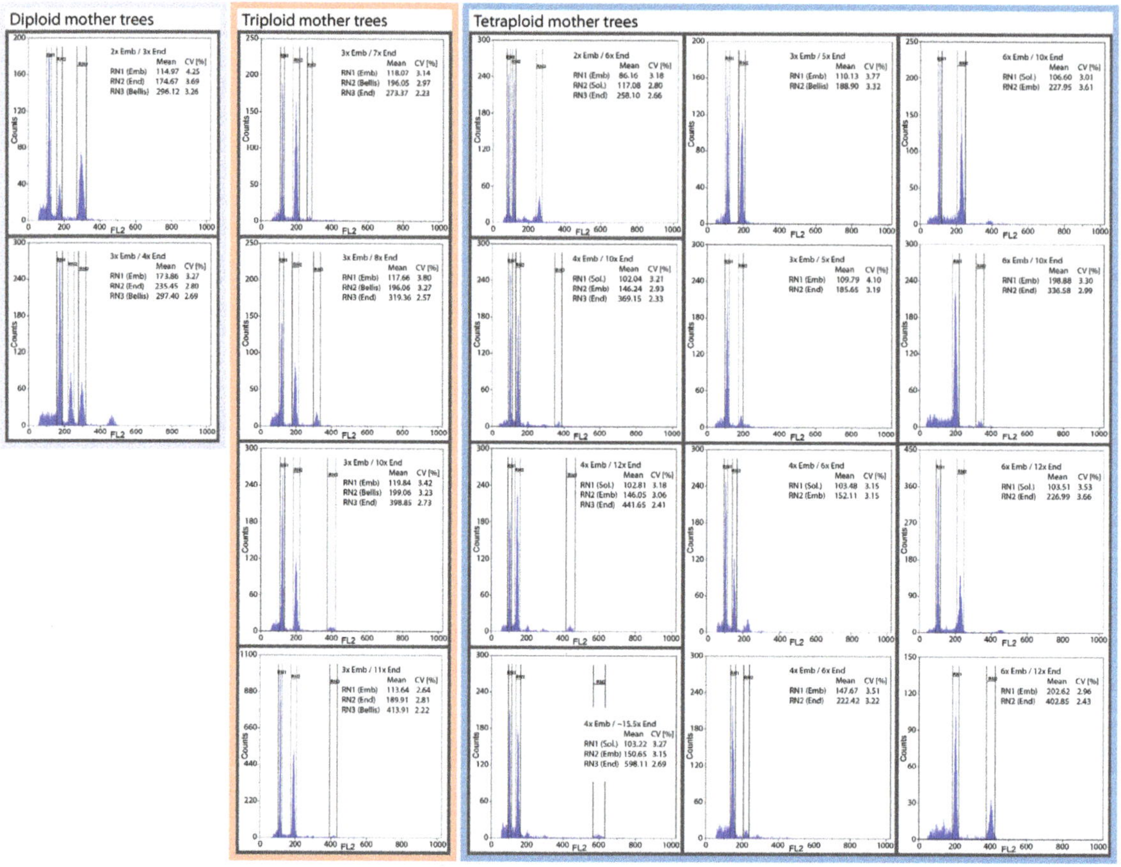

Figure 4. Examples of flow cytometric seed screen (FCSS) histograms and relative fluorescence (*FL2*) vs. counts of nuclei (*Counts*) for most frequent seed types in diploids, triploids, and tetraploids. Each analyzed seed is outlined in dark grey. Two separate measurements, embryo + reference standard and embryo + endosperm, were necessary to clarify the reproduction mode of some seeds. Mean value of fluorescence intensity of nuclei (*Mean*) delimited in peaks (*RN1–RN3*) and corresponding coefficient of variation (*CV*) are reported. *Emb*–embryo, *End*–endosperm, *Bellis*–reference standard *Bellis perennis*, *Sol.*–reference standard *Solanum lycopersicum*.

2.5. Reproduction Modes in *S. amici-petri*, *S. dolomiticola*, and *S. hornadensis*

All 34 seeds analyzed in diploid *S. aria* and *S. torminalis* were found to be of sexual origin. Of these, 33 seeds were $2x_{emb}/3x_{end}$. These can be interpreted as the result of meiotic reduced embryo sacs double-fertilized with reduced sperm cells. One $3x_{emb}/4x_{end}$ originated from a meiotically reduced embryo sac but was double-fertilized with 2x sperm cells.

The triploid *S. dolomiticola* produced 60 (65.93%) and 16 (17.58%) exclusively asexual seeds of categories $3x_{emb}/8x_{end}$ and $3x_{emb}/10x_{end}$, respectively (Table 1). Embryos and endosperms of these seeds developed from apomeiotic unreduced embryo sacs (3x egg cell + 6x central cell). Embryos developed parthenogenetically from unreduced egg cells and the endosperm origin was produced via pseudogamy, indicating that central cell fertilization is necessary for the successful development of parthenogenetic embryos. The central cell (6x) was fertilized by one 2x or two 1x sperm cells (origin of 8x endosperms) or two 2x sperm cells (10x endosperms). Other seed categories were rarely represented (see Table 1). The data indicated that the species can be considered an obligate apomict.

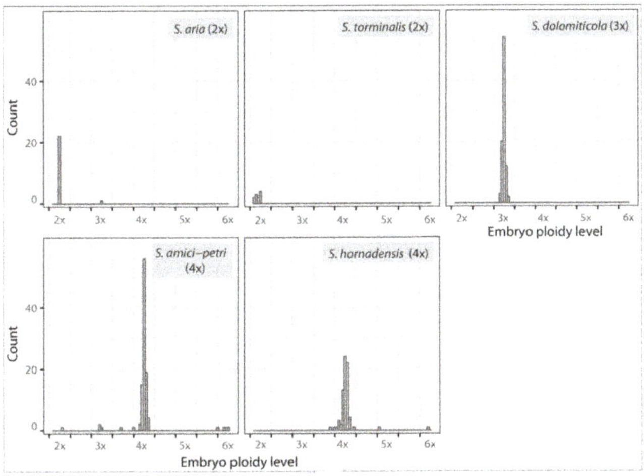

Figure 5. Calculated embryo ploidy levels of 305 seeds of *Sorbus* analyzed in the present study. The estimated euploid and aneuploid ploidy level categories are separated by vertical lines.

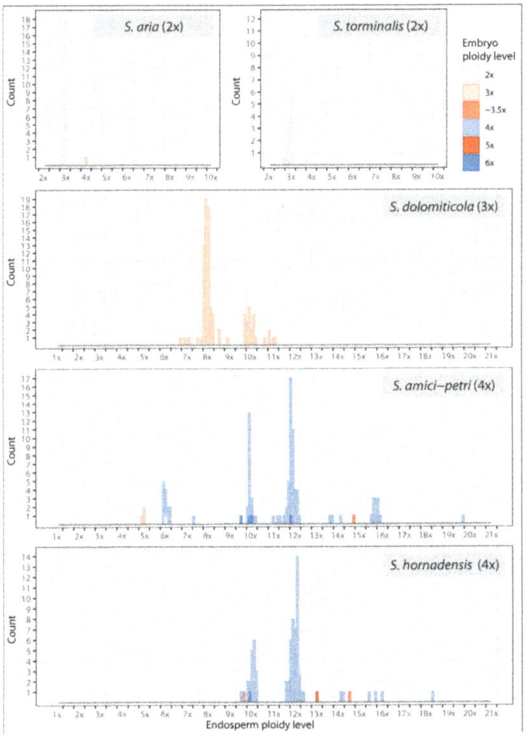

Figure 6. Calculated endosperm ploidy levels of 305 seeds of *Sorbus* analyzed in the present study. The estimated euploid and aneuploid ploidy level categories are separated by vertical lines. The embryo ploidy level associated with endosperm is depicted using sequential and qualitative coloring for increasing ploidy level and for even vs. odd ploidy, respectively.

Table 1. Frequency of seed categories in *Sorbus amici-petri*, *S. dolomiticola*, and *S. hornadensis*. Embryo ploidy levels are listed in rows and endosperm ploidy levels are listed in columns.

Embryo ploidy level	**Endosperm Ploidy Level**																								
S. amici-petri	3x	4x	5x	6x	~6.5x	7x	~7.5x	8x	~8.5x	9x	~9.5x	10x	~10.5x	11x	~11.5x	12x	~12.5x	13x	~13.5x	14x	~14.5x	~15.5x	16x	~18.5x	~19.5x
2x	.	.	.	1	.	.	.	.	.	.	.	.	.	.	.	.	.	.	.	.	.	.	.	.	.
3x	.	.	3	.	.	.	.	.	.	.	.	.	.	.	.	.	.	.	.	.	.	.	.	.	.
~3.5x	.	.	.	.	.	.	.	.	.	.	.	.	.	.	.	.	.	.	.	.	1	.	.	.	.
4x	.	.	.	13	.	.	1	.	.	.	.	20	.	2	9	37	.	.	2	1	.	7	4	.	1
6x	.	.	.	.	.	.	.	.	.	.	1	2	.	.	.	1	.	.	.	.	.	.	.	.	.
S. dolomiticola	3x	4x	5x	6x	~6.5x	7x	~7.5x	8x	~8.5x	9x	~9.5x	10x	~10.5x	11x	~11.5x	12x	~12.5x	13x	~13.5x	14x	~14.5x	~15.5x	16x	~18.5x	~19.5x
3x	.	.	.	.	1	2	1	60	2	1	3	16	1	4	.	.	.	.	.	.	.	.	.	.	.
S. hornadensis	3x	4x	5x	6x	~6.5x	7x	~7.5x	8x	~8.5x	9x	~9.5x	10x	~10.5x	11x	~11.5x	12x	~12.5x	13x	~13.5x	14x	~14.5x	~15.5x	16x	~18.5x	~19.5x
~3.5x	.	.	.	.	.	.	.	.	.	.	1	.	.	.	.	.	.	.	.	.	1	.	.	.	.
4x	.	.	.	.	.	.	.	.	.	.	1	18	.	.	4	38	2	1	.	2	.	2	1	1	.
5x	.	.	.	.	.	.	.	.	.	.	.	.	.	.	.	.	.	1	.	.	.	.	.	.	.
6x	.	.	.	.	.	.	.	.	.	.	.	1	.	.	.	.	.	.	.	.	.	.	.	.	.

The tetraploid *S. amici-petri* showed the highest variation in reproduction modes among the investigated endemics (Table 1). Most of the analyzed seeds were 4x, but rare 2x, 3x, ~3.5x, and 6x seeds were also recorded. The categories of $4x_{emb}/10x_{end}$ and $4x_{emb}/12x_{end}$ were most frequent, and 20 (18.87% in *S. amici-petri*) and 37 seeds (34.91%), respectively, were found. The embryo sacs of these seeds were apomeiotic and unreduced (4x egg cell + 8x central cell), and embryos developed parthenogenetically. The pseudogamous origin of endosperms was further inferred, with the one or two 2x sperm cells' participation. Other seeds had 4x embryos associated with 11x, ~11.5x, ~13.5x, 14x, ~15.5x, 16x, and ~19.5 endosperms (26 seeds, 24.53%), and altogether suggested common pseudogamous apomixis in *S. amici-petri*. A considerable number of seeds (18, 16.98%) originated from meiotically reduced embryo sacs, namely $3x_{emb}/5x_{end}$ and $4x_{emb}/6x_{end}$ (possibly also $4x_{emb}/{\sim}7.5x_{end}$). The only recorded seed of $2x_{emb}/6x_{end}$ represented a rare event of reduced parthenogenesis. Finally, unreduced egg cells may also be fertilized (origin of B_{III} hybrids), as found in seed categories $6x_{emb}/{\sim}9.5x_{end}$, $6x_{emb}/10x_{end}$, and $6x_{emb}/12x_{end}$.

The presence of sexually derived seeds in *S. amici-petri* has a non-random distribution across sampled trees. We found nine trees that almost exclusively produced asexually derived seeds (>80%), with only three sexual seeds out of 78 in total. The other two trees had higher proportions of sexual seeds (37.5% and 60%), with 15 sexual seeds out of 28 in total.

All endosperms in the morphologically distinct tetraploid species *S. hornadensis* showed a ploidy level higher than 8x, which indicates that the species can be considered an obligate apomict (Table 1). Most of the analyzed seeds were 4x, but ~3.5x, 5x, and 6x seeds were also occasionally recorded. The categories of $4x_{emb}/10x_{end}$ and $4x_{emb}/12x_{end}$ were most frequent, with 18 (24.32% in *S. hornadensis*) and 38 seeds (51.35%), respectively. The embryo sacs of these seeds were apomeiotic and unreduced (4x egg cell + 8x central cell), as in *S. amici-petri*, and the embryos developed parthenogenetically. The endosperms in these seeds were of pseudogamous origin, with an 8x central cell fertilized by one or two 2x sperm cells. Furthermore, 14 seeds (18.92%) containing 4x embryos had endosperms of various ploidy levels, namely ~9.5x, ~11.5x, ~12.5x, 13x, 14x, ~15.5x, 16x, and ~18.5x. Rare B_{III} hybrids may also appear in *S. hornadensis*, which was evidenced by the presence of two seeds of the categories $5x_{emb}/13x_{end}$ and $6x_{emb}/10x_{end}$. Additionally, we recorded two seeds with decreased embryo and endosperm DNA content, namely ${\sim}3.5x_{emb}/{\sim}9.5x_{end}$ and ${\sim}3.5x_{emb}/{\sim}14.5x_{end}$, for which the possible route for such DNA contents is difficult to infer. In both of these seeds, a possible partial reduction of DNA content may be explained by the loss of some chromosomes.

3. Discussion

3.1. Reproduction Characteristics and Variation of Tetraploid S. amici-petri and S. hornadensis

Studies of reproduction modes in apomictic plants depend on the precise ploidy level determination of the egg and the central cells (embryo sac), as well as the sperm cells (pollen grain), which interact during fertilization [40–42]. The direct observation of gametophyte development and the fertilization process by cyto-embryological microscopic techniques in sufficiently high numbers is very time consuming and laborious [43]. In most cases, the ploidy levels of eggs and central cells can be determined based on FCSS, thereby enabling the sexual or apomictic origin of embryos to be precisely distinguished [20,39,42]. However, the employment of FCSS is limited in the estimation of the number of sperm cells contributing to endosperm [42]. For instance, the fertilization process for the seed origin with $3x_{emb}/8x_{end}$ could be explained either by the participation of two 1x sperm cells or one 2x sperm cell in the endosperm origin, which cannot be distinguished based on FCSS. Nevertheless, FCSS is one of the most informative techniques in cyto-embryological studies, as it enables the rapid screening of hundreds of seeds in a short time period [44].

In the present study, we applied FCSS to study the reproduction modes of the stenoendemic species *S. amici-petri*, *S. dolomiticola*, and *S. hornadensis* for the first time. We confirmed their cytotypic differentiation; while *S. amici-petri* and *S. hornadensis* are tetraploids, *S. dolomiticola* is triploid species. Both tetraploids *S. amici-petri* and *S. hornadensis* are pseudogamous apomicts that predominantly reproduce clonally, with sexual reproduction being rare (see below). In most cases, we documented $4x_{emb}/10x_{end}$ and $4x_{emb}/12x_{end}$ seeds. Endosperms usually originated from pseudogamy, the fertilization of unreduced central cells with reduced 2x/2x + 2x sperm cells in both species. These are likely their own pollen grains because pollen's self-compatibility is often restored in polyploids [45] and has also been documented among the tetraploid *Sorbus* in pollination and molecular studies [17]. We documented only 18 sexually derived seeds in *S. amici-petri*. Despite its rarity, sexual reproduction may effectively modify genetic makeup and increase the evolutionary potential of apomictic species [14].

We assumed a high frequency of 2x pollen in the studied locality (the presence of many tetraploids). In our study, we documented a higher proportion of $3x_{emb}/8x_{end}$ seeds than $3x_{emb}/10x_{end}$ seeds in triploids, which may suggest more frequent one-sperm-fertilized central cells ($1 \times 2x$ over $2 \times 2x$ sperm cells from tetraploids); however, the opposite pattern was observed in tetraploids (more $4x_{emb}/12x_{end}$ than $4x_{emb}/10x_{end}$ seeds). In the case of triploids, we could not confirm the ploidy levels of sperm cells, as both 1x pollen from diploids and 2x pollen from tetraploids may possibly be utilized in the pollination of triploid. Therefore, frequent 8x endosperms may possibly be the result of the higher pollen compatibility of triploids with 1x pollen from diploids, and in this case, the two-sperm-fertilization of the central cell is equally frequent in triploids as it is in tetraploids. In the case of tetraploids, which are likely self-compatible [17], the central cell is likely more frequently fertilized by the two sperm cells than only by the one sperm cell. Both possibilities are apparently possible. The question of the biological role of one-sperm- vs. two-sperm-fertilization of central cells has been debated. The genus *Sorbus* belongs to the subtribe Malinae Reveal within Rosaceae that includes other genera whose reproduction modes are similar to those of *Sorbus* [39,46–49]. The predominance of two sperm cells over one sperm cell in the fertilization of the central cell is typical in *Amelanchier* Medik. [47] and has been reported in other studies of *Sorbus* as well [18,20]. In contrast, the one-sperm fertilization of the central cell is more frequent in *Cotoneaster* Medik. [48,49]. For *Crataegus* L. species, both alternatives were predominant in different species [39,46,50]. Therefore, this question remains to be resolved for rosaceous genera, but reproductive modes may differ even within plants of a single genus.

Sorbus amici-petri and *S. hornadensis* include seeds with 4x embryos and ~16x endosperms (~15.5x, 16x). We interred that such seeds are more frequent in *S. amici-petri*. The origin of endosperm in this case may involve trinucleated central cells (12x) fertilized with 2x + 2x sperm cells or the endopolyploidization of autonomous endosperms (8x $\rightarrow$ 16x).

Trinucleated central cells have been frequently inferred in seed origins of several roseaceous genera, including *Amelanchier*, *Cotoneaster*, *Crataegus*, and *Sorbus* [18,20,39,46–49]. An alternative explanation that involves endopolyploidization has been proposed, for instance, in some triploids of *Crataegus* [50], which are obligate apomicts with 6x central cells, and rarely produce seed types of $6x_{emb}/{>}12x_{end}$. The origin of such seeds may be better explained as a consequence of the endopolyploidization of unreduced embryo sacs or newly originated embryos and endosperms simultaneously. Generally, species may differ either in various pathways of embryo and endosperm origin, and even in their variation within the species.

Sorbus amici-petri can produce asexual as well as sexual seeds. The presence of seeds with $2x_{emb}/6x_{end}$, $3x_{emb}/5x_{end}$, and $4x_{emb}/6x_{end}$ indicates rare regular meiosis in *S. amici-petri*. Most of these seeds contain embryos produced by the sperm-cell fertilization of reduced egg cells. Even if embryo sacs are of apomeiotic origin, egg cells are frequently fertilized and B_{III} hybrids originate ($6x_{emb}/10x_{end}$). We recorded almost exclusively the 2x sperm cell contribution to the endosperm origin, irrespective of origin from the fertilization of reduced or unreduced central cells. Rarely, we found that, for the 1x sperm cells contribution ($3x_{emb}/5x_{end}$), the pollen donor was likely diploid *S. aria* or *S. torminalis* (they predominate in populations). The frequency of both reproduction modes (sexual or apomictic) varied between individuals of *S. amici-petri*. Meiotically derived embryos were found on certain trees with a frequency of 0–60%. Our sample size of this local endemic does not allow for general comparison. However, some apomicts of the genus *Boechera* Á. Löve et D. Löve showed a bimodal distribution of the frequency of apomeiosis [51]. This means that facultative apomicts produced asexual seeds at very low or very high frequencies, and those with intermediate frequencies were absent. This pattern has yet to be fully explained, but is likely a result of interplay between environmental effects during gamete development, gene flow dynamics in the populations, and regulatory changes associated with apomeiosis [51].

3.2. Dependence of Seed Production of Apomict S. dolomiticola on Heterospecific Pollen and Conservation Implications

We have shown that *S. dolomiticola* is an apomictic species, and we have exclusively documented the unreduced embryo sac formation and parthenogenetic embryo origins. Nevertheless, sperm cells are important in the reproduction of *S. dolomiticola*. Although necessary for the origin of functional endosperms, they are excluded from egg cell fertilization, and thus do not affect the genotype of the embryo. Contributions of 2x (or 1x + 1x) and two 2x, sperm cells are the most frequent among reconstructed endosperm origins. Species of the genus *Sorbus* produce reduced pollen grains [20], as do related genera, e.g., *Crataegus* [52], and although we did not perform ploidy analyses of pollen cells, our pollen investigations revealed that *S. dolomiticola* has irregular pollen grains, which suggested an unbalanced reduced pollen grain formation [53]. However, contributions of unbalanced ~1.5x sperm cells, as manifested by ~7.5x or 9x (~9.5x) endosperms, are very rare in *S. dolomiticola*. This is consistent with results of another study [20], but opposite patterns, including the frequent fertilization of unreduced central cells of triploids by unbalanced reduced ~1.5x sperm cells (origin of $3x_{emb}/9x_{end}$ and $3x_{emb}/{\sim}9.5x_{end}$), were documented in *Sorbus* [18]. Additionally, this is the case for *Crataegus* L. [50,52]. The self-incompatibility observed in triploid *Sorbus* [17] may also be the reason for the low frequency of ~1.5x pollinations. Because *S. dolomiticola* is an apomictic species and forms clonal populations, pollen incompatibility is very likely effective, even among different individuals (supposedly genetically identical clones).

Our results show that contributions of heterospecific pollen from diploids (1x/1x + 1x) or tetraploids (2x/2x + 2x) is necessary for the origin of nutritive tissue endosperm (pseudogamy), and regular seed production in *S. dolomiticola*. This latter example suggests that the evolutionary success of self-incompatible or sterile pseudogamous apomicts therefore depends on other related species. This has an important implication for conservation strategies of such endemics. Seed formation in *S. dolomiticola* trees is conditioned by the sympatric occurrence of cross-compatible congeners. Mating cytotypically differentiated partners must be maintained in sympatry with *S. dolomiticola* to ensure the production of well-developed seeds and the long-term survival of the species [38]. Unless precise inter-species compatibilities are known, conservation priorities should also focus on the preservation of population heterogeneity, with the presence of several cytotypically variable *Sorbus* species [54]. Absence of congeners at sites may potentially lead to *S. dolomiticola* extinction. Several similar biological associations have implications in conservation management. Conservation efforts for rare parasitic plants, which may be limited due to host availability or host preference, should consider the needs of the host [55], and should appropriately adjust conservation management accordingly. Conservation implications should also be considered in cases of plant-insect interactions. The conservation of dependent species has been proposed for threatened herbivorous insects that depend on specific host plants [56]. The dependent species are often lost due to extinction before their host [55,56]; conservation planning should not ignore such biotic interactions.

3.3. Microevolutionary Dynamics in Mixed-Cytotype Populations in Stredné Pohornádie Valley

The origin of parthenogenetic 2x, allotriploid 3x, and sexual 4x individuals, as well as B_{III} 6x hybrids, represents a palette of various cytotypes with microevolutionary potential. Mature plants with the ability to produce functional pollen and seeds can interact with co-occurring species. Mixed cytotype populations of pseudogamous and facultative apomicts are, therefore, considered natural evolutionary laboratories. Cytotypic diversity in populations results in either a higher proportion of observed 3x offspring on diploid plants [18] or a larger proportion of hybrid plants. They are likely to have originated because of a higher proportion of cytotypically diverse pollen available for pollination [20,39]. For instance, either the reduced 2x pollen from 4x plants or the unreduced 2x pollen from 2x plants may give rise to 3x embryos in 2x trees. However, FCM does not allow interploidy to be distinguished from within-ploidy crossings.

The presence of various endemics of *Sorbus* in Stredné Pohornádie valley is not surprising. Heterogeneous orography and mosaics of forests, xerothermous forest-steppe-like habitats, and rocky terraces allow several *Sorbus* species to co-exist in relatively close proximity [34–36]. The diploids *S. aria*, *S. aucuparia*, and *S. torminalis* may be found together with several other polyploid taxa, such as *S. danubialis* s.l or *S. thaiszii* s.l. Their mutual reproductive interactions and between-cytotype crosses may result in locally originating, hybrid triploids [57] that can serve as a triploid bridge to the origin of new genotypes. Gametophytic apomixis is likely a process that can stabilize new genotypes and lead to the evolution of new species [27,45]. Stredné Pohornádie valley may be considered a region with several evolutionarily significant units, with the ongoing speciation of *Sorbus*. Presumed combinations of parental species *S. danubialis* s.l., *S. thaiszii* s.l., *S. aucuparia*, and *S. torminalis* for the hybridogenous origin of three stenoendemics were proposed based on morphological variations and they need to be tested as probable hypotheses using artificial pollination techniques and molecular and cytogenomic research tools.

4. Materials and Methods

4.1. Plant Material

Three species of the genus *Sorbus*: *S. amici-petri*, *S. dolomiticola*, and *S. hornadensis* (Figure 1) are narrow endemics, which are known only from a few sites in the close vicinity of the Kysak, Obišovce, and Trebejov villages (Stredné Pohornádie phytogeographical district, eastern Slovakia, Central Europe, [58]), which are located approximately 15 km north of Košice city. In this study, trees were sampled from two xerothermous steep S-W oriented slopes (N: 48°51′52″ E: 21°13′00″, alt. ~320 m a.s.l. and N: 48°50′46″, E: 21°14′10″, ~320 m a.s.l). Floral buds were collected in the spring (May) of 2022 and mature fruits in the autumn (October November) of 2020 and 2022; all seeds originated from open pollination. Details of the sampled material are listed in Table 2. Additionally, fruits of two diploid species, *S. aria* and *S. torminalis*, which grow sympatrically with the investigated polyploids, were sampled for comparison and the correct interpretation of flow cytometry (FCM) data. Voucher specimens of the trees were deposited in the herbarium of the Botanical Garden of Pavol Jozef Šafárik University in Košice [59].

Table 2. Summary of plant material belonging to the genus *Sorbus* analyzed in the present study. The number of investigated individuals and number of pollen grains, fruits, and seeds used for pollen analyses (*Pollen Stainability* and *Pollen Grain Size*), seed production rate determination (*Fruits*), flow cytometry determination of ploidy level of mature plants (*FCP*), and flow cytometric seed screen (*FCSS*) are reported in parenthesis. For FCSS, the number of analyzed seeds and number of successfully analyzed seeds are mentioned before and after the forward slash, respectively.

Species	Pollen Stainability	Pollen Grain Size	Fruits	FCP	FCSS
Sorbus amici-petri	2 (401)	2 (192)	8 (327)	15	11 (107/106)
S. dolomiticola	3 (550)	3 (246)	4 (46)	15	9 (93/91)
S. hornadensis	3 (409)	3 (159)	7 (176)	12	10 (82/74)
SUM	8 (1360)	8 (597)	19 (549)	42	30 (282/271)

4.2. Pollen Stainability and Pollen Grain Size Determination

Male meiosis may be irregular in polyploids; therefore, pollen viability and the shape of pollen grains may be indicative of the disruption of male gamete development. In the present study, pollen viability was assessed indirectly using a staining technique [60]. We used a simplified method for the differential staining of aborted and non-aborted pollen grains [61]. Floral buds in the balloon stage (larger buds close to anthesis [17]) or open flowers were sampled and immediately placed in a solution of ethanol (96%) and acetic acid (98%) in a ratio of 3:1. Four to five anthers were dissected from one randomly selected bud (or flower) per tree and placed in 60 μL of staining solution prepared according to Ross et al. [61]. After incubating for ~30 min and heating for 5 min (moving the glass slide above a flame), anthers were thoroughly crushed on a glass slide until pollen grains were released into the staining solution. Prepared slides were photographed and analyzed immediately using a Leica DM 2500 microscope equipped with a DFC 290 HD camera (Leica Microsystems GmbH, Wetzlar, Germany) and the Leica application suite software (ver. 3.5.0, Leica Microsystems GmbH, Wetzlar, Germany). The slides were then observed under 400× magnification. We scored at least 80 pollen grains per sample. The proportion of stained pollen grains was calculated as a percentage: 100 × (number of stained pollen grains)/(number of all pollen grains).

The size and shape of pollen grains were determined via image analysis using TpsDig2 software (ver. 2.26, [62]). Pollen size was scored as the area of the 2D projection of the stained pollen grain (in μm^2).

4.3. Seed Production Rate

Fruits were transversely halved and seeds with well-developed embryos were counted. We expressed the seed production rate as the ratio between the number well-developed seeds (Ns) and the number of examined fruits (Nf).

4.4. Flow Cytometric Determination of the Ploidy Level of Mother Plant, Embryo, and Endosperm

FCM was used to determine the ploidy level of investigated mother individuals (flow cytometry ploidy level determination; FCP [63]) and their seed tissues, embryo, and endosperm (flow cytometric seed screen; FCSS [40,41]). The ploidy level was determined based on DNA content determinations of nuclei in maternal (mother DNA content/ploidy level), embryo (embryo DNA content/ploidy level), and endosperm tissue (endosperm DNA content/ploidy level). We used terminal or subterminal buds on short brachyblasts of *Sorbus* trees as a source material for nuclear isolation and subsequent FCP. Seeds were halved, and embryos and endosperms were used to isolate nuclei for FCSS. We conducted a two-step FCM procedure [64]. Samples were prepared using the following internal reference standards [65,66]: *Solanum lycopersicum* L. 'Stupické polní tyčkové rané', 1.96 pg DNA [67] or *Bellis perennis* L. wild population, 3.61 pg DNA [68] (note, we recalculated the genome size of *B. perennis* against *S. lycopersicum*). We used a modified sample preparation protocol [52,69]. Approximately 0.5 cm^2 of the leaf of the reference standard and two or three buds of the *Sorbus* tree and halves of seeds were co-chopped using a razor blade in a petri dish containing 1 mL of general purpose buffer for FCP and FCSS, respectively [69]. A suspension containing isolated nuclei was then filtered through a 42-μm nylon filter. Samples were supplemented with 4 μL of β-mercaptoethanol, 50 μL RNAse (1 mg mL^{-1}), and 50 μL of the intercalating dye propidium iodide (1 mg mL^{-1}). After 10 min of incubation at 4 °C, the fluorescence intensity of stained nuclei was measured using a Partec CyFlow ML flow cytometer (Partec GmbH, Münster, Germany) and FloMax software (ver. 2.7, Partec GmbH, Münster, Germany). We recorded at least 4000 measured nuclei via FCM (min. 3780 but often more than 5000 nuclei), usually with at least 1000 nuclei for both the *Sorbus* and reference standard. The coefficient of variation of the nuclei was <5% in FCM peaks of *Sorbus* (maternal tissues or embryos) and the reference standard, and <8% in peaks of endosperms (mostly because of less abundant endosperms in *Sorbus* seeds).

The DNA content of mother plants was calculated based on the following formula, where a deviation of 1.1% from flow cytometry machine linearity was incorporated:

$$\text{DNA quantity of sample} = \text{DNA quantity of standard} \times [(G_0/G_1 \text{ peak mean of } \textit{Sorbus})/\ 1.011 \times (G_0/G_1 \text{ peak mean of standard})]$$

where G_0/G_1 refers to the population of nuclei (FCM peak) in G_0 or G_1 phases of the cell cycle. The DNA quantity and ploidy level determination were interpreted based on the previous inference of relatively stable and distinct DNA content differences observed between 2x, 3x, 4x, and 5x individuals [70].

In FCSS analyses, we first determined the embryo DNA content and then calculated the endosperm DNA content as follows:

$$\text{DNA quantity of endosperm} = \text{DNA quantity of embryo} \times [(G_0/G_1 \text{ peak mean of endosperm})/\ 1.011 \times (G_0/G_1 \text{ peak mean of embryo})]$$

where G_0/G_1 refers to the population of nuclei (FCM peak) in G_0 or G_1 phases of the cell cycle. This is useful when separate measurements, such as the embryo + reference standard and embryo + endosperm, are performed. We determined embryo and endosperm ploidy levels based on the inference of their DNA contents, adjusted for the mean DNA content of embryos in diploids (those of the same 2x ploidy level as their mother trees averaged per mother tree) or of parthenogenetic embryos in polyploids (averaged per mother tree). We analyzed 282 seeds of triploids and tetraploids, and 34 seeds of diploids for comparison.

4.5. Examination of Reproduction Modes

We inferred the occurrence of meiosis/apomeiosis, the autonomous/pseudogamous endosperm formation, the embryo origin (parthenogenesis/sexual origin/B_{III} hybrids), and the contribution of sperm cells in the embryo and the endosperm origin, as per recent studies [40–42]. The reconstruction of the embryo and endosperm origin in seeds is crucial for the determination of reproduction modes in the FCSS method [40–42]. Possible interpretations of the seed origin are summarized in Supplementary Material Table S1.

4.6. Statistical Analyses

All summary statistics and plots were performed using the R environment (ver. 3.5.3 [71]) and the ggplot2 package [72]. One-way analysis of variance (ANOVA) and Tukey's HSD pairwise multiple comparison test were employed to test for differences between the means of the seed production rate of different species. Data were log transformed to meet assumptions of ANOVA, data normality and homoscedasticity.

5. Conclusions

In the present study, we documented the reproduction modes of triploid *S. dolomiticola* and tetraploid *S. amici-petri* and *S. hornadensis*, three stenoendemics of the Stredné Pohornádie phytogeographical district in eastern Slovakia. All three species reproduced predominantly asexually, and offspring were identical clones of mother trees. Embryos developed parthenogenetically from unreduced egg cells, but pseudogamy, the fertilization of the unreduced central cells of embryo sacs, is necessary for regular seed development. While tetraploids produced regular pollen grains, and these were involved in the pseudogamic origin of endosperm, triploid *S. dolomiticola* formed irregular pollen grains and appeared to be dependent on pollen availability from different cytotypes (diploids and tetraploids). Rare, sexually originated seeds in tetraploids, mostly in *S. amici-petri*, may increase the genetic and also cytotypic diversity of populations and may further accelerate within-population microevolutionary dynamics; this may potentially lead to the origin of new species. We conclude that determining the modes of reproduction in endangered pseudogamous apomicts may help to set up appropriate conservation strategies. Conservation planning focused on triploids, as in the case of *S. dolomiticola*, should include strategies that maintain or even increase the species and cytotypic diversity of *Sorbus* populations.

Supplementary Materials: The following supporting information can be downloaded at: https://www.mdpi.com/article/10.3390/plants12020373/s1, Table S1: Survey of seed types recorded in the present study.

Author Contributions: Conceptualization, V.K. and V.M.; methodology, V.K., M.M. and V.M.; validation, V.K.; investigation, V.K. and M.M.; resources, V.K. and V.M.; data curation, V.K.; writing—original draft preparation, V.K., M.M. and V.M.; writing—review and editing, V.K. and V.M.; supervision, V.K.; project administration, V.K.; funding acquisition, V.K. All authors have read and agreed to the published version of the manuscript.

Funding: This research was funded by the grant agency Vedecká Grantová Agentúra MŠVVaŠ SR a SAV (VEGA), grant no: 1/0741/19.

Institutional Review Board Statement: Not applicable.

Informed Consent Statement: Not applicable.

Data Availability Statement: Not applicable.

Acknowledgments: Kristína Kolarčiková and Zuzana Košturiaková are acknowledged for help with plant material sampling. We thank Jaroslav Doležel (Olomouc, Czech Republic) for the kind provision of some FCM reference standards.

Conflicts of Interest: The authors declare no conflict of interest.

References

1. Faure, J.-E. Double Fertilization in Flowering Plants: Discovery, Study Methods and Mechanisms. *Comptes Rendus De L'académie Des Sci. -Ser. III-Sci. De La Vie* **2001**, *324*, 551–558. [CrossRef]
2. Raghavan, V. Some Reflections on Double Fertilization, from Its Discovery to the Present. *New Phytol.* **2003**, *159*, 565–583. [CrossRef] [PubMed]
3. Barrett, S.C.H. Major Evolutionary Transitions in Flowering Plant Reproduction: An Overview. *Int. J. Plant Sci.* **2008**, *169*, 1–5. [CrossRef]
4. Maheswari, P. *An Introduction to the Embryology of Angiosperms*; McGraw-Hill: New York, NY, USA, 1950.
5. Kolarčik, V.; Kocová, V.; Vašková, D. Flow Cytometric Seed Screen Data Are Consistent with Models of Chromosome Inheritance in Asymmetrically Compensating Allopolyploids. *Cytom. Part A* **2018**, *93*, 737–748. [CrossRef] [PubMed]
6. Savidan, Y. Gametophytic Apomixis. In *Current Trends in the Embryology of Angiosperms*; Bhojwani, S.S., Soh, W.-Y., Eds.; Springer Netherlands: Dordrecht, The Netherlands, 2001; pp. 419–433.
7. Normark, B.B. 13-Unusual Gametic and Genetic Systems. In *Sperm Biology an Evolutionary Perspective*; Birkhead, T.R., Hosken, D.J., Pitnick, S., Eds.; Academic Press: London, UK, 2009; pp. 507–538. ISBN 978-0-12-372568-4.
8. Vallejo-Marín, M.; Dorken, M.; Barrett, S.C.H. The Ecological and Evolutionary Consequences of Clonality for Plant Mating. *Annu. Rev. Ecol. Evol. Syst.* **2010**, *41*, 193–213. [CrossRef]
9. Barrett, S.C.H. Influences of Clonality on Plant Sexual Reproduction. *Proc. Natl. Acad. Sci. USA* **2015**, *112*, 8859–8866. [CrossRef]

10. De Meeûs, T.; Prugnolle, F.; Agnew, P. Asexual Reproduction: Genetics and Evolutionary Aspects. *Cell. Mol. Life Sci.* **2007**, *64*, 1355–1372. [CrossRef] [PubMed]
11. Ozias-Akins, P. Apomixis: Developmental Characteristics and Genetics. *Crit. Rev. Plant Sci.* **2006**, *25*, 199–214. [CrossRef]
12. Vozárová, R.; Herklotz, V.; Kovařík, A.; Tynkevich, Y.; Volkov, R.; Ritz, C.; Lunerová, J. Ancient Origin of Two 5S RDNA Families Dominating in the Genus *Rosa* and Their Behavior in the Canina-Type Meiosis. *Front. Plant Sci.* **2021**, *12*, 643548. [CrossRef] [PubMed]
13. Pellino, M.; Hojsgaard, D.; Schmutzer, T.; Scholz, U.; Hörandl, E.; Vogel, H.; Sharbel, T. Asexual Genome Evolution in the Apomictic *Ranunculus auricomus* Complex: Examining the Effects of Hybridization and Mutation Accumulation. *Mol. Ecol.* **2013**, *22*, 5908–5921. [CrossRef] [PubMed]
14. Hojsgaard, D.; Hörandl, E. A Little Bit of Sex Matters for Genome Evolution in Asexual Plants. *Front. Plant Sci.* **2015**, *6*, 82. [CrossRef] [PubMed]
15. Hojsgaard, D.; Klatt, S.; Baier, R.; Carman, J.; Hörandl, E. Taxonomy and Biogeography of Apomixis in Angiosperms and Associated Biodiversity Characteristics. *Crit. Rev. Plant Sci.* **2014**, *33*, 414–427. [CrossRef] [PubMed]
16. Sennikov, A.; Kurtto, A. A Phylogenetic Checklist of *Sorbus* s.l. (Rosaceae) in Europe. *Memo. Soc. Pro Fauna Flora Fenn.* **2017**, *93*, 1–78.
17. Ludwig, S.; Robertson, A.; Rich, T.; Đorđević, M.; Cerović, R.; Houston, L.; Harris, S.; Hiscock, S. Breeding Systems, Hybridization and Continuing Evolution in Avon Gorge *Sorbus*. *Ann. Bot.* **2013**, *111*, 563–575. [CrossRef] [PubMed]
18. Hajrudinović, A.; Siljak-Yakovlev, S.; Brown, S.C.; Pustahija, F.; Bourge, M.; Ballian, D.; Bogunić, F. When Sexual Meets Apomict: Genome Size, Ploidy Level and Reproductive Mode Variation of *Sorbus aria* s.l. and *S. austriaca* (Rosaceae) in Bosnia and Herzegovina. *Ann. Bot.* **2015**, *116*, 301–312. [CrossRef] [PubMed]
19. Lepší, M.; Lepší, P.; Koutecký, P.; Bílá, J.; Vít, P. Taxonomic Revision of *Sorbus* Subgenus *Aria* Occurring in the Czech Republic. *Preslia* **2015**, *87*, 109–162.
20. Lepší, M.; Koutecký, P.; Nosková, J.; Lepší, P.; Urfus, T.; Rich, T.C.G. Versatility of Reproductive Modes and Ploidy Level Interactions in *Sorbus* s.l. (Malinae, Rosaceae). *Bot. J. Linn. Soc.* **2019**, *191*, 502–522. [CrossRef]
21. Robertson, A.; Rich, T.C.G.; Allen, A.M.; Houston, L.; Roberts, C.A.T.; Bridle, J.O.N.R.; Harris, S.A.; Hiscock, S.J. Hybridization and Polyploidy as Drivers of Continuing Evolution and Speciation in *Sorbus*. *Mol. Ecol.* **2010**, *19*, 1675–1690. [CrossRef] [PubMed]
22. Uhrinová, V.; Zozomová-Lihová, J.; Bernátová, D.; Paule, J.; Paule, L.; Gömöry, D. Origin and Genetic Differentiation of Pink-Flowered *Sorbus* Hybrids in the Western Carpathians. *Ann. Bot.* **2017**, *120*, 271–284. [CrossRef]
23. Chester, M.; Cowan, R.S.; Fay, M.F.; Rich, T.C.G. Parentage of Endemic *Sorbus* L. (Rosaceae) Species in the British Isles: Evidence from Plastid DNA. *Bot. J. Linn. Soc.* **2007**, *154*, 291–304. [CrossRef]
24. Liljefors, A. Studies on Propagation, Embryology, and Pollination in *Sorbus*. *Acta Horti Bergiani* **1953**, *16*, 277–329.
25. Jankun, A.; Kovanda, M. Apomixis in *Sorbus sudetica* (Embryological Studies in *Sorbus* 1). *Preslia* **1986**, *58*, 7–19.
26. Jankun, A.; Kovanda, M. Apomixis and Origin of *Sorbus bohemica* (Embryological Studies in *Sorbus* 2). *Preslia* **1987**, *59*, 97–116.
27. Hojsgaard, D.; Greilhuber, J.; Pellino, M.; Paun, O.; Sharbel, T.F.; Hörandl, E. Emergence of Apospory and Bypass of Meiosis via Apomixis after Sexual Hybridisation and Polyploidisation. *New Phytol.* **2014**, *204*, 1000–1012. [CrossRef]
28. Vít, P.; Lepší, M.; Lepší, P. There Is No Diploid Apomict among Czech *Sorbus* Species: A Biosystematic Revision of *S. eximia* and Discovery of *S. barrandienica*. *Preslia* **2013**, *84*, 71–96.
29. Somlyay, L.; Sennikov, A. Atlas Florae Europaeae Notes 23. The Typification and Revised Taxonomic Circumscription of *Sorbus bakonyensis* (Rosaceae), with a Description of *Sorbus udvardyana*, a New Apomictic Species Endemic to Hungary. *Phytotaxa* **2014**, *164*, 265–275. [CrossRef]
30. Hajrudinović, A.; Frajman, B.; Schönswetter, P.; Silajdžić, E.; Siljak-Yakovlev, S.; Bogunić, F. Towards a Better Understanding of Polyploid *Sorbus* (Rosaceae) from Bosnia and Herzegovina (Balkan Peninsula), Including Description of a Novel, Tetraploid Apomictic Species. *Bot. J. Linn. Soc.* **2015**, *178*, 670–685. [CrossRef]
31. Somlyay, L.; Lisztes-Szabó, Z.; Sennikov, A.N. Atlas Florae Europaeae Notes 29. Two New Species of *Sorbus* (Rosaceae) Endemic to Hungary, Previously Confused with *S. subdanubialis*. *Ann. Bot. Fenn.* **2016**, *53*, 361–372. [CrossRef]
32. Levin, J.; Fay, M.F.; Pellicer, J.; Hedrén, M. Multiple Independent Origins of Intermediate Species between *Sorbus aucuparia* and *S. hybrida* (Rosaceae) in the Baltic Region. *Nord. J. Bot.* **2018**, *36*, e02035. [CrossRef]
33. Velebil, J.; Lepší, M.; Nosková, J.; Lepší, P. Taxonomic Assessment of *Sorbus* Subgenus *Aria* in the Malé Karpaty Mountains. *Preslia* **2022**, *94*, 305–334. [CrossRef]
34. Mikoláš, V. *Sorbus dolomiticola* Mikoláš, a New Hybridogenous Species of the Genus *Sorbus* s.l. from Eastern Slovakia. *Thaiszia J. Bot.* **1996**, *6*, 1–12.
35. Mikoláš, V. *Sorbus amici-petri* Mikoláš, a New Hybridogenous Species of the Genus *Sorbus* s.l. from Eastern Slovakia. *Thaiszia J. Bot.* **2003**, *13*, 127–133.
36. Mikoláš, V. *Sorbus hornadensis* Mikoláš (Rosaceae, Pyreae), a New Hybridogenous Species from Eastern Slovakia. *Thaiszia J. Bot.* **2015**, *25*, 21–27.
37. Rivers, M.; Beech, E.; Bazos, I.; Bogunić, F.; Buira, A.; Caković, D.; Carapeto, A.; Carta, A.; Cornier, B.; Fenu, G.; et al. *European Red List of Trees*; IUCN: Cambridge, UK; Brussels, Belgium, 2019.

38. Hamston, T.J.; Wilson, R.J.; de Vere, N.; Rich, T.C.G.; Stevens, J.R.; Cresswell, J.E. Breeding System and Spatial Isolation from Congeners Strongly Constrain Seed Set in an Insect-Pollinated Apomictic Tree: *Sorbus subcuneata* (Rosaceae). *Sci. Rep.* **2017**, *7*, 45122. [CrossRef] [PubMed]
39. Talent, N.; Dickinson, T. Endosperm Formation in Aposporous *Crataegus* (Rosaceae, Spiraeoideae, Tribe Pyreae): Parallels to Ranunculaceae and Poaceae. *New Phytol.* **2007**, *173*, 231–249. [CrossRef] [PubMed]
40. Matzk, F.; Meister, A.; Schubert, I. An Efficient Screen for Reproductive Pathways Using Mature Seeds of Monocots and Dicots. *Plant J.* **2000**, *21*, 97–108. [CrossRef]
41. Matzk, F.; Meister, A.; Brutovská, R.; Schubert, I. Reconstruction of Reproductive Diversity in *Hypericum perforatum* L. Opens Novel Strategies to Manage Apomixis. *Plant J.* **2001**, *26*, 275–282. [CrossRef]
42. Dobeš, C.; Lückl, A.; Hülber, K.; Paule, J. Prospects and Limits of the Flow Cytometric Seed Screen—Insights from *Potentilla* Sensu Lato (Potentilleae, Rosaceae). *New Phytol.* **2013**, *198*, 605–616. [CrossRef]
43. Leblanc, O.; Mazzucato, A. Screening Procedures to Identify and Quantify Apomixis. In *The Flowering of Apomixis: From Mechanisms to Genetic Engineering*; Savidan, Y., Carman, J.G., Dresselhaus, T., Eds.; CIMMYT, IRD, European Commission DG VI (FAIR): Mexico City, Mexico, 2001; pp. 121–136.
44. Matzk, F. Reproduction Mode Screening. In *Flow Cytometry with Plant Cells*; Doležel, J., Greilhuber, J., Suda, J., Eds.; Wiley-VCH Verlag: Weinheim, Germany, 2007; pp. 131–152. ISBN 9783527610921.
45. Hojsgaard, D.; Hörandl, E. The Rise of Apomixis in Natural Plant Populations. *Front. Plant Sci.* **2019**, *10*, 358. [CrossRef] [PubMed]
46. Talent, N.; Dickinson, T.A. The Potential for Ploidy Level Increases and Decreases in *Crataegus* (Rosaceae, Spiraeoideae, Tribe Pyreae). *Can. J. Bot.* **2007**, *85*, 570–584. [CrossRef]
47. Burgess, M.B.; Cushman, K.R.; Doucette, E.T.; Talent, N.; Frye, C.T.; Campbell, C.S. Effects of Apomixis and Polyploidy on Diversification and Geographic Distribution in *Amelanchier* (Rosaceae). *Am. J. Bot.* **2014**, *101*, 1375–1387. [CrossRef] [PubMed]
48. Macková, L.; Nosková, J.; Ďurišová, L.; Urfus, T. Insights into the Cytotype and Reproductive Puzzle of *Cotoneaster* integerrimus in the Western Carpathians. *Plant Syst. Evol.* **2020**, *306*, 58. [CrossRef]
49. Bogunić, F.; Siljak-Yakovlev, S.; Mahmutović-Dizdarević, I.; Hajrudinović-Bogunić, A.; Bourge, M.; Brown, S.C.; Muratović, E. Genome Size, Cytotype Diversity and Reproductive Mode Variation of *Cotoneaster integerrimus* (Rosaceae) from the Balkans. *Plants* **2021**, *10*, 2798. [CrossRef]
50. Kolarčik, V.; Kocová, V.; Mikoláš, V.; Mártonfiová, L.; Hajdučeková, N.; Mártonfi, P. Variability of Reproduction Pathways in the Central-European Populations of Hawthorns with Emphasis on Triploids. *Plants* **2022**, *11*, 3497. [CrossRef]
51. Aliyu, O.M.; Schranz, M.E.; Sharbel, T.F. Quantitative Variation for Apomictic Reproduction in the Genus *Boechera* (Brassicaceae). *Am. J. Bot.* **2010**, *97*, 1719–1731. [CrossRef]
52. Vašková, D.; Kolarčik, V. Breeding Systems in Diploid and Polyploid Hawthorns (*Crataegus*): Evidence from Experimental Pollinations of *C. monogyna*, *C. subsphaerica*, and Natural Hybrids. *Forests* **2019**, *10*, 1059. [CrossRef]
53. Comai, L. The Advantages and Disadvantages of Being Polyploid. *Nat. Rev. Genet.* **2005**, *6*, 836–846. [CrossRef] [PubMed]
54. Hamston, T.; de Vere, N.; King, A.; Pellicer, J.; Fay, M.; Cresswell, J.; Stevens, J. Apomixis and Hybridization Drives Reticulate Evolution and Phyletic Differentiation in *Sorbus* L.: Implications for Conservation. *Front. Plant Sci.* **2018**, *9*, 1796. [CrossRef] [PubMed]
55. Marvier, M.A.; Smith, D.L. Conservation Implications of Host Use for Rare Parasitic Plants. *Conserv. Biol.* **1997**, *11*, 839–848. [CrossRef]
56. Moir, M.; Coates, D.; Kennington, W.; Barrett, S.R.; Taylor, G. Concordance in Evolutionary History of Threatened Plant and Insect Populations Warrant Unified Conservation Management Approaches. *Biol. Conserv.* **2016**, *198*, 135–144. [CrossRef]
57. Kolarčik, V.; (Pavol Jozef Šafárik University, Košice, Slovakia); Mikoláš, V.; (independent researcher, Košice, Slovakia). Microevolutionary Processes in Mixed-Cytotype Population of Sexuals and Pseudogamous Apomicts: A Case Study of the Genus *Sorbus*. 2022, *unpublished work*.
58. Futák, J. Fytogeografické Členenie Slovenska. In *Flóra Slovenska IV/1*; Bertová, L., Ed.; Veda: Bratislava, Slovakia, 1984; pp. 418–420.
59. Thiers, B. *Index Herbariorum: A Global Directory of Public Herbaria and Associated Staff*; New York Botanical Garden's Virtual Herbarium: New York, NY, USA, 2022.
60. Kearns, C.A.; Inouye, D.W. *Techniques for Pollination Biologists*; University Press of Colorado: Niwot, CO, USA, 1993.
61. Ross, P.; Slovin, J.P.; Chen, C. A Simplifed Method for Differential Staining of Aborted and Non-Aborted Pollen Grains. *Int. J. Plant Biol.* **2010**, *1*, e13. [CrossRef]
62. Rohlf, F.J. The Tps Series of Software. *Hystrix Ital. J. Mammal.* **2015**, *26*, 9–12. [CrossRef]
63. Suda, J.; Kron, P.; Husband, B.C.; Trávníček, P. Flow Cytometry and Ploidy: Applications in Plant Systematics, Ecology and Evolutionary Biology. In *Flow Cytometry with Plant Cells*; Doležel, J., Greilhuber, J., Suda, J., Eds.; Wiley-VCH Verlag: Weinheim, Germany, 2007; pp. 103–130.
64. Doležel, J.; Göhde, W. Sex Determination in Dioecious Plants *Melandrium album* and *M. rubrum* Using High-Resolution Flow Cytometry. *Cytometry* **1995**, *19*, 103–106. [CrossRef] [PubMed]
65. Doležel, J. Flow Cytometric Analysis of Nuclear DNA Content in Higher Plants. *Phytochem. Anal.* **1991**, *2*, 143–154. [CrossRef]
66. Greilhuber, J.; Temsch, E.M.; Loureiro, J.C.M. Nuclear DNA Content Measurement. In *Flow Cytometry with Plant Cells*; Doležel, J., Greilhuber, J., Suda, J., Eds.; Wiley-VCH Verlag: Weinheim, Germany, 2007; pp. 67–101.
67. Doležel, J.; Sgorbati, S.; Lucretti, S. Comparison of Three DNA Fluorochromes for Flow Cytometric Estimation of Nuclear DNA Content in Plants. *Physiol. Plant.* **1992**, *85*, 625–631. [CrossRef]

68. Schönswetter, P.; Suda, J.; Popp, M.; Weiss-Schneeweiss, H.; Brochmann, C. Circumpolar Phylogeography of *Juncus biglumis* (Juncaceae) Inferred from AFLP Fingerprints, CpDNA Sequences, Nuclear DNA Content and Chromosome Numbers. *Mol. Phylogenet. Evol.* **2007**, *42*, 92–103. [CrossRef] [PubMed]
69. Loureiro, J.; Rodriguez, E.; Doležel, J.; Santos, C. Two New Nuclear Isolation Buffers for Plant DNA Flow Cytometry: A Test with 37 Species. *Ann. Bot.* **2007**, *100*, 875–888. [CrossRef]
70. Pellicer, J.; Clermont, S.; Houston, L.; Rich, T.C.G.; Fay, M.F. Cytotype Diversity in the *Sorbus* Complex (Rosaceae) in Britain: Sorting out the Puzzle. *Ann. Bot.* **2012**, *110*, 1185–1193. [CrossRef] [PubMed]
71. R Core Team. *R: A Language and Environment for Statistical Computing*; R Foundation for Statistical Computing: Vienna, Austria, 2019.
72. Wickham, H. *Ggplot2: Elegant Graphics for Data Analysis*; Springer: New York, NY, USA, 2016.

plants

MDPI

Article

Flower Size as an Honest Signal in Royal Irises (*Iris* Section *Oncocyclus*, Iridaceae)

Sissi Lozada-Gobilard [1,2,*], Nadine Nielsen [1] and Yuval Sapir [1]

[1] The Botanical Garden, School of Plant Sciences and Food Security, G.S. Wise Faculty of Life Science, Tel Aviv University, Tel Aviv 69978, Israel; nadinen@mail.tau.ac.il (N.N.); sapiry@tauex.tau.ac.il (Y.S.)
[2] Biodiversity Unit, Department of Biology, Lund University, 223 62 Lund, Sweden
* Correspondence: sissi.lozada_gobilard@biol.lu.se

Abstract: Flower traits, such as flower size or color changes, can act as honest signals indicating greater rewards such as nectar; however, nothing is known about shelter-rewarding systems. Large flowers of Royal irises offer overnight shelter as a reward to *Eucera* bees. A black patch might signal the entrance to the tunnel (shelter) and, together with the flower size, these might act as honest signals. We hypothesize that larger flowers and black patches indicate larger tunnels, and larger tunnels will increase pollinator visits, enhancing the plants' reproductive success. We measured seven species in a controlled environment and two species from three natural populations varying in flower size. Fruit and seed sets were assessed in these natural populations. We found a positive correlation between the flower, patch size, and tunnel volume, suggesting that the flowers and patch size act as honest signals, both under controlled conditions and in the wild. However, in natural populations, this positive relationship and its effect on fitness was population-specific. Flower size increased the fitness in YER *I. petrana*, and interactions between flower/patch size and tunnel size increased the fitness in YER and *I. atropurpurea* NET populations. This suggests that the honesty of the signal is positively selected in these two populations. This study supports the hypothesis that pollinator-mediated selection leads to the honest signaling of flower advertisement.

Keywords: honest signal; floral traits; flower size; color signal; shelter reward; fitness; Royal irises; *Oncocyclus*; endemic plant species; morphometrics

Citation: Lozada-Gobilard, S.; Nielsen, N.; Sapir, Y. Flower Size as an Honest Signal in Royal Irises (*Iris* Section *Oncocyclus*, Iridaceae). *Plants* **2023**, *12*, 2978. https://doi.org/10.3390/plants12162978

Academic Editors: Brenda Molano-Flores and James Cohen

Received: 29 June 2023
Revised: 6 August 2023
Accepted: 15 August 2023
Published: 18 August 2023

1. Introduction

Plant–pollinator interactions are a typical example of a mutualistic relationship. Plants use flower traits for advertisement and the attraction of pollinators [1]. Plants benefit from pollinators transporting pollen from flower to flower to assure their reproduction [2], and in return, pollinators receive a reward, commonly food (e.g., nectar or pollen). When a flower trait is positive correlated to the reward, it is known as honest signaling [3,4]. Honest signaling, which is a positive relationship between the signal and reward (reviewed in [5]), has been found within species [6,7], communities [8], and across ecosystems [9]. They are mostly found between floral advertisement (e.g., flower size, color, etc.) and nectar reward [10,11], but have also been found in pollen [12,13] and resin rewarding systems [14].

Plants changing color once they have been pollinated is the most obvious example of an honest signal [3,15]. Color changes in nectar guides were shown to be reliable signals to pollinators, enhancing the plants' reproductive success [16]. Larger nectar guides can increase pollen deposition [17] and be positively selected by pollinators [18]. However, the size and shape of nectar guides could cause disruptive selection when extreme phenotypes are selected by distinct pollinator groups (i.e., insects and birds) [19]. In some nectarless species such as *Oncocyclus* irises, dark-colored flowers mimic shelters [20,21], and a darker spot at the entrance of the shelter may serve as guide for pollinators [22], functioning similarly to nectar guides.

The flower size can be a visual signal of higher-energy rewards such as nectar or pollen, and therefore, are another example of an honest signal [10,23–25]. Larger flowers are easier to detect from a distance [26], and thus, the flower size might be positively selected by pollinators [27–30]. As larger flowers attract naïve pollinators, the positive feedback between the visual cue and the reward enhances pollinator learning [26,31,32].

Royal irises (*Iris* section *Oncocyclus*) possess exceptionally large, beautiful flowers up to 12 cm in diameter, making them among the largest flowers in the flora of the Middle East [33,34]. These irises are endemic to the Middle East and serve as models for the evolutionary processes of speciation, pollination, and ecology [34–38]. They typically produce one large flower per stem, and an individual plant can have from one to hundreds of stems growing in well-defined patches [21]. The complex flower morphology consists of three outer tepals that fall downwards ("falls") and three inner upright tepals ("standards"). At the base of each fall tepal, there is a dark mark, a cluster of specialized black cells that appear to be black regardless of the flower's color, which will be referred to herein as the "black patch". Behind this black patch on the fall tepal, a petaloid style forms a tunnel-shaped space together with the base of the fall. Each flower has three black patches and three tunnels (Figure 1).

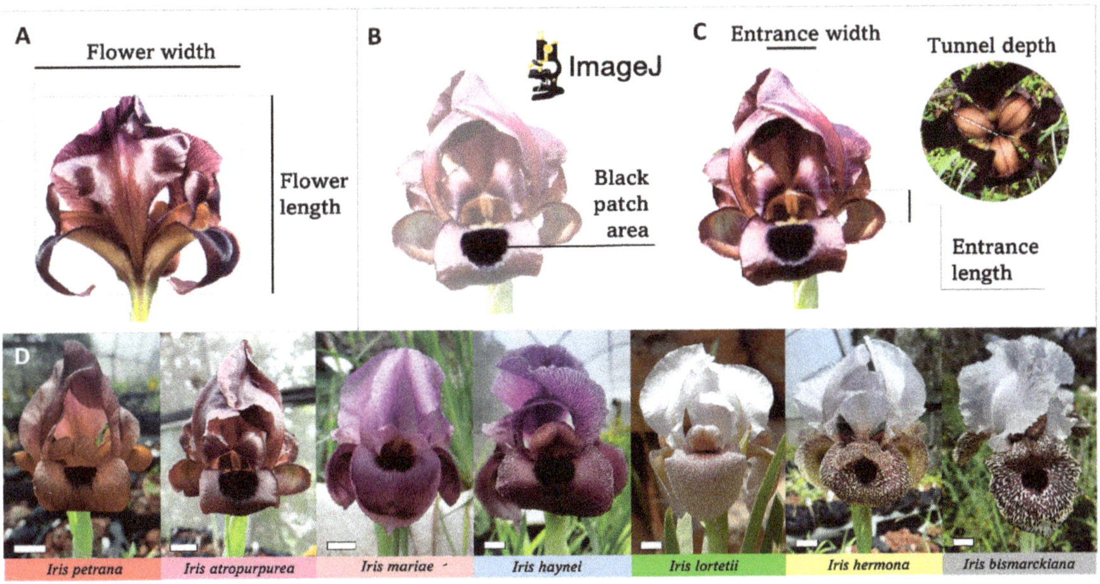

Figure 1. Flower size, black patch size, and tunnel volume measurements. Flower size was calculated by multiplying the width and length (**A**). Black patch size was calculated from digital photographs using ImageJ (**B**). Tunnel volume was calculated by multiplying the width and length of the entrance and the tunnel depth (**C**). Examples of flowers corresponding to each species measured under controlled conditions of the TAUBG (**D**). Scale bar in photographs = 1 cm. Measurements (**A**–**C**) are demonstrated on *Iris atropurpurea*.

Royal irises are self-incompatible and rely on pollinators for their reproduction [21]. A bee transporting pollen and sheltering in at least one tunnel might perform fertilization and fruit development [21]. The seeds are dispersed by ants and are carried up to 60 m away from the plant [39]. These flowers do not produce any nectar, but they offer shelter as a reward to *Eucera* male bees that use the tunnels of the flowers as shelters overnight [21]. Bees that shelter in flower tunnels emerge earlier in the morning than bees sheltering on the bare ground [40]. This observation is associated with the increased temperature within the tunnels compared with the ambient air temperature in the first 60–90 min after sunrise [40]. Thus, a heat reward has been suggested to be associated with the night-sheltering reward system.

Previous studies have shown a positive-mediated selection on flower size in *Iris atropurpurea* [38]. The high correlations between rainfall and flower size, but not with black patch size in *Iris petrana*, suggest that flower size might be selected both by water availability and pollinators, while the size of the black patch might be selected by pollinators only [22]. However, whether the size of the tunnel is related to the flower size and black patch size, and its covariation with fitness, remain unclear.

Here, we tested whether flower and black patch sizes act as honest signals for a shelter size (tunnel volume) reward in Royal irises. Larger flowers might increase the plant's attractiveness to pollinators from a distance, while at closer range, the black patch might signal the location of the reward (i.e., entrance to the shelter). If flower and black patch sizes are honest signals for the shelter reward, they are predicted to be in close correlation with the tunnel size, and this correlation would be under selection. This prediction assumes that more bees could shelter in larger tunnels, increasing the likelihood of a pollination event and the probability of a flower becoming a fruit and producing seeds. We measured the flowers of seven species of Royal irises from the collection of the Tel Aviv University Botanical Garden (TAUBG) to test the relationship between flower size, black patch size, and tunnel volume. In addition, we tested the effects of flower size, black patch size, and tunnel size, and their correlations on fitness, in three natural populations of *Iris atropurpurea* and *Iris petrana* that differ in terms of their flower sizes.

2. Results

Flower traits including flower size, black patch size, and tunnel volume (Figure 1A–C) were measured in seven species of Royal irises (Figure 1D) from a controlled environment at the Tel Aviv University Botanical Garden (TAUBG). Additionally, the flower traits were measured in three natural populations of two species: *Iris petrana* in Yeruham (YER), and *Iris atropurpurea* in Netanya (NET) and Yavne-Kur (KUR). Hereafter, these populations are referred to as YER, NET, and KUR, respectively.

2.1. Controlled Environment (TAUBG)

Flower size at TAUBG collection did not differ between 2021 and 2022 (LM, $F_{1,163} = 1.64$, $p = 0.20$, Supplementary Figure S1). When all species were analyzed together, we found positive correlations between all three flower traits. Flower size was positively correlated with black patch size (Pearson´s $r = 0.72$, $p < 0.001$, Figure 2A) and tunnel volume (Pearson´s $r = 0.66$, $p < 0.001$, Figure 2B). Black patch size was also positively correlated with tunnel volume (Pearson´s $r = 0.54$, $p < 0.001$, Figure 2C). When the species were analyzed separately, significant positive relationships for all combinations were found in *Iris atropurpurea* only (Figure 2D–F), except for *I. mariae*, which showed a positive significant relationship between black patch size and flower size (Supplementary Figure S2).

2.2. Natural Populations

In the wild, flower size differed among populations, with larger flowers found in NET, followed by KUR and YER (LM, $F_{2,116} = 87.51$, $p < 0.001$, Figure 3A). A similar tendency was found in black patch size (LM, $F_{2,116} = 89.70$, $p < 0.001$, Figure 3B). Tunnel volume was the largest in NET but similar in YER and KUR (LM, $F_{2,116} = 20.13$, $p < 0.001$, Figure 3C). Significant positive correlations for all flower traits were found in YER (Figure 3D–F), as well as a significant positive relationship between black patch size and tunnel volume in the *I. atropurpurea* NET population (Supplementary Figure S3).

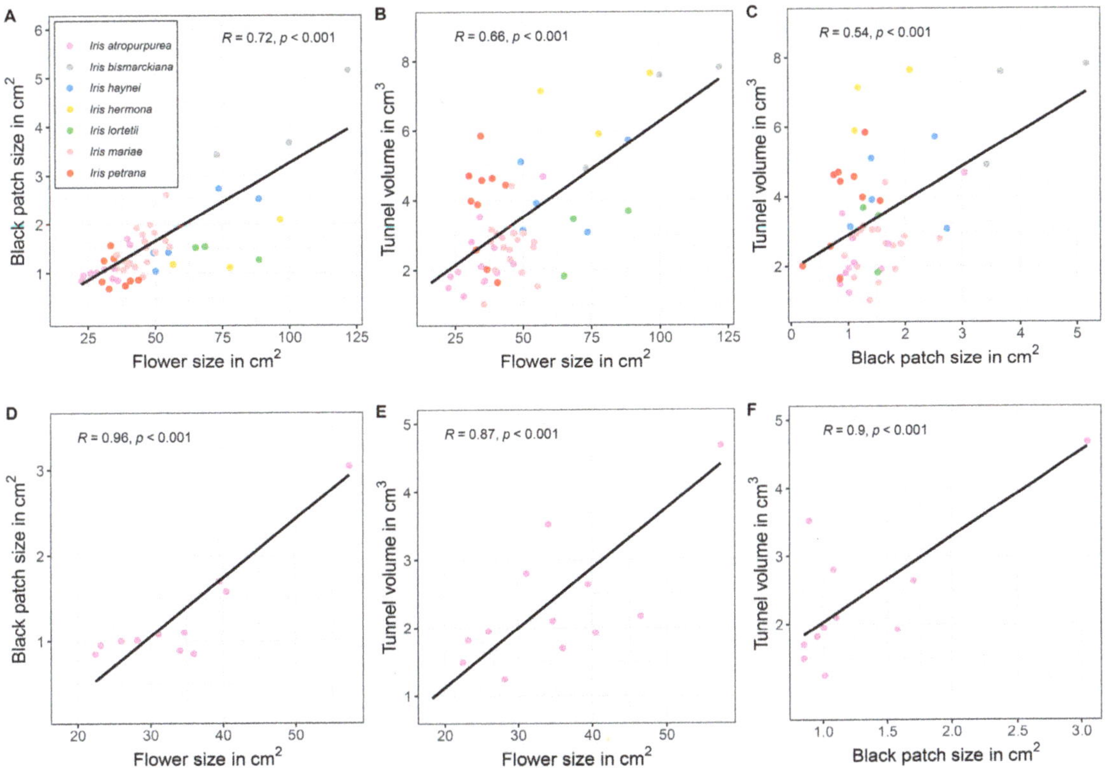

Figure 2. Flower size, black patch size, and tunnel volume in seven species of Royal irises from the TAUBG collection. When all species were analyzed together, all flower traits exhibited positive relationships (**A–C**). When species were analyzed separately, significant correlations of all flower traits were only found in *Iris atropurpurea* (**D–F**).

2.3. Fitness from Natural Populations

From a total of 93 flowers marked, 51 developed into fruits, but not all of them bore seeds (Table 1). In the YER population, 75% of fruits had seeds, followed by the KUR population with 62%, and the NET population with 40%. The mean number of seeds per fruit did not differ significantly between YER and KUR (~18 seeds/fruit); the NET population had an average of 5 seeds/fruit (Table 1, Supplementary Figure S4).

Table 1. Summary of fitness measurements and sample sizes in the natural populations. Populations KUR and NET correspond to *Iris atropurpurea*, while YER corresponds to *I. petrana*.

Pop	N Plants	Flowers Measured for Traits	Flowers Marked for Fitness	Flowers That Became Fruits (%)	Fruits with Seeds (%)	Seeds per Fruit (Mean ± SD)
KUR	15	48	35	16 (46%)	10 (62%)	17.6 ± 2.5
NET	13	30	27	15 (56%)	6 (40%)	4.6 ± 1.3
YER	23	41	31	20 (65%)	15 (75%)	18.2 ± 3.2

Fruit set, measured as the probability of a flower to become a fruit, significantly increased with flower size and tunnel volume in the *Iris petrana* YER population (Table 2, Figure 4A). Similarly, seed set, measured as the number of seeds per fruit, significantly increased with the flower size and tunnel volume in this population (Table 2, Figure 4B,D). The size of the black patch did not affect the seed set, but its interaction with tunnel volume did (Table 2,

Figure 4C,E). No relationships between fruit and seed set with flower size, black patch size, tunnel volume, or their interactions were found in the KUR population (Table 2). Fruit set did not show any effect on single flower traits in the NET population, but their interactions had a significant effect on seed set (Table 2, Supplementary Figure S5).

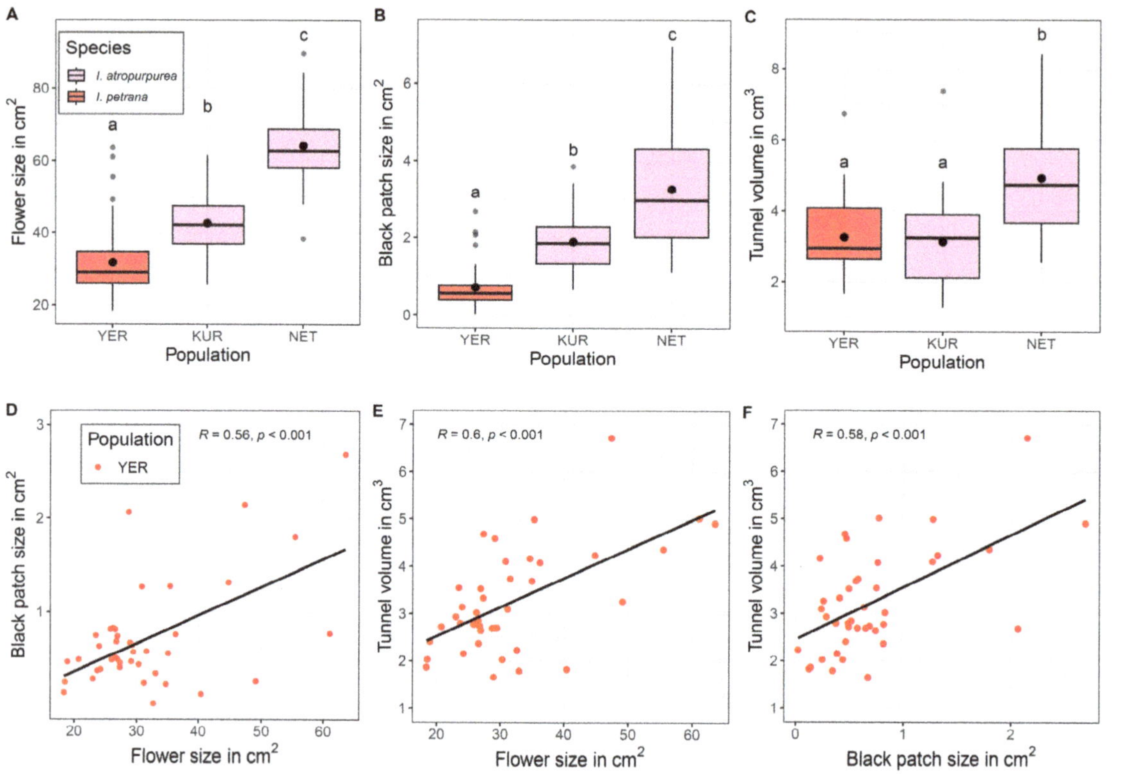

Figure 3. Flower size, black patch size, and tunnel volume of *Iris atropurpurea* (KUR and NET) and *Iris petrana* (YER). Flower size (**A**), black patch size (**B**), and tunnel volume (**C**) varied between species and populations. When populations were analyzed separately, there was only a positive relationship between tunnel, black patch size, and flower size in the *Iris petrana* YER population (**D–F**). Letters in (**A–C**) indicate significant differences between populations; black circles indicate mean values.

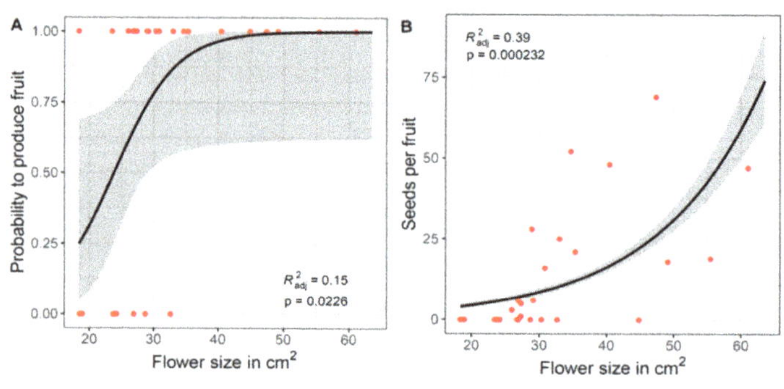

Figure 4. *Cont.*

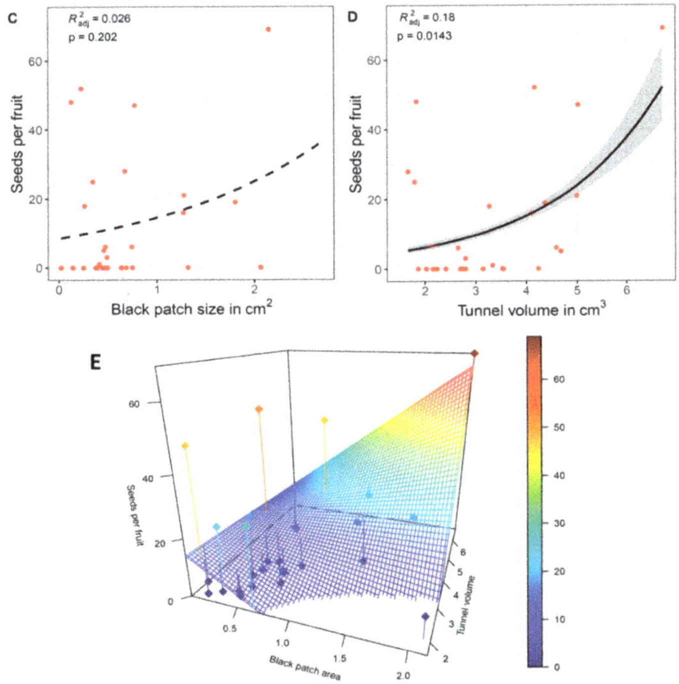

Figure 4. Effect of flower size, black patch size, and tunnel volume on fitness in the YER population. Flower size increased the probability of producing fruits (**A**) and the number of seeds per fruit (**B**). Black patch size alone did not affect the seed set (**C**), while tunnel volume (**D**) and black patch size × tunnel volume interactions (**E**) had a positive effect on the seed set. Dashed lines indicate no significance.

Table 2. Effects of flower size, black patch size, and tunnel volume on the fruit set and seed set in three populations of Royal irises. Significant effects are in bold. Asterisks represent significance levels: ** $p < 0.01$, * $p < 0.05$, (*) shows values that are marginally significant ~0.05.

Species	Pop	Predictors		Fruit Set		Seed Set	
			Df	**F Value**	***p* Value**	**F Value**	***p* Value**
I. atropurpurea	KUR	Flower size	1	0.16	0.686	0.33	0.568
		Black patch size	1	0.00	0.980	0.07	0.786
		Tunnel volume	1	0.10	0.752	0.45	0.511
		Flower size × Black patch size	1	0.66	0.428	0.62	0.441
		Flower size × Tunnel volume	1	0.86	0.368	0.55	0.465
		Black patch size × Tunnel volume	1	0.57	0.458	0.13	0.715
	NET	Flower size	1	1.15	0.301	1.13	0.030
		Black patch size	1	1.49	0.210	2.26	0.154
		Tunnel volume	1	2.24	0.156	0.07	0.793
		Flower size × Black patch size	1	1.08	0.314	**10.6**	**<0.01 ****
		Flower size × Tunnel volume	1	1.80	0.200	**7.87**	**<0.05 ***
		Black patch size × Tunnel volume	1	**3.71**	**0.07 (*)**	**5.40**	**<0.08 (*)**
I. petrana	YER	Flower size	1	**8.62**	**<0.01 ****	**8.48**	**<0.01 ****
		Black patch size	1	0.30	0.586	0.23	0.636
		Tunnel volume	1	**3.41**	**0.07 (*)**	**5.01**	**<0.05 ***
		Flower size × Black patch size	1	1.78	0.193	0.40	0.532
		Flower size × Tunnel volume	1	0.64	0.429	0.54	0.470
		Black patch size × Tunnel volume	1	1.17	0.289	**5.12**	**<0.05 ***

3. Discussion

Honest signaling in plant–pollinator systems is key in the coevolution of flowers and pollinating animals. Flower size and coloration signals have scarcely been studied in this context. In this study, we tested whether two visual signals, flower size and black patch size, act as honest signals in the Royal irises. The flowers of Royal irises offer a tunnel-shaped shelter as a reward, where bees shelter during the night. Here, we tested the relationship between flower size and black patch size with tunnel volume (as an indicator for the extent of the shelter reward) and the possible selection acting on these advertisements and the reward.

We found evidence for honest signaling in a controlled environment, expressed as a positive relationship between either flower size or black patch size and tunnel volume (Figure 2). However, in the natural populations, this positive relationship was found only in *Iris petrana* (Figure 3). In addition, the interaction between traits, i.e., the correlation between flower/patch size and tunnel size, increased fitness in *I. petrana* YER and *I. atropurpurea* NET populations (Figure 4, Table 2). This suggests a positive selection on the honesty of the signal, and a correlation between advertisement and reward in these populations.

3.1. Honest Signals of Flower and Black Patch Are Species- and Population-Specific

Although we found a positive relationship between the visual signals (i.e., flower size and black patch size) and the reward (i.e., tunnel volume), this honest signaling was not present in all species and/or populations, evident only in *Iris atropurpurea* in the controlled environment, and in *Iris petrana* YER population in the wild, as well as in the *Iris atropurpurea* NET population. These results suggest that flower size and black patch size are honest signals for the size of the reward, but this honest signaling varies across species and populations.

Population-specific selection on flower traits mediated by pollinators was previously documented in other species (e.g., [41–45]). In Royal irises, pollinator-mediated selection on flower size was shown in *Iris atropurpurea*, but not in *Iris haynei* [38]; our results suggest that selection mediated by pollinators varies between the two populations of *Iris atropurpurea*. *Iris petrana* did not show any significant correlation under controlled conditions, but the natural population of YER did, as well as a clear positive effect of flower size and tunnel volume on fitness. These results highlight the differences among populations, but whether these relationships are also present in other populations of *I. petrana* remains to be tested. More important than the effect of single flower traits (i.e., flower size and black patch size) on fitness was the effect of their interaction with tunnel volume. All interactions were significant in the NET population, although in the YER population, only black patch size $\times$ tunnel volume had an effect on fitness (Table 2). These results suggest selection of the honest signal, which varies between populations as well.

3.2. Are Larger Shelters Better?

In Royal irises, shelter is the reward [20,21]. Larger tunnels might offer more space for bees to shelter. *Eucera* bees tend to sleep in aggregations [46–48], which help them maintain higher body temperature [49], or as a dilution effect against predators [48]. Previous observations in *I. atropurpurea* showed an average of 2 male bees sheltering in a single tunnel (mode = 1), and an exponential decrease in the frequency of number of bees, up to a single case of 22 bees in one tunnel (Y. Sapir unpublished). Therefore, aggregations (at least 2 bees) seem to be common, likely increasing pollen deposition and pollen import.

Pollination success highly depends on the pollinator visitation rate and pollinator preference [50–52]. Interactions between flower/patch size and tunnel size, and increased fitness in *I. petrana* YER and *I. atropurpurea* NET populations, suggest that larger flowers and patches might be preferred by pollinators in these populations [22,38]. However, pollen limitation was previously identified in the NET population [38], which might explain the high fraction (60%) of fruits with zero seeds observed in this study. In the *I. atropurpurea* KUR population, a high proportion of fruits with seeds (75%) and a high number of seeds per fruit (~18, similar to YER; Table 1, Supplementary Figure S4), indicate that pollination events do occur there. However, lack of evidence for an honest signal or its effect on fitness in this population (Table 2) suggests that pollinators might not be driven by flower or black patch size. More studies including direct observations on pollinators' behavior are needed.

Larger tunnels may provide better shelters as a reward. Previously, temperature increase within the tunnels after sunrise was argued to be the reward to night-sheltering pollinators [40]. However, the relationship between tunnel size and this heat reward remains unclear. We hypothesize that larger tunnels provide a better microclimate that heats up faster than smaller tunnels in the morning. Nonetheless, this hypothesis is still not resolved.

3.3. Abiotic Factors Can Affect the Selection of Flower Size

Resource availability is an important limiting factor on flower traits, including flower size [53]. In Royal irises, flower size relates to the north–south aridity gradient of Israel, decreasing towards the desert, as an adaptation to drought [34]. In arid habitats where water and nutrients are scarce, large flowers can be very costly to produce while small ones are favored [54,55], causing changes in plant–pollinator interactions [56,57].

The decrease in flower size of the seven Royal irises measured under controlled conditions matches their natural distribution from north to south (Figure 1, Supplementary Figure S6A), following the aridity gradient (for the natural occurrence of the species, see Figure 1 in Ref. [58]). Black patch size and tunnel volume did not exhibit such a decrease, and seem to be less variable among species (Supplementary Figure S6B,C), suggesting that the climate gradient does not affect these traits as much as it affects flower size.

The natural occurrence of *Iris petrana* is located in the Negev desert, an arid environment with low water availability. In the YER population, flower size was found to highly depend on rainfall over the years, while black patch size remained constant [22], suggesting that flower size is a very plastic trait highly dependent on water availability and costly for the plant to produce. In addition, we found a direct positive effect of flower size on fitness in this population, which suggests that pollinators select larger flowers with larger tunnels. This might cause a selection conflict where the climate might select for smaller flowers, while pollinators select for larger flowers.

In the *Iris atropurpurea* NET population, no direct effect of flower size on fitness was found, but interactions of flower size and black patch size with tunnel volume increased the seed set (Table 2). These results suggest that both traits together might be important and likely selected by pollinators. Indeed, pollinator-mediated selection on flower size was previously found in this population [38]. Whether pollinators still prefer larger flowers/patches occurring in populations under more favorable conditions (i.e., populations in the north) remains unclear. More studies to test the effect of water availability on reproductive success and the effect of pollinators are needed [59].

3.4. Is There an Indirect Selection of Black Patch Size?

We found a positive relationship between flower and black patch size, and between black patch size and tunnel volume, both in controlled conditions and in wild populations (Figures 2 and 3). However, in the natural populations, black patch size did not affect fruit and seed sets (Table 1). In the *Iris petrana* YER population, larger flowers with larger tunnels significantly increased fitness, while in *Iris atropurpurea*, the interactions of flower size × black patch size and black patch size × tunnel volume increased fitness, although this effect was population-specific (in the NET population only). There was no direct effect of black patch size on the fitness component of either species or populations, but the significant effect of tunnel volume × black patch size interaction on fitness suggests a complex synergistic effect of the signal and the reward, which may hint for an indirect selection on this signal.

While the size of the black patch might not be under selection per se, the selection may act indirectly through the black patch in gaining heat and transferring the energy to the tunnel, playing a role on the heat reward (Y. Sapir and R. Heliczer, unpublished). An ongoing study is comparing temperature increase within the tunnels in these natural populations and the role of the black patch as an underlying mechanism of flower heating (Lozada-Gobilard et al., in preparation).

4. Materials and Methods

4.1. Flower Size, Black Patch Size, and Tunnel Volume Measurements

To determine the size of the flower, we measured the length and the width using calipers and calculated a flower size by multiplying them together (Figure 1A). Black patch area was estimated from digital photographs using ImageJ [60]. Each black patch was carefully, manually encircled, and its area was calculated using the standardized measuring protocol implemented in ImageJ (Figure 1B). Tunnel volume was calculated by multiplying the length and the width of the tunnel entrance with the depth of the tunnel (Figure 1C). For black patch size and tunnel volume, one tunnel and fall tepal out of the three were selected randomly. Flower and black patch size are presented in cm^2, while the tunnel volume is in cm^3.

4.2. Sampling of Plant Material

The flower measurements were collected from the Royal irises collection, and maintained in a nethouse at the Tel Aviv University Botanical Garden (TAUBG; 32°06′ N, 34°48′ E), a controlled isolated environment, permissible to light but impermissible to pollinators (Supplementary Figure S7). The collection was established in 2008–2009 by transplanting rhizomes from natural populations throughout Israel. Rhizomes were transplanted to individual bags with garden soil and tuff (50:50) and watered regularly (~2 L per week) with an automatic irrigation system. Rhizomes of the plants are replanted in new soil approximately every five years.

Seven species of Royal irises were measured, namely, *Iris bismarckiana, I. hermona, I. lortetii, I haynei, I. mariae, I. atropurpurea*, and *I. petrana*, representing species endemic or sub-endemic to Israel (Figure 1D). We measured 59 flowers in total: *I. petrana* (n = 10), *I. atropurpurea* (n = 16), *I. mariae* (n = 19), *I. haynei* (n = 5), *I. lortetii* (n = 3), *I. hermona* (n = 3), and *I. bismarckiana* (n = 3). In their natural environment, these species are distributed along the north–south aridity gradient of Israel [34] and are eco-geographically isolated with different levels of pre- and post-zygotic reproductive barriers [58,61]. For an overview of the natural distributions of these species in Israel and Palestine, see Figure 1 in Ref. [58].

Additionally, flower traits were measured in three natural populations of two species: *Iris petrana* in Yeruham (YER), and *Iris atropurpurea* in Netanya (NET) and Yavne-Kur (KUR). Sample sizes: YER (n = 41), KUR (n = 48), and NET (n = 30). Both KUR and NET are located in the Mediterranean coastal region of Israel. The NET population is at the Netanya Iris reserve, located 26 km north of Tel Aviv (32°170 N, 34°500 E, altitude 37 m), and KUR is 15 km south of Tel Aviv (31°53′25.4 N, 34°42′35.60 E, altitude 15 m). The YER population is in the Yeruham Iris Nature Reserve, located in the arid region of Israel (31°01′14.46 N, 34°58′21.4 E, altitude 549 m). These three populations occur along a latitudinal gradient from north to south, receiving 600, 500, and 200 mm of mean annual precipitation, respectively. The average temperature from February to April in NET is 18–23 °C, 19–25 °C in KUR, and 18–26 °C in YER. The average temperature from February to April in NET varies from 18 to 23 °C, from 19 to 25 °C in KUR, and from 18 to 26 °C in YER.

Data from both TAUBG and natural populations were collected during the Royal irises flowering season, between February and April 2022. *Iris atropurpurea* exhibits wide within-species variation in flower size; thus, to account for most of the variation, we selected two extreme populations with large (NET) and small (KUR) flowers, while *Iris petrana* corresponds to the smallest range among all species (see the average flower sizes per population in Supplementary Table S1). Collection of data in the wild for these two species was possible since their flowering times do not overlap (*I. atropurpurea* = January until mid-March; *I. petrana* = March to April). Populations of larger flowers such as *I bismarckiana* or *I hermona* flower simultaneously with *I. petrana*, making their data collection logistically difficult.

It was previously shown that flower size highly depends on water availability (i.e., rainfall) [22]. To ensure that flower size did not change under the controlled conditions of the TAUBG collection, we measured flower sizes in two consecutive years (2021 and 2022) and tested whether there were differences between the two years in flower size. Tunnel and black patch size data were collected in 2022 only. Since the environmental conditions in the natural populations vary, we tested the relationships between flower, black patch size, and tunnel volume by population.

4.3. Fitness Estimates in Natural Populations

In the three natural populations of NET, KUR, and YER, we randomly marked flowers that were opened for at least 5 days and bagged them to preserve the fruits and seeds for later collection. In total, 119 flowers were measured for flower traits (YER = 41, KUR = 48, and NET = 30), and 93 were marked for fruit development. In YER, 31/41 flowers were marked, 28 were later recovered, and of these, 20 (71%) flowers developed into fruits. In KUR, 35/48 flowers were marked, 23 recovered, and 16 (69%) developed into fruits. In NET, 27/30 were marked, 19 recovered, and 15 (79%) developed into fruits (Table 1). Since these species are self-incompatible, at least one efficient visit to the tunnel of an *Eucera* bee is needed for fertilization. Moreover, because all three stigma lobes are merged into one style, pollen deposition on a single stigma is sufficient to fertilize seeds in all three carpels of the ovary (Y. Sapir, unpublished). We recorded whether the marked flowers developed into a fruit (Yes = 1, No = 0). In YER, a total of 31 flowers were marked, corresponding to 15 individuals. In NET, 57 flowers were bagged, corresponding to 24 individuals, while in KUR, 35 flowers corresponding to 13 individuals were marked. Due to a high variation in flower size within genotype [22] and some cases with a lack of clear separation between individuals (S. Lozada-Gobilard pers. obs.), each flower was analyzed independently; hence, the analysis considers the ecological interaction of the single flower, rather than the fitness of the genotype (individual plant). From the collected fruits, we recorded whether they developed seeds and counted the number of seeds per fruit. Flowers that did not set fruits, and fruits that did not contain seeds, were recorded as zero seeds.

4.4. Statistical Analyses

The TAUBG and natural populations datasets were analyzed separately. Flower size and tunnel volume from the TAUBG dataset followed a normal distribution, while the black patch size was log-transformed to improve normality. All three variables were log-transformed from the dataset of the natural population to improve normality. To evaluate whether flower size changed in two years at the TAUBG, we applied a linear model using the *"lm"* and *"Anova"* functions. To develop a general overview about the relationships between flower size, black patch size, and tunnel volume in Royal irises, we applied Pearson's correlations, including all species in the TAUBG dataset (N = 59), and separately by species for those with N > 5 (i.e., *I. atropurpurea*, *I. petrana*, and *I. mariae*). Only the flower size measurements from TAUBG corresponding to 2022 were used.

To compare populations from the natural populations dataset, we applied a linear model using *"lm"* from the "stats" package and the *"Anova"* function from the "car" package, and performed pairwise comparison using Tukey post hoc tests with the *"TukeyHSD"* function from the "stats" package. Correlations within populations (KUR, NET, and YER) were analyzed separately using Pearson's tests. To test whether flower size, black patch size, or tunnel volume influenced fruit set, we applied a generalized linear model (GLM) with a binomial family distribution. Finally, to test the effect on seed set between populations, we applied log-linear regression using GLM with a Poisson family distribution suitable for counting data. All statistical tests were performed using R software version 4.2.2 [62].

5. Conclusions

In this study, we tested the hypothesis that flower size and black patch size are honest signals for the shelter reward (tunnel volume) in Royal irises. Our results showed that flower size and black patch size could act as an honest signal, where large flowers/patches indicate larger tunnels (where pollinators shelter), increasing the probability of fruits and seeds. Under controlled conditions, evidence of honest signaling was found in an entire group of Royal irises, but only in *Iris atropurpurea*. However, in the wild, this positive relationship was only found in YER *Iris petrana* and NET *Iris atropurpurea* populations. These results suggest that the positive relationship between flower/patch size and tunnel volume might be a common trait in Royal irises, but its effect on fitness might be species- or population-specific. In addition, flower size showed a direct positive effect on fitness in the YER *I. petrana* population; correlation between flower/patch size and tunnel size (i.e., interactions between traits) increased fitness in YER *I. petrana* and NET *I. atropurpurea* populations, suggesting a positive selection on the honesty of the signal. These results suggest that flower size might act as an honest signal for larger tunnels putatively attracting pollinators from a distance, whereas the size of the black patch at closer range might not be as important as the size of the flower. More studies are needed to evaluate pollinator preferences and selection on flower size, black patch size, and tunnel size in more species of Royal irises with an extended size range. Understanding how flower traits attract pollinators is essential for plant fitness, in particular for those that fully depend on pollinators to reproduce. Changes in flower traits due to various factors, including pollinators, directly affect plant reproduction success, which is crucial for conservation purposes, especially of endemic plant species such as the Royal irises.

Supplementary Materials: The following supporting information can be downloaded at: https://www.mdpi.com/article/10.3390/plants12162978/s1, Figure S1. Comparison between 2021 and 2022 of flower size measurements in the TAUBG collection. Figure S2. Relationship between flower size and black patch size per species n > 5 from the TAUBG. Figure S3. Relationship between flower size, black patch size, and tunnel volume per population. Figure S4. Proportion of fruits that developed seeds and the mean number of seeds per fruit per population. Figure S5. Single and interactive effects of flower size, black patch size, and tunnel volume on seed set in the NET population. Figure S6. Flower traits measured in seven species of Royal irises under controlled conditions of the TAUBG. Figure S7. Royal irises collection at the Tel Aviv University Botanical Garden. Table S1. Sampling overview of individuals from the TAUBG and natural populations.

Author Contributions: Conceptualization, S.L.-G. and Y.S.; methodology, S.L.-G. and Y.S.; data acquisition, S.L.-G. and N.N.; formal analysis, S.L.-G.; resources, Y.S.; data curation, S.L.-G. and N.N.; writing—original draft preparation, S.L.-G.; writing—review and editing, Y.S.; visualization, S.L.-G.; supervision, Y.S.; project administration, Y.S.; funding acquisition, S.L.-G. and Y.S. All authors have read and agreed to the published version of the manuscript.

Funding: S.L.-G. was supported by a postdoctoral fellowship from the University of Potsdam—Tel Aviv University collaboration program. The study was partially supported by a grant from the Israel Science Foundation (336/16) to Y.S.

Institutional Review Board Statement: Not applicable.

Informed Consent Statement: Not applicable.

Data Availability Statement: Data and R script code are available on Figshare at https://doi.org/10.6084/m9.figshare.23596698 (accessed on 16 October 2023).

Acknowledgments: We thank Valentina Elettra Alberti, and Daniele Garzoni for their help in the field, and Yamit Bar-Lev and Rony Gafny for their essential help with the collection of data in the field and in the lab. We thank two anonymous reviewers and editors whose comments and suggestions improved the original version of the manuscript. S.L.-G. acknowledges Mahua Ghara for helpful discussions.

Conflicts of Interest: The authors declare no conflict of interest.

References

1. Chittka, L.; Raine, N.E. Recognition of Flowers by Pollinators. *Curr. Opin. Plant Biol.* **2006**, *9*, 428–435. [CrossRef] [PubMed]
2. Kearns, C.A.; Inouye, D.W.; Waser, N.M. Endangered Mutualisms: The Conservation of Plant-Pollinator Interactions. *Annu. Rev. Ecol. Syst.* **1998**, *29*, 83–112. [CrossRef]
3. Schaefer, H.M.; Schaefer, V.; Levey, D.J. How Plant-Animal Interactions Signal New Insights in Communication. *Trends Ecol. Evol.* **2004**, *19*, 577–584. [CrossRef]
4. Knauer, A.C.; Schiestl, F.P. Bees Use Honest Floral Signals as Indicators of Reward When Visiting Flowers. *Ecol. Lett.* **2015**, *18*, 135–143. [CrossRef]
5. van der Kooi, C.J.; Reuvers, L.; Spaethe, J. Honesty, Reliability, and Information Content of Floral Signals. *iScience* **2023**, *26*, 107093. [CrossRef] [PubMed]
6. Armbruster, W.S.; Antonsen, L.; Pélabon, C. Phenotypic Selection on Dalechampia Blossoms: Honest Signaling Affects Pollination Success. *Ecology* **2005**, *86*, 3323–3333. [CrossRef]
7. Fenster, C.B.; Cheely, G.; Dudash, M.R.; Reynolds, R.J. Nectar Reward and Advertisement in Hummingbird-Pollinated *Silene virginica* (Caryophyllaceae). *Am. J. Bot.* **2006**, *93*, 1800–1807. [CrossRef]
8. Ortiz, P.L.; Fernández-Díaz, P.; Pareja, D.; Escudero, M.; Arista, M. Do Visual Traits Honestly Signal Floral Rewards at Community Level? *Funct. Ecol.* **2021**, *35*, 369–383. [CrossRef]
9. Ornelas, J.F.; Ordano, M.; De-Nova, A.J.; Quintero, M.E.; Garland, T. Phylogenetic Analysis of Interspecific Variation in Nectar of Hummingbird-Visited Plants. *J. Evol. Biol.* **2007**, *20*, 1904–1917. [CrossRef]
10. Tavares, D.C.; Freitas, L.; Gaglianone, M.C. Nectar Volume Is Positively Correlated with Flower Size in Hummingbird-Visited Flowers in the Brazilian Atlantic Forest. *J. Trop. Ecol.* **2016**, *32*, 335–339. [CrossRef]
11. Parachnowitsch, A.L.; Manson, J.S.; Sletvold, N. Evolutionary Ecology of Nectar. *Ann. Bot.* **2019**, *123*, 247–261. [CrossRef]
12. Choteau, M.; Barabé, D.; Gibernau, M. A Comparative Study of Inflorescence Characters and Pollen—Ovule Ratios among the Genera *Philodendron* and *Anthurium* (Araceae). *Int. J. Plant Sci.* **2006**, *167*, 817–829. [CrossRef]
13. Stanton, M.L.; Preston, R.E. Ecological Consequences and Phenotypic Correlates of Petal Size Variation in Wild Radish, *Raphanus sativus* (Brassicaceae). *Am. J. Bot.* **1988**, *75*, 528–539. [CrossRef]
14. Pélabon, C.; Thöne, P.; Hansen, T.F.; Armbruster, W.S. Signal Honesty and Cost of Pollinator Rewards in *Dalechampia scandens* (Euphorbiaceae). *Ann. Bot.* **2012**, *109*, 1331–1339. [CrossRef] [PubMed]
15. Weiss, M.R. Floral Colour Changes as Cues for Pollinators. *Nature* **1991**, *354*, 227–229. [CrossRef]
16. Zhang, C.; Vereecken, N.J.; Wang, L.; Tian, B.; Dafni, A.; Yang, Y.; Duan, Y. Are Nectar Guide Colour Changes a Reliable Signal to Pollinators That Enhances Reproductive Success? *Plant Ecol. Divers.* **2017**, *10*, 89–96. [CrossRef]
17. Peach, K.; Liu, J.W.; Klitgaard, K.N.; Mazer, S.J. Sex-Specific Floral Attraction Traits in a Sequentially Hermaphroditic Species. *Ecol. Evol.* **2020**, *10*, 1856–1875. [CrossRef]
18. Wang, L.L.; Zhang, C.; Tian, B.; Sun, X.D.; Guo, W.; Zhang, T.F.; Yang, Y.P.; Duan, Y.W. Reproductive Isolation Is Mediated by Pollen Incompatibility in Sympatric Populations of Two Arnebia Species. *Ecol. Evol.* **2015**, *5*, 5838–5846. [CrossRef]
19. Medel, R.; Botto-Mahan, C.; Kalin-Arroyo, M. Pollinator-Mediated Selection on the Nectar Guide Phenotype in the Andean Monkey Flower, Mimulus Luteus. *Ecology* **2003**, *84*, 1721–1732. [CrossRef]
20. Vereecken, N.J.; Dorchin, A.; Dafni, A.; Hötling, S.; Schulz, S.; Watts, S. A Pollinators' Eye View of a Shelter Mimicry System. *Ann. Bot.* **2013**, *111*, 1155–1165. [CrossRef]
21. Sapir, Y.; Shmida, A.; Ne'eman, G. Pollination of *Oncocyclus* irises (*Iris*: Iridaceae) by Night-Sheltering Male Bees. *Plant Biol.* **2005**, *7*, 417–424. [CrossRef]
22. Lozada-Gobilard, S.; Motter, A.; Sapir, Y. Among-years Rain Variation Is Associated with Flower Size, but Not with Signal Patch Size in Iris Petrana. *Ecology* **2023**, *104*, e3839. [CrossRef] [PubMed]
23. Galen, C. Measuring Pollinator-Mediated Selection on Morphometric Floral Traits: Bumblebees and the Alpine Sky Pilot, *Polemonium viscosum*. *Evolution* **1989**, *43*, 882–890. [CrossRef] [PubMed]
24. Spaethe, J.; Tautz, J.; Chittka, L. Visual Constraints in Foraging Bumblebees: Flower Size and Color Affect Search Time and Flight Behavior. *Proc. Natl. Acad. Sci. USA* **2001**, *98*, 3898–3903. [CrossRef] [PubMed]
25. Arista, M.; Ortiz, P.L. Differential Gender Selection on Floral Size: An Experimental Approach Using Cistus Salvifolius. *J. Ecol.* **2007**, *95*, 973–982. [CrossRef]
26. Hempel De Ibarra, N.; Langridge, K.V.; Vorobyev, M. More than Colour Attraction: Behavioural Functions of Flower Patterns. *Curr. Opin. Insect Sci.* **2015**, *12*, 64–70. [CrossRef]
27. Galen, C.; Newport, M.E.A. Bumble Bee Behavior and Selection on Flower Size in the Sky Pilot, *Polemonium viscosum*. *Oecologia* **1987**, *74*, 20–23. [CrossRef]
28. Parachnowitsch, A.L.; Kessler, A. Pollinators Exert Natural Selection on Flower Size and Floral Display in Penstemon Digitalis. *New Phytol.* **2010**, *188*, 393–402. [CrossRef]
29. Lebel, M.; Obolski, U.; Hadany, L.; Sapir, Y. Pollinator-Mediated Selection on Floral Size and Tube Color in Linum Pubescens: Can Differential Behavior and Preference in Different Times of the Day Maintain Dimorphism? *Ecol. Evol.* **2018**, *8*, 1096–1106. [CrossRef]
30. Ne'eman, G.; Ne'eman, R. Factors Determining Visual Detection Distance to Real. *J. Pollinat. Ecol.* **2017**, *20*, 1–12. [CrossRef]

31. Blarer, A.; Keasar, T.; Shmida, A. Possible Mechanisms for the Formation of Flower Size Preferences by Foraging Bumblebees. *Isr. J. Zool.* **2002**, *46*, 159. [CrossRef]
32. Biernaskie, J.M.; Walker, S.C.; Gegear, R.J. Bumblebees Learn to Forage like Bayesians. *Am. Nat.* **2009**, *174*, 413–423. [CrossRef] [PubMed]
33. Monty, A.; Saad, L.; Mahy, G. Bimodal Pollination System in Rare Endemic *Oncocyclus* Irises (Iridaceae) of Lebanon. *Can. J. Bot.* **2006**, *84*, 1327–1338. [CrossRef]
34. Sapir, Y.; Shmida, A.; Fragman, O.; Comes, H.P. Morphological Variation of the *Oncocyclus* Irises (*Iris*: Iridaceae) in the Southern Levant. *Bot. J. Linn. Soc.* **2002**, *139*, 369–382. [CrossRef]
35. Sapir, Y.; Shmida, A. Species Concepts and Ecogeographical Divergence of *Oncocyclus* Irises. *Isr. J. Plant Sci.* **2002**, *50*, 119–127. [CrossRef]
36. Volis, S.; Zhang, Y.H.; Deng, T.; Dorman, M.; Blecher, M.; Abbott, R.J. Divergence and Reproductive Isolation between Two Closely Related Allopatric Iris Species. *Biol. J. Linn. Soc.* **2019**, *127*, 377–389. [CrossRef]
37. Saad, L.; Mahy, G. Molecular and Morphological Variation of Rare Endemic *Oncocyclus* Irises (Iridaceae) of Lebanon. *Bot. J. Linn. Soc.* **2009**, *159*, 123–135. [CrossRef]
38. Lavi, R.; Sapir, Y. Are Pollinators the Agents of Selection for the Extreme Large Size and Dark Color in *Oncocyclus* Irises? *New Phytol.* **2015**, *205*, 369–377. [CrossRef]
39. Sapir, Y.; Mazzucco, R. Post-Zygotic Reproductive Isolation among Populations of Iris Atropurpurea: The Effect of Spatial Distance among Crosses and the Role of Inbreeding and Outbreeding Depression in Determining Niche Width. *Evol. Ecol. Res.* **2012**, *14*, 425–445.
40. Sapir, Y.; Shmida, A.; Ne'eman, G. Morning Floral Heat as a Reward to the Pollinators of the *Oncocyclus* Irises. *Oecologia* **2006**, *147*, 53–59. [CrossRef]
41. Boberg, E.; Ågren, J. Despite Their Apparent Integration, Spur Length but Not Perianth Size Affects Reproductive Success in the Moth-Pollinated Orchid Platanthera Bifolia. *Funct. Ecol.* **2009**, *23*, 1022–1028. [CrossRef]
42. Chapurlat, E.; Ågren, J.; Sletvold, N. Spatial Variation in Pollinator-Mediated Selection on Phenology, Floral Display and Spur Length in the Orchid Gymnadenia Conopsea. *New Phytol.* **2015**, *208*, 1264–1275. [CrossRef] [PubMed]
43. Sletvold, N.; Ågren, J. Pollinator-Mediated Selection on Floral Display and Spur Length in the Orchid Gymnadenia Conopsea. *Int. J. Plant Sci.* **2010**, *171*, 999–1009. [CrossRef]
44. Boberg, E.; Alexandersson, R.; Jonsson, M.; Maad, J.; Ågren, J.; Nilsson, L.A. Pollinator Shifts and the Evolution of Spur Length in the Moth-Pollinated Orchid Platanthera Bifolia. *Ann. Bot.* **2014**, *113*, 267–275. [CrossRef] [PubMed]
45. Robertson, J.L.; Wyatt, R. Evidence for Pollination Ecotypes in the Yellow-Fringed Orchid, Platanthera Ciliaris. *Evolution* **1990**, *44*, 121–133. [CrossRef] [PubMed]
46. Mahlmann, T.; Hipólito, J.; de Oliveira, F.F. Male Sleeping Aggregation of Multiple Eucerini Bee Genera (Hymenoptera: Apidae) in Chapada Diamantina, Bahia, Brazil. *Biodivers. Data J.* **2014**, *2*, e1556. [CrossRef]
47. Shimron, O.; Hefetz, A. Mating Behavior and Sex Attraction of Eucera Palestinae Friese (Hymenoptera: Anthophoridae). *J. Kansas Entomol. Soc.* **1985**, *58*, 526–531.
48. Alcock, J. Sleeping Aggregations of the Bee Idiomelissodes Duplocincta (Cockerell) (Hymenoptera: Anthophorini) and Their Possible Function. *J. Kansas Entomol. Soc.* **1998**, *71*, 74–84.
49. Linsley, E.G.; Cazier, M. Diurnal and Seasonal Behavior Patterns among Adults of Protoxaea Gloriosa (Hymenoptera, Oxaeidae). *Am. Museum Novit.* **1972**, *25*, 1–25.
50. Burd, M. Bateman's Principle and Plant Reproduction: The Role of Pollen Limitation in Fruit and Seed Set. *Bot. Rev.* **1994**, *60*, 373–425. [CrossRef]
51. Wesselingh, R.A. Pollen Limitation Meets Resource Allocation: Towards a Comprehensive Methodology: Research Review. *New Phytol.* **2007**, *174*, 26–37. [CrossRef] [PubMed]
52. Sletvold, N.; Agren, J. There Is More to Pollinator-Mediated Selection than Polllen Limitation. *Evolution* **2014**, *68*, 1907–1918. [CrossRef] [PubMed]
53. Caruso, C.M.; Eisen, K.E.; Martin, R.A.; Sletvold, N. A Meta-Analysis of the Agents of Selection on Floral Traits. *Evolution* **2018**, *73*, 4–14. [CrossRef]
54. Galen, C. High and Dry: Drought Stress, Sex-Allocation Trade-Offs, and Selection on Flower Size in the Alpine Wildflower *Polemonium viscosum* (Polemoniaceae). *Am. Nat.* **2000**, *156*, 72–83. [CrossRef]
55. Teixido, A.L.; Barrio, M.; Valladares, F. Size Matters: Understanding the Conflict Faced by Large Flowers in Mediterranean Environments. *Bot. Rev.* **2016**, *82*, 204–228. [CrossRef]
56. Gallagher, M.K.; Campbell, D.R. Shifts in Water Availability Mediate Plant–Pollinator Interactions. *New Phytol.* **2017**, *215*, 792–802. [CrossRef]
57. Phillips, B.B.; Shaw, R.F.; Holland, M.J.; Fry, E.L.; Bardgett, R.D.; Bullock, J.M.; Osborne, J.L. Drought Reduces Floral Resources for Pollinators. *Glob. Chang. Biol.* **2018**, *24*, 3226–3235. [CrossRef]
58. Osmolovsky, I.; Shifrin, M.; Gamliel, I.; Belmaker, J.; Sapir, Y. Eco-Geography and Phenology Are the Major Drivers of Reproductive Isolation in the Royal Irises, a Species Complex in the Course of Speciation. *Plants* **2022**, *11*, 3306. [CrossRef]
59. Sapir, Y.; Ghara, M. The (Relative) Importance of Pollinator-Mediated Selection for Evolution of Flowers. *Am. J. Bot.* **2017**, *104*, 1787–1789. [CrossRef]

60. Schneider, C.A.; Rasband, W.S.; Eliceiri, K.W. NIH Image to ImageJ: 25 Years of Image Analysis. *Nat. Methods* **2012**, *9*, 671–675. [CrossRef]
61. Volis, S.; Zhang, Y.H.; Dorman, M.; Abbott, R.J. Incipient Speciation in *Oncocyclus* Irises: Eco-Geographic Isolation and Genetic Divergence with No Reproductive Isolation? *Flora Morphol. Distrib. Funct. Ecol. Plants* **2021**, *275*, 151746. [CrossRef]
62. R Core Team R Core Team. *R: A Language and Environment for Statistical Computing;* R Foundation for Statistical Computing: Vienna, Austria, 2019; Available online: http//www.R-project.org (accessed on 28 June 2023).

plants

MDPI

Article

Phenology and Floral Biology of *Diospyros sericea* A. DC. (Ebenaceae): Inconstant Males May Be behind an Enigma of Dioecy

Bárbara Ramaldes [1], Renata Santos [2], André Rodrigo Rech [3] and Michellia Soares [4,*]

1 Botany Laboratory, Alegre Campus, Federal University of Espírito Santos, Alegre 29500-000, ES, Brazil
2 Laboratory of Forest Seeds, Department of Silviculture and Forest Genetics, Federal University of Lavras, Lavras 37200-900, MG, Brazil
3 Centre of Advanced Studies on Ecological Systems and Interactions, JK Campus, Federal University of Jequitinhonha and Mucuri Valley, Diamantina 39100-000, MG, Brazil
4 Laboratory of Plant Ecology and Systematics, Federal Institute of Northern Minas Gerais, Salinas Campus, Salinas 39560-000, MG, Brazil
* Correspondence: michellia.soares@ifnmg.edu.br

Abstract: *Diospyros sericea* is a tree/shrub species considered dioecious and broadly distributed in Brazil. Despite its importance for niche composition in a range of ecosystems, there is little knowledge about this species, and so far no study has analyzed its sexual system. We aimed to investigate dioecy expression in *D. sericea* through sexual dimorphisms in its phenology and floral biology. We analyzed the phenological events over a year and studied floral biology traits (morphology, flower development, floral resource, floral attractants supply, viability of pollen, and stigma receptivity) in both male and female plants. *D. sericea* presents typical features of dioecious plants like well-established primary and secondary dimorphisms that contribute to its reproductive success. However, we also identified fruit development in what should be structurally male individuals. We suggest that the evolutionary pathway leading to the observed phenomenon may be the existence of subdioecious populations with "inconstant males". Although our data prevented us from making further assumptions about the origin of this trait, the study contributes to future analyses towards unraveling the enigma of dioecy not only in *D. sericea* but in other *Diospyros* species.

Keywords: genus *Diospyros*; sexual dimorphism; reproductive system; evolution

Citation: Ramaldes, B.; Santos, R.; Rech, A.R.; Soares, M. Phenology and Floral Biology of *Diospyros sericea* A. DC. (Ebenaceae): Inconstant Males May Be behind an Enigma of Dioecy. *Plants* **2022**, *11*, 2535. https://doi.org/10.3390/plants11192535

Academic Editors: Brenda Molano-Flores and James Cohen

Received: 16 June 2022
Accepted: 23 September 2022
Published: 27 September 2022

1. Introduction

Angiosperms are currently the most diverse plant group on earth. Their origin and the causes of their great expansion were considered by Darwin as an abominable mystery [1]. As sessile organisms, these plants depend on pollinating agents for pollen transfer [2], establishing a coevolutionary and exclusive relationship with animals [3,4]. Floral visitors directly affect the pollen flow [5]; therefore, investing in attraction resources may increase the number of fertilized ovules [6] and favor outcrossing.

Outcrossing is essential to genetic diversity [7], while selfing may pose harmful effects to subsequent generations [8,9]. Therefore, mechanisms that reduce self-pollination and allow outcrossing may have directly influenced floral evolution [2], contributing to the diversity of floral morphologies currently observed among angiosperms.

Functional hermaphroditism is the most common sexual system among angiosperms [10,11]. In cosexual plants, it may appear in hermaphroditic plants bearing only perfect flowers (about 90% of angiosperm species) and monoecious plants in which a single individual bears both male and female unisexual flowers [2,12–14], among others. Dioecy, with unisexual male and female flowers in different individuals (reviewed by [7,12,13,15,16]), appears at a lower frequency of around 6 to 7% of the species [16,17].

Many hypotheses have been proposed to explain the establishment of dioecy in angiosperms. It has been demonstrated across multiple groups that dioecy has had multiple origins across evolutionary time [2,13,18]. One possibility is the evolution of dioecy from self-compatible non-dioecious ancestors [14,19], avoiding inbreeding and optimizing resource allocation. This would have involved at least two types of mutation: one that caused male infertility, producing individuals with structurally female flowers, and one (or more) mutations that suppressed female fertility, producing individuals with structurally male flowers [20–23] that occurred successively, not simultaneously. This would have involved the existence of populations with intermediate sexual types composed of some individuals bearing perfect flowers and other individuals bearing flowers with one of the organs (stamen/carpel) sterile based on the mutation, characterizing androdioecy and gynodioecy [20].

Current research has attempted to trace the evolutionary pathways to the existence of sex chromosomes in plants and, for this effect, sex determination has been widely studied [14,24,25]. Genetic mutations causing the sterility of reproductive functions are expected to have antagonistic and pleiotropic effects [14,26]. For example, the female sexual organs were not expressed, and/or male functions were enhanced in hermaphroditic ancestors by re-allocating reproductive resources from female to male functions, thus resulting in subdioecious populations [14]. Although normally considered dioecious, with only occasional monoecious individuals, male plants are regarded as "inconstant" due to their ability of occasional seed production [14,21]. Conversely, during this hypothetical evolutionary pathway between monoecy and dioecy, intermediate paradioecious populations could have appeared, in which a given plant, mostly bearing unisexual flowers of one sex, could bear flowers from the opposite sex [12].

The differences between female and male individuals in dioecious species are related to sexual dimorphism [10], which can occur in morphology, physiology, and life history [10,27]. Primary sexual dimorphism refers to sexual differences between the androecium and the gynoecium, whereas secondary sexual dimorphism produces morphological, physiological, and phenological differences between the sexes [12,28].

Dioecy is generally related to some ecological traits [15,29–31]. These include the occurrence at high elevations [16], generalized pollination [32], fleshy fruit production [15,30,31], and animal seed dispersal [30,33]. Molecular phylogenetic analyses have shown an association of these ecological traits with the diversification of dioecious lineages [34,35].

Dioecious species have been found to compose up to 12% of the endemic flora of the Brazilian Cerrado [30]. These species occur in open areas, are brevi-deciduous, and show irregular and barely evident reproductive phenology. The angiosperm family Ebenaceae is among the exclusively dioecious families that occur in the Cerrado. Hence, Ebenaceae is considered a pantropical family with a center of diversity in South America [36]. It is mainly composed of woody plants with a tree, shrub, or sub-shrub habit [36–38]. Encompassing more than 500 species, *Diospyros* L. is the most diverse genus within Ebenaceae [36,39,40]. Sixty-two species of this genus occur in Brazil and twenty-nine are endemic to the country [41].

Species of *Diospyros* have been used as models in molecular research aiming to elucidate dioecy evolution and expression. Although sex-determining genes among *Diospyros* species have been identified [7,14], studies on its reproductive biology are scarce [10]. This scarcity can be confirmed with *D. sericea* A. DC. This species is broadly distributed in South America, including Venezuela, Colombia, and Brazil [42]. However, *D. sericea* reproductive biology is completely unknown, and, although considered a dioecious species, no study has yet attempted to compare individuals of *D. sericea* of different sexes.

Thus, here we aim to analyze dioecy expression through sexual dimorphisms in *D. sericea* by describing its phenological events and floral biology. We hope to produce data that contribute to the conservation of *D. sericea* and the interdependent ecological balance associated with it. Our work may provide a basis for future analyses on this species towards unraveling the enigma of the origin of dioecy and its establishment in the genus and angiosperms in general.

2. Results

2.1. Phenological Analyses

2.1.1. Vegetative Phenology

We observed leafing and defoliation in all individuals (13 female and 17 male) of *D. sericea* during the monitoring months. These phenological events could be considered highly synchronous but occurred at a low intensity.

Regardless of climatic variation, leafing remained constant in both male and female individuals throughout the monitored period, with ~25% intensity. Defoliation happened similarly: between 25 and 35% intensity and with little variation during the observed year (Figure 1A,B). One exception to this trend was the significant increase in defoliation that exceeded leaf production among female individuals between September and December (Figure 1B).

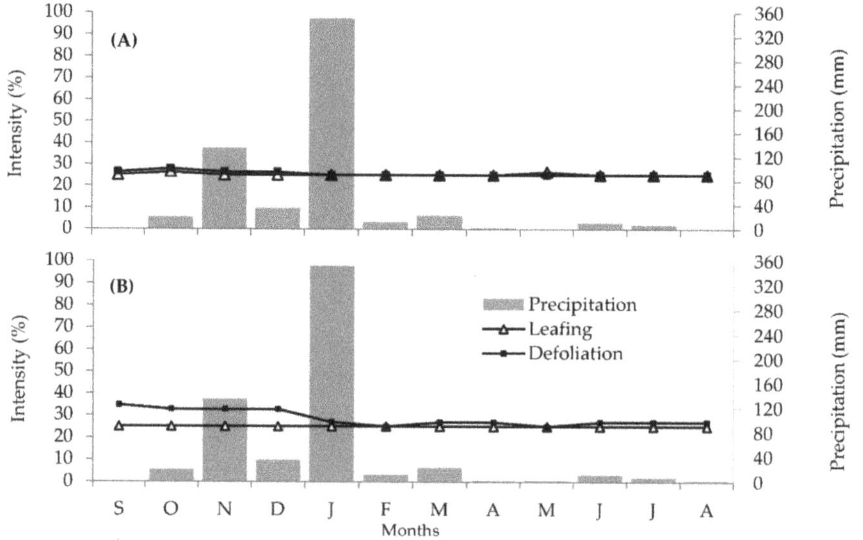

Figure 1. Fournier intensity of the vegetative phenology of Diospyros sericea and climatic variables in the area surrounding the Serra Nova e Talhado State Park (PESNT), district of Rio Pardo de Minas, Minas Gerais state, Brazil. Data recorded between September 2015(S) and August 2016(A). (**A**) Male individuals. (**B**) Female individuals.

The Spearman correlation analysis between the vegetative phenophases and the climatic variables revealed a significant correlation between defoliation and climatic variables (temperature and humidity). Defoliation was positively correlated with temperature, peaking in the period with the highest temperature averages, and strongly negatively correlated with humidity, peaking in the periods with the lowest relative air humidity (Table 1).

Our results support the classification of *D. sericea* as an evergreen species, characterized by continuous growth (ECG), prolonged leaf production, and an absence of evident deciduousness.

Table 1. Spearman correlation between the climatic variables (average temperature (°C), relative air humidity (% UR), and precipitation (mm); recorded between September 2015 and August 2016) and the intensity of vegetative (leafing and defoliation) and reproductive (buds, flowers, immature and mature fruits) phenophases of individuals of *D. sericea*.

Phenology	Individuals	Phenophases	Average Temperature	Relative Humidity	Precipitation
Vegetative	Male	Leafing	-	-	−0.16
		Defoliation	0.75 *	−0.81 *	0.21
	Female	Leafing	-	-	-
		Defoliation	0.56 *	−0.75 *	0.16
Reproductive	Male	Initial flowering	−0.27	-	0.002
		Established flowering	−0.08	−0.02	0.18
		Initial fruiting	0.31	−0.15	0.50
	Female	Initial flowering	−0.40	0.04	−0.02
		Established flowering	−0.05	0.03	0.27
		Initial fruiting	0.52	−0.43	0.16
		Established fruiting	0.91 *	−0.60 *	0.47

Initial flowering (flower bud); established flowering (flowers in anthesis); initial fruiting (immature fruit); established fruiting (mature fruit). * Indicates significant differences ($p \leq 0.05$) according to the Spearman correlation analysis (rs).

2.1.2. Reproductive Phenology

Flowering was observed from initial flowering (floral bud) to establishment (anthesis flowers) in *D. sericea* during the whole monitoring year (Figure 2A,B). However, the synchronicity of these events behaved differently among individuals of the same sex. Flowering was highly synchronous among male individuals. Among female individuals, flowering was only highly synchronous during the peak months (November, January, July to August) and remained hardly synchronous during the rest of the year.

Flowering intensity also differed between male and female populations. Flowering in male populations was twice as intense as in female populations. This trait may be connected with the number of flowers per flowering branch since male plants presented more flowers than female ones.

We observed two flowering peaks at different times. The first peak, with the highest flowering intensity, happened during the rainy season: it was first recorded among the male individuals between October and December and, later, among female individuals from November to January. The second peak, observed during the dry season between May and August, followed a similar pattern whereby flowering intensity was initially higher among the male population and was later followed by the female population between June and August. We highlight that the production of flower buds and flowers varied gradually over time among the male population, with some periods of stability. Conversely, among the female population, flowering intensity peaked rapidly and abruptly.

As with the other phenophases, fruiting was also observed throughout the monitoring period and was, for the most part, highly synchronous. However, only a single fruiting peak was observed in the transition between the dry and the rainy seasons (Figure 2C,D). The highest intensity of initial fruiting (immature fruits) was recorded in September and of mature fruits in October. We highlight that some male individuals of *D. sericea* also produced fruits (Figure 2C), although these were smaller in size, produced in lower numbers, and failed to reach maturity.

The Spearman correlation analysis revealed a general weak correlation between flowering and the climatic variables analyzed. The established fruiting (mature fruit) phenophase was the only one to show a significant correlation with climate. Mature fruit production was positively correlated with temperature and negatively with humidity (Table 1), suggesting higher fruit production under higher temperatures and lower relative air humidity.

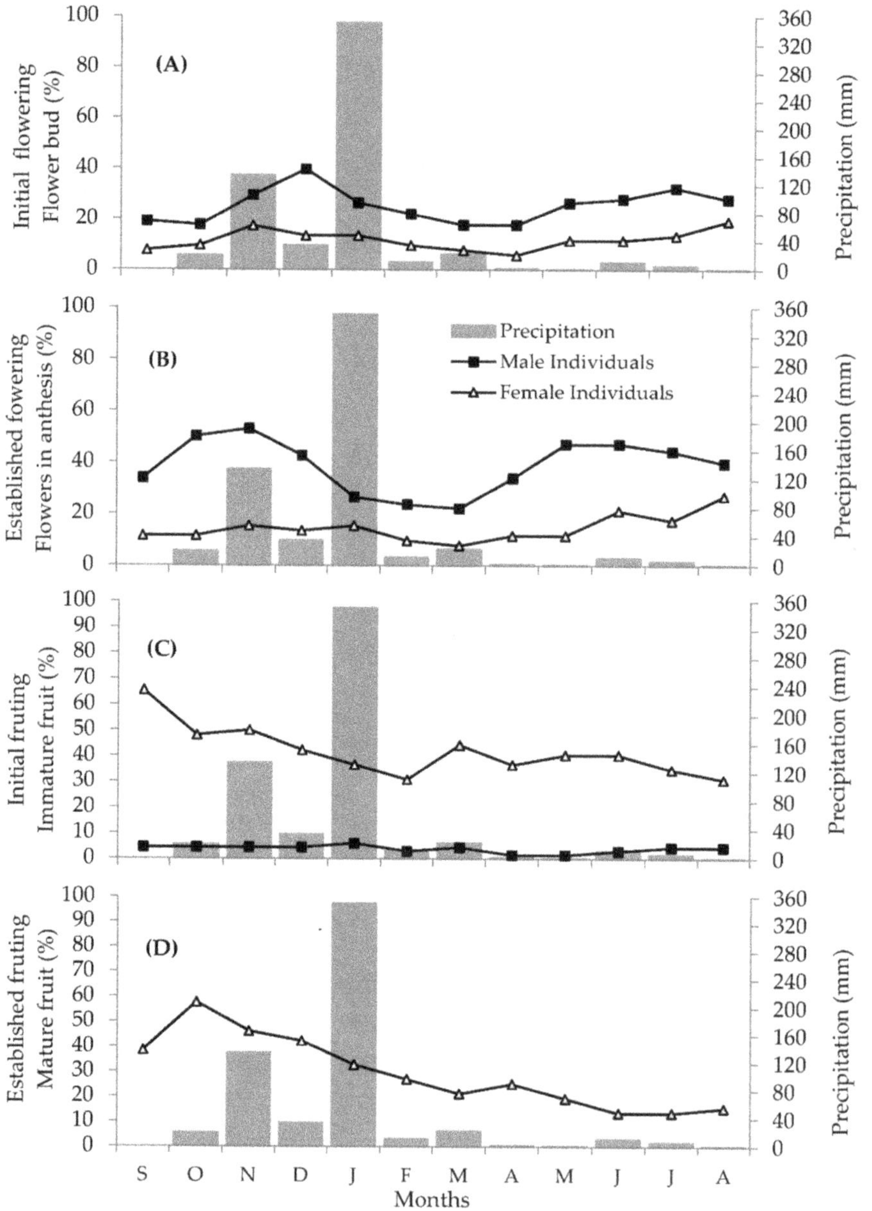

Figure 2. Fournier intensity of the reproductive phenology of male and female plants of *D. sericea* and accumulated precipitation (mm). Data recorded between September 2015 and August 2016. (**A**) Initial flowering intensity (flower bud); (**B**) established flowering intensity (flowers in anthesis); (**C**) initial fruiting intensity (immature fruits); (**D**) established fruiting intensity (mature fruits).

2.2. Floral Biology

2.2.1. Morphological Traits and Sexual Expression

As suggested by its name, *D. sericea* has yellow and golden pilosity that provides it a sericeous aspect. On its vegetative structures, trichomes are found along the branches and abaxial leaf surfaces (Figure 3A). On its flowers, trichomes are found on the entirely pilous sepals and the central region of the outer surface of the petals (Figure 3B).

Figure 3. Floral biology traits of *D. sericea* located in the surroundings of the PESNT. (**A**) Sericeous aspect of the vegetative structures and (**B**) flowers, indicated by the arrow, bar = 1 cm; (**C**) multiple buds per axil in a male individual, bar = 1 cm; (**D**) solitary bud per axil in a female individual, and bracts indicated by the arrow, bar = 1 cm; (**E**) female flower with a bottle-shaped pistil, bar = 1 cm; (**F**) male flower, bar = 1 cm; (**G**) polyadelphous stamens fused at the base with free anthers, bar = 0.2 cm; (**H**) sericeous pistillode in a structurally male flower, bar = 0.5 cm; (**I**) staminodes in a structurally female flower indicated by the arrow, bar = 0.5 cm; (**J**) structurally male flower with a well-developed pistillode, with stigma indicated by the arrow, bar = 0.5 cm; (**K**) structurally female flower with well-developed staminodes indicated by the arrow, bar = 0.5 cm; (**L**) structurally male individual with a developed fruit at the branch base indicated by the arrow, bar = 3 cm; (**M**) structurally female individual with a fruit-bearing branch, bar = 3 cm.

Its sessile and downward-oriented flowers are inserted in leaf axils along the branches, covered by the abaxial surface of leaves (Figure 3A,B). Structurally male individuals have more than one flower bud per axil, forming a cymose inflorescence (Figure 3C). Female individuals showsolitary flowers (Figure 3D).

One or two small green bracts are also observed in the floral axils. These bracts are formed along with the flower buds but quickly dehydrate (Figure 3D) and may persist after flower abscission.

Morphological similarities were observed between the male and female flowers, including color, form, symmetry, and number of perianth parts (see the comparative description of floral size below). Both the pistillate and staminate flowers are dichlamydeous, heterochlamydeous, actinomorphic, and pentamerous (Figure 3E,F), and occasionally tetramerous and hexamerous.

Both male and female flowers are inconspicuous, with a gamosepalous calyx with greenish sepals and an opaque cream-colored corolla with cyclically arranged free petals. Due to their morphological similarities, male and female flowers are only distinguishable by their reproductive whorls.

The structurally male flowers are polystaminate with a heterodynamous androecium (Figure 3F,H). The stamens are epipetalous and organized in polyadelphous bundles, each composed of three stamens attached by the filament base, with anthers pivoting freely (Figure 3G). Male flowers also have a sericeous vestigial pistillode (Figure 3H).

Structurally female flowers have a syncarpous gynoecium composed of three carpels with terminal styles that form a bottle-shaped pistil, which does not surpass the corolla height (Figure 3I). The ovary is superior and trilocular, with two ovules in each cavity. Female flowers generally have vestigial staminodes (Figure 3I).

The pistillodes and staminodes are occasionally similar in size to those of functional pistils and stamens (Figure 3J,K). Some predominantly male individuals, with occasional hermaphroditic flowers, produced fruits. In these cases, a single individual developed several fruits, usually at the base of the branches (Figure 3L). However, these fruits failed to reach maturity and were considerably smaller than those originating from ovary development in structurally female flowers (Figure 3L,M).

2.2.2. Morphometrics

Both pistillate and staminate flowers had no significant differences in the calyx and corolla lengths between the sexes, but significant differences were found in the diameter of the perianth, pistil length and diameter, and stamen length (Table 2).

Table 2. Mean, standard deviation and variation in the morphometric measurements of floral whorls (centimeters), the number of stamens/staminodes, and the number of buds per axil (units) of individuals of *D. sericea* PESNT. $\overline{x}$ = mean; SD = standard deviation.

Structures Measured	Male		Female		*t*-Test or *U* Test	*p*
	$\overline{x} \pm$ SD	Min-Max Range	$\overline{x} \pm$ SD	Min-Max Range		
Calyx						
Length	0.77 ± 0.077	0.59–0.94	0.80 ± 0.053	0.71–0.90	$t = -1.96$	0.0549
Diameter	0.45 ± 0.046	0.32–0.53	0.53 ± 0.038	0.44–0.59	$t = -6.42$	<0.0001 *
Corolla						
Length	1.13 ± 0.107	0.98–1.35	1.09 ± 0.075	0.95–1.21	$t = 1.88$	0.0655
Diameter	0.43 ± 0.054	0.30–0.55	0.48 ± 0.050	0.34–0.59	$t = -3.30$	0.0018 *
Pistil/pistillodes						
Length	0.12 ± 0.087	0.06–0.52	0.68 ± 0.083	0.39–0.79	$t = -23.62$	<0.0001 *
Diameter	0.13 ± 0.043	0.08–0.23	0.33 ± 0.022	0.30–0.41	$U = 0.00$	<0.0001 *
Stamens/staminodes						
Length	0.62 ± 0.056	0.49–0.72	0.43 ± 0.042	0.35–0.50	$t = 13.47$	<0.0001 *
Number	43.07 ± 4.811	34–54	9.35 ± 3.908	4–15	$U = 0.00$	<0.0001 *
Number of buds *per* axil	2.18 ± 0.566	1–7	1	1	-	-

* Indicates significant differences ($p \leq 0.05$) according to a *t*-test or a Mann–Whitney test.

The number and length of the staminodes present in the female flowers differed significantly from that of the functional stamens present in the male flowers. Among the male flowers, this was also observed in the length and diameter of the pistillodes, which differed significantly from the functional gynoecium of the female flowers. These data revealed significant structural differences in the reproductive whorls and vestigial reproductive organs of female and male flowers in D. sericea (Table 2).

The sexual dimorphism observed in the reproductive whorls reflects the significant differences found in calyx and corolla diameters between pistillate and staminate flowers. The ovary in female flowers leads to broader sepal and petal diameters. However, this was not observed among the structurally male flowers, where the pistillodes were approximately three times smaller than the pistils of structurally female flowers (Table 2).

The number of flower buds per axil is higher in structurally male individuals (1–7 buds per axil), on average twice as many as in female individuals.

2.2.3. Floral Development

Floral development begins with the formation of buds, which are initially completely covered by the sepals (Figure 4A,G). The petals display spiral flowering (Figure 4B,H). The anthesis period may last from two to three days until complete flower opening (Figure 4C,I).

Figure 4. Floral development of *D. sericea* observed in the surroundings of the PESNT. Structurally male individuals: (**A**) flower buds; (**B**) flower in anthesis; (**C**) blooming flower; (**D**) flower at the onset of senescence with brownish anthers and petal lobes; (**E**) oxidation of the floral features (corolla and stamens); (**F**) completely oxidated flower. Structurally female individuals: (**G**) buds; (**H**) flower in anthesis; (**I**) blooming flower; (**J**) immature fruits; (**K**) mature fruits; (**L**) seeds. (**A–G**,**I**) scale bar = 1 cm; (**H**) bar = 0.5 cm; (**J–L**) bar = 3 cm.

Structurally male flowers remain completely open during the fourth and fifth day of floral development, when senescence begins. This process can be observed through the oxidation of the floral whorls, beginning in the anthers and petal lobes (Figure 4D), which become progressively brown (Figure 4E). From the sixth to the seventh day, the flowers become completely oxidated and dry (Figure 4F) and may suffer abscission.

In structurally female flowers, fruit development begins early, during anthesis, before the flowers are completely open. Initially, *D. sericea* berries are densely sericeous with a greenish color (Figure 4J). During maturation, the fruits become orange and some of their pilosity is lost (Figure 4K). The seeds are elongated and brown (Figure 4L). The flowers that fail to develop fruits before the completion of anthesis undergo an oxidative process that culminates in flower senescence.

2.2.4. Pollen Integrity and Stigma Receptivity

In individuals with structurally male flowers, the flower buds, flowers in pre-anthesis, and completely open flowers had an expressive number of intact pollens. This trait was identified by the acetocarmine reaction. Pollen is spherical, whitish, and arranged in monads, with a powdery appearance (Figure 5A).

Pollen integrity exceeded 90% in the three stages of floral development (Table 3; Figure 5A). In structurally male flowers, the percentage of pollen integrity was maintained throughout floral development. According to the analysis of variance (ANOVA), there were no significant differences in the pollen integrity between flower buds, flowers in pre-anthesis, and completely open flowers from different and random individuals ($F = 0.7627$; $p = 0.5198$, Table 3). On the other hand, in flowers and flower buds of female individuals, no pollen grains were detected in the staminodes.

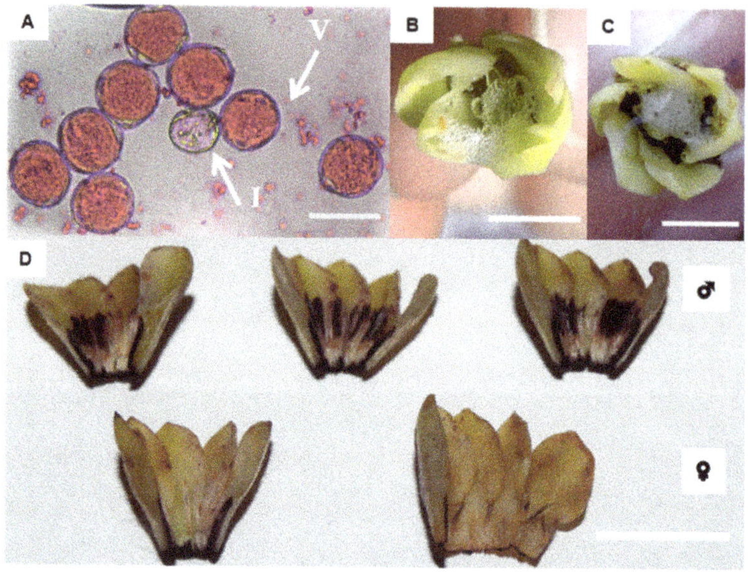

Figure 5. Results of the pollen integrity and stigma receptivity tests and the presence of osmophores in flowers of *D. sericea* collected in the surroundings of the PESNT. (**A**) Acetocarmine-stained pollen, V = viable, and non-stained pollen, I = inviable, bar = 50 µm; (**B**) sign of stigma receptivity from a reaction to hydrogen peroxide in a pre-anthetic flower and (**C**) in a blooming flower, bar = 0.5 cm; (**D**) neutral red-positive coloration of scent glands in the petals and stamens or staminodes: above, male flowers (♂), below, female flowers (♀), bar = 1 cm.

Table 3. Pollen integrity (%) tested in flower buds, flowers in pre-anthesis, and open flowers of *D. sericea* from individuals located in the surroundings of PESNT.

Floral Development Stage	Min–Max Range	$\overline{x} \pm$ SD	Pollen Integrity (%)
Flower buds	163–197	182.4 ± 12.72	91.2%
Flowers in pre-anthesis	167–197	186.0 ± 9.043	93%
Open flowers	177–197	187.7 ± 6.667	94%

The stigma receptivity test conducted in flower buds, flowers in pre-anthesis, and open flowers showed positive results (Figure 5B,C). The open flowers displayed a stronger reaction than the other stages, with greater formation of bubbles than in the other stages. These data could either indicate a greater receptivity during this stage or a reaction to necrosis enzymes. We also tested receptivity in the pistillodes of structurally male flowers, but the results were always negative.

2.2.5. Floral Attractants

The scent exhaled by *D. sericea* flowers is sweet and, by the human sense of smell, it is considerably more intense in male flowers than in female flowers. Scent glands (osmophores) were identified at the corolla margins and in the anther region of the stamens. The structurally male flowers showed a larger red-brownish-colored region than the female flowers (Figure 5D).

3. Discussion

D. sericea proved to be structurally unisexual at the individual level in the analyzed populations due to the primary sexual dimorphism identified. This characteristic has been previously observed and described in other studies—e.g., [42,43]. Moreover, we identified the presence of vestigial reproductive organs of the opposite sex in both female (staminodes) and male (pistillodes) flowers. This had also been described in the species [38,42,43], as well as in several species of the genus *Diospyros* [44–46] and of the Ebenaceae family [7,36]. This evidence confirms that *D. sericea* is structurally dioecious. However, we also identified variations, such as fruit development in what should be structurally male individuals. These individuals are predominantly male with occasional hermaphroditic flowers.

In terms of secondary dimorphisms, the species has a wide variation in phenology between its areas of occurrence. In Brazil, flowers and fruits were recorded throughout the year, varying according to the region of occurrence [42]. Flowering synchronicity between the sexes is essential for the reproductive success of dioecious species [47]. However, male and female plant strategies may differ from one another to favor the pollen flow [10,15,48]. For example, we observed that, in *D. sericea*, female inflorescences are reduced to a single flower which probably also allows a better resource supply. Another common characteristic is that the flowering period of female individuals tends to be shorter, allowing greater resource allocation and energy provision for fruit and seed development [15,48,49].

The flowering of male individuals tends to happen earlier [50–52], over a longer period, and with a higher flower production [53,54]. Flowering intensity among the male flowers was nearly twice that observed among the female flowers, which could be explained by the higher number of flowers per axil found in male individuals. This trait, frequent among other dioecious species [47], has been previously identified in *D. sericea*—e.g., [37,55]—as well as the whole family Ebenaceae [36].

The flowering phenophases occur in *D. sericea* throughout the whole year, which possibly induces constant pollinator visits [56,57]. Since the majority of sexually dimorphic dioecious plants are pollinated by animals [15], their reproductive success relies on these pollinating agents [15,58]. Thus, as well as depending on a certain degree of flowering synchronicity between the sexes, dioecious species (including *D. sericea*) also depend on the phenology of floral visitors. This relationship is particularly advantageous if the flowering peaks coincide with periods of higher pollinator abundance [15]. Asynchronous flowering peaks between the sexes, which we observed in *D. sericea*, may favor outcrossing for reducing competition for flower visitors [57,59]. Altogether, these data suggest that the reproductive success of this dioecious species lies beyond its own conditions and phenological traits, depending directly on the pollinator's presence.

In terms of floral morphology, male flowers tend to be smaller than female ones as a common pattern in dioecious species, which also occurs in the family Ebenaceae and the genus *Diospyros* [36,38]. However, this characteristic is not uniform, as we can see in *D. sericea*. We observed morphological similarities in size and general physical aspects between pistillate and staminate flowers. Both female and male flowers were inconspicuous and barely specialized, which are frequent traits among dioecious species [15,30,32,60]. These morphological similarities, along with differences in resource availability offered by male and female flowers (e.g., pollen), may enable pollination by deceit [61]. When male and female flowers are similar and a given resource is available in only one of the sexes, pollinators may visit both staminate and pistillate flowers [61–64]. Thus, considering the morphological similarities between its unisexual flowers and the differences in resource supply between the sexes, pollination by deceit may be common in *D. sericea* flowers.

In species whose female flowers have an anticipated flowering peak among the male flowers, this works as a strategy for pollinator deceit. Pollinators visit the staminate flowers in search for pollen and eventually visit pistillate flowers, where they fail to find this resource [61]. This strategy seems successful in *D. sericea*, where female flowers do not produce pollen but are visited for deceit probably due to the morphological similarity between female and male flowers.

The presence of scent in flowers of both sexes is also important for the reproduction of dioecious species, given that olfaction is one of the most used senses by insects [65,66]. In this case, scent plays the role of a distance attraction resource [67–69], a mechanism that also contributes to automimicry and which has been reported in other dioecious species [15,70,71]. The typical scent of *D. sericea*, although also present in female flowers, is stronger in male flowers, as also observed in other dioecious species—e.g., [72]. This may promote greater efficiency in pollen transport due to higher attractiveness to pollinators.

Other than the pollinator attraction strategies, *D. sericea* also seems to have adapted to the environmental conditions towards maximizing the success of its reproductive cycle until its final stage, which is seed dispersal. Female individuals have the highest flower production at the end of the dry season, as also described in *D. lasiocalyx* [73]. This phenological behavior does not seem ideal for these individuals, given that female reproduction is costly, and water is limited during this period. However, according to [74], plants can overcome limitations in water availability through mechanisms that avoid the effects of drought on reproduction. One of these mechanisms was observed in *D. sericea*, whereby defoliation in female individuals only increased significantly between the dry and the rainy seasons, which coincided with the flowering peak and the most intense fruiting phase. This trait suggests nutrient relocation from senescing leaves to fruit formation [75]. Another favorable phenological adaptation is fruit development during more humid months, a common feature among animal-dispersed species [76] (birds and primates consume fruits and disperse seeds of *D. sericea* [42]). An advantage associated with this is that water availability favors the production of fleshy fruits, which become more conspicuous to seed dispersers [77], increasing the likelihood of reproductive success.

The onset of fruiting in female flowers before anthesis completion may indicate high efficiency in male reproductive strategies for providing large amounts of pollen, coupled with pollinator efficiency. However, in this case, apomixis also cannot be ignored. Considered a type of asexual reproduction, apomixis is the formation of seeds without fertilization, which can result in embryos developed from the ovule tissues [78,79]. The facultative apomictic species reproduce sexually but may produce seeds without fertilization as an alternative path to reproduction [78]. *D. sericea* fruiting traits and dependency on pollinators, coupled with the common observation of apomixis in its genus [80,81], could suggest the possibility of the species showing facultative apomixis, however we have not found evidence for apomixis in the studied species.

The analysis of *D. sericea* floral biology calls the evolutionary process of its reproductive system into consideration. A possible evolutionary pathway to dioecy involves gynodioecy, reviewed by Ashman [82]. Since a gynodioecious population is composed of female plants and monoecious ones [12] or plants with female and hermaphroditic flowers [20,83], autogamy (i.e., self-pollination) could be possible, not depending directly on pollinating agents. Despite fertilization taking place before anthesis completion in *D. sericea*, no pollen was found in female flowers' staminodes; thus, autogamy is unlikely unless the female flowers showed occasional well-developed staminodes with viable pollen, which we did not find.

Fruit development in predominantly male individuals with occasional hermaphroditic flowers of *D. sericea* suggests a direct link with the presence of vestigial female reproductive organs in its flowers. According to Wallnöfer [36], structurally male flowers in the Ebenaceae family bear pistillodes that are seldom absent and sporadically well-developed. Considering this trait, an evolutionary pathway that passes through androdioecy is also possible. Androdioecy is considered an intermediate sexual type in the evolutionary process towards dioecy, defined by the occurrence of populations formed by monoecious individuals [12] or by individuals with perfect flowers (i.e., hermaphroditic) and others with sterile female sexual functions [20,83]. Thus, the fact that its reproductive organs are highly viable from the beginning of floral development may allow self-fertilization in male flowers bearing pistillodes that, for a random evolutionary transition cause, are occasionally functional, leading to fruit formation in male individuals.

Another evolutionary pathway would be the inexpression of the female sex and/or intensification of the male functions in hermaphroditic ancestors, allowing resource relocation from female to male functions to evolve and resulting in subdioecious populations [14]. These are normally considered dioecious, with occasional occurrence of monoecious individuals [14,21]. In subdioecious populations, male individuals are considered "inconstant" due to their ability of occasional seed production. Thus, with occasional manifestations of monoecy in *D. sericea* populations, some individuals that appear predominantly male would be capable of producing fruits, even if at a low frequency. This could be the most probable evolutionary pathway for dioecy in *D. sericea*.

4. Materials and Methods

4.1. Study Site and Model Plant Species

The study was performed in the Serra Nova district, Rio Pardo de Minas municipality, Minas Gerais state, Brazil. The data were collected in the surroundings of the Parque Estadual de Serra Nova e Talhado (PESNT) (coordinates 42°57′30″ W and 16°1′30″ S), which has a total area of 49,890 hectares.

The climate in the region falls in the category BSh, according to the Köppen climate classification (hot, semi-arid climate), with average precipitation between 380 and 760 mm and an average annual temperature above 18 °C. According to the climate data collected in the municipality of Rio Pardo de Minas between 1983 and 2005 [84], the average temperature in the region is 22.5 °C, oscillating between 19.3 and 24 °C. The rainy season occurs between November and March, representing 85.4% of total annual precipitation, and peaks in December (200 mm). The dry season occurs between May and September, representing 3.84% of total annual precipitation, and peaks in August (2 mm).

The PESNT is located in a transitional region between the Cerrado and Caatinga phytogeographic domains in a landscape intersected by rocky outcrops from the northern Espinhaço mountain range. Grasslands, rocky grasslands (Campos Rupestres), cerrado savannas, cerrado grasslands, semideciduous seasonal forests (gallery forests), and deciduous seasonal forests (dry forests) compose the PESNT vegetation [84].

The selected plant species was *Diospyros sericea*, a widely distributed species in Brazil, occurring in the Amazon, Caatinga, Cerrado, Atlantic Forest, and Pantanal phytogeographic domains [85]. It has a tree or shrub habit, and its main human use is for building houses [43,86]. Although edible, its fruits are seldom consumed by humans [87] but are highly appreciated by birds and primates. The latter, particularly, play an essential role in *D. sericea* seed dispersal. *D. sericea* is well distributed in the surroundings of the PESNT, especially in abandoned pastures and monocultures under natural regeneration. Wood extraction is frequent and indiscriminate on the site.

4.2. Phenology

4.2.1. Data Collection

We selected 17 adult male and 13 adult female individuals (at reproductive age) of *D. sericea* with completely visible canopies to monitor phenology. When possible, we ensured a distance of 5 m between individuals aiming to capture greater genetic variability and lower relatedness among them [88].

We recorded phenological data every month for one year between September 2015 and August 2016. To identify the vegetative phenophases (leafing and defoliation), we adapted the methodology proposed by [89]. Leafing was identified as the period between initial bud development and the formation of young leaves. Defoliation was identified as a change in leaf color and subsequent fall. The reproductive phenological events that we observed were flowering and fruiting. We subdivided the flowering phase into initial flowering (flower buds) and established flowering (flowers in anthesis). We subdivided the fruiting phase into initial fruiting (immature fruits) and established fruiting (mature fruits).

4.2.2. Data Analysis

We observed the crown of each individual to record the phenological observations, estimating the total manifestation of each phenological event. We followed Fournier [90] to analyze phenological data. We also estimated phenological synchronicity in the population, following Morellato et al. [91].

We used our observations on the development and persistence of crown leaves to classify the species into 4 phenological groups (according to [92]): (1) evergreen with continuous growth (ECG), without evident deciduousness and leaf production during long periods; (2) evergreen with seasonal growth (ESG), without total deciduousness but with leaf replacement in the transition between the dry and rainy periods; (3) brevi-deciduous (BDC), with total deciduousness in the dry season for a period shorter than two weeks; and (4) deciduous (DEC), with total deciduousness in the dry season for a period longer than two weeks.

4.3. Floral Biology

For the floral biology analyses, we collected one flower from 10 male and 10 female individuals. These analyses were conducted in the Laboratory of Plant Ecology and Systematics at the Federal Institute of Northern Minas Gerais—Salinas campus.

We assessed floral development in situ by marking and observing one pre-anthetic flower of ten different individuals (n = 10) during the morning, afternoon, and night for seven days (totaling 54 h). Pre-anthesis was considered as the period when flowers were fully formed but with a closed corolla.

4.3.1. Morphometric Analyses

We measured the floral features of randomly selected male (n = 26) and female (n = 26) flowers. Two or three flowers were collected from ten different plants of each sex. We used a digital caliper with a precision of 0.1 mm to measure the studied traits. The floral features analyzed were the length and diameter of the calyx, corolla, and pistil, and the number and length of stamens. We also characterized the distribution and number of buds per axil in male and female individuals.

4.3.2. Pollen Integrity and Stigma Receptivity

Pollen release was assessed in situ from flowers (n = 10) in pre-anthesis, open flowers (n = 10), and senescing flowers (n = 10). The pollen viability test was done in pre-anthetic flowers (n = 10) and open flowers (n = 10) belonging to ten different structurally male individuals. For this, the anthers were macerated and stained with acetic carmine (1.2%) in a semipermanent slide [93]. The slides were analyzed under a light microscope (Nova Optical Systems, 180i) with an objective lens of 10× magnification, and the first 200 pollen grains observed were manually counted. We considered viable the pollen grains that changed color to reddish pink and inviable the pollen grains that did not change color at all. Pollen quantity was expressed in percentage terms.

Stigma receptivity was tested in situ in flower buds (n = 10), pre-anthesis flowers (n = 10), and open flowers (n = 10). We applied hydrogen peroxide at 3% (H_2O_2) to the stigmas [94] and, with a magnifying glass (10×), interpreted the formation of bubbles as a sign of respiratory activity and therefore stigma receptivity.

4.3.3. Scent

We assessed the presence and type of floral scent by placing recently opened flowers inside glass jars that were later sealed for 20 min [94]. We used separate containers for male and female flowers. Then, we analyzed the odor intensity by the human sense of smell from three different individuals. To detect floral scent glands (osmophores), we submerged the flowers for five minutes in neutral red solution (1:1000) and washed them in running water [70].

4.4. Statistical Analyses

All data were analyzed in BioEstat 5.0. We compared the morphometric measurements between male and female flowers (floral features and number of flower buds per leaf axil) with a Student's *t*-test ($p \leq 0.05$). In a few cases, when the requirements for the *t*-test were not met, the Mann–Whitney U test was used ($p \leq 0.05$). The results of the pollen integrity test at different floral developmental stages were analyzed with a one-way ANOVA.

We calculated Spearman correlations (*rs*; 5% of significance) between the phenological phases of leafing, defoliation, initial flowering (flower buds), established flowering (flowers in anthesis), initial fruiting (immature fruits), and established fruiting (mature fruits), and the climatic variables during the sampling months (average temperature, relative humidity, and accumulated precipitation). The climate data (Figures 1 and 2) were drawn from the website of Brazil's National Institute of Meteorology—INMET [95], specifically from the automatic weather station from Rio Pardo de Minas, Minas Gerais state

5. Conclusions

D. sericea exhibits the common features of a dioecious species in all aspects analyzed. Primary sexual dimorphism is evident, whereas secondary sexual dimorphism mainly appears in phenological traits and floral attractants and resources. However, female and male flowers are morphologically similar.

Importantly, vestigial reproductive organs were generally found in flowers of both sexes, which is common among *Diospyros* species. This trait may be associated with occasional fruit development in predominantly male individuals with occasional hermaphroditic flowers, characterizing them as inconstant males

The strong morphological similarity between male and female flowers and the existence of staminodes/pistillodes together with the occasional fruiting of male flowers suggests that dioecy in this species is of recent origin. Although our data do not explain the causes of fruit formation in predominantly male individuals with occasional hermaphroditic flowers, our descriptions may be a preliminary step for future analyses on the evolution of dioecy in *D. sericea*. The production of seeds exclusively by females was already pointed out as a handicap of dioecy [96]. Due to the general absence of fruit production in males, female individuals should produce twice as much seeds [20], a problem that may be attenuated in *D. sericea* due to fruit production in males. In this case, seed production is adding to other traits associated with the success of dioecious species such as precocious reproduction and animal-dispersed fleshy fruits, among others [97]. The promising information presented in this study calls attention to the need for a phylogenetic-based comparative approach to compare species in the *Diospyros* genus. Thus, perhaps it will be possible to unveil the enigma of the origin and establishment of dioecy in *D. sericea* and in the genus *Diospyros*.

The comprehension of dioecious species' sexual system, such as we found in *D. sericea*, is essential to its conservation and the entire ecosystem in which it interacts with pollinating and seed-dispersing animals. In this particular case, the variations in the expression of the sexual system requires that, for conserving a dioecious species, beyond focusing on male and female individuals, it would also be essential to identify matrices that represent the sexual transition, such as inconstant males. Thus, in a broader perspective, these studies may contribute to the conservation of this species and dioecious species in general. By taking into account the morphological variation, the conservation will be preserving not only a species but its ecological and evolutionary history.

Author Contributions: Conceptualization, M.S.; methodology, B.R., R.S. and M.S.; validation, B.R., R.S. and M.S.; formal analysis, B.R., R.S. and M.S.; investigation, B.R., R.S. and M.S.; resources, M.S.; data curation, B.R. and R.S.; writing—original draft preparation, B.R., R.S., A.R.R. and M.S.; writing—review and editing, B.R., R.S., A.R.R. and M.S. All authors have read and agreed to the published version of the manuscript.

Funding: Instituto Federal do Norte de Minas Gerais (IFNMG)—campus Salinas: logistical support and funding part of this research through the Programa de Apoio à Pesquisa (PROAPE); and the technical support from: Fundação de Amparo à Pesquisa do Estado de Minas Gerais (FAPEMIG-APQ-00932-21) and Conselho Nacional de Pesquisas Científicas CNPq (Processo 423939/2021-1).

Institutional Review Board Statement: Not applicable.

Acknowledgments: We thank the collaborators from the Parque Estadual de Serra Nova e Talhado (PESNT) for lodging and support in the collection sites. We also thank UFVJM through the PAP—Program for the funding support.

Conflicts of Interest: The authors declare no conflict of interest.

References

1. Crepet, W.L. The Abominable Mystery. *Science* **1998**, *282*, 1653–1654. [CrossRef]
2. Barrett, S.C.H. The evolution of mating strategies in flowering plants. *Trends Plant Sci.* **1998**, *3*, 335–341. [CrossRef]
3. Lloyd, D.G.; Barrett, S.C.H. (Eds.) *Floral Biology: Studies on Floral Evolution in Animal-Pollinated Plants*; Chapman & Hall: New York, NY, USA, 1996.
4. Crepet, W.L. Progress in understanding angiosperm history, success, and relationships: Darwin's abominably "perplexing phenomenon". *Proc. Natl. Acad. Sci. USA* **2000**, *97*, 12939–12941. [CrossRef] [PubMed]
5. Harder, L.D.; Barret, S.C.H. Pollen dispersal and mating patterns in animal-pollinated plants. In *Floral Biology, Studies on Floral Evolution in Animal-Pollinated Plants*; Chapman & Hall: New York, NY, USA, 1996; pp. 140–190.
6. Cohen, D.; Shmida, A. The evolution of flower display and reward. *Evol. Biol.* **1993**, *27*, 197–243.
7. Akagi, T.; Henry, I.M.; Tao, R.; Comai, L. A Y-chromosome-encoded small RNA acts as a sex determinant in persimmons. *Science* **2014**, *346*, 646–650. [CrossRef]
8. Knight, T.A. XII. An account of some experiments on the fecundation of vegetables. In a letter from Thomas Andrew Knight, Esq. to the Right Hon. Sir Joseph Banks, K. B. P. R. S. *Philos. Trans. R. Soc. Lond.* **1799**, *89*, 195–204. [CrossRef]
9. Ralls, K.; Frankham, R.; Ballou, J.D. Inbreeding and Outbreeding. *Encycl. Biodivers.* **2013**, *2*, 245–252. [CrossRef]
10. Barrett, S.C.; Hough, J. Sexual dimorphism in flowering plants. *J. Exp. Bot.* **2012**, *64*, 67–82. [CrossRef]
11. Barrett, S.C. Plant sex: Best to be bisexual when mates are scarce. *Curr. Biol.* **2021**, *31*, R298–R300. [CrossRef]
12. Sakai, A.K.; Weller, S.G. Gender and sexual dimorphism in flowering plants: A review of terminology, biogeographic patterns, ecological correlates, and phylogenetic approaches. In *Sexual and Gender Dimorphism in Flowering Plants*; Geber, M.A., Dawson, T.E., Delph, L.F., Eds.; Springer: Berlin/Heidelberg, Germany, 1999; pp. 1–31.
13. Ainsworth, C. Boys and Girls Come Out to Play: The Molecular Biology of Dioecious Plants. *Ann. Bot.* **2000**, *86*, 211–221. [CrossRef]
14. Akagi, T.; Charlesworth, D. Pleiotropic effects of sex-determining genes in the evolution of dioecy in two plant species. *Proc. R. Soc. Boil. Sci.* **2019**, *286*, 20191805. [CrossRef]
15. Bawa, K.S. Evolution of dioecy in flowering plants. *Annu. Rev. Ecol. Syst.* **1980**, *11*, 15–39. [CrossRef]
16. Renner, S.S.; Ricklefs, R.E. Dioecy and its correlates in the flowering plants. *Am. J. Bot.* **1995**, *82*, 596–606. [CrossRef]
17. Renner, S.S. The relative and absolute frequencies of angiosperm sexual systems: Dioecy, monoecy, gynodioecy, and an updated online database. *Am. J. Bot.* **2014**, *101*, 1588–1596. [CrossRef]
18. Charlesworth, D. Plant sex determination and sex chromosomes. *Heredity* **2002**, *88*, 94–101. [CrossRef]
19. Barrett, S.C.H. The evolution of plant reproductive systems: How often are transitions irreversible? *Proc. R. Soc. Boil. Sci.* **2013**, *280*, 20130913. [CrossRef]
20. Charlesworth, B.; Charlesworth, D. A Model for the Evolution of Dioecy and Gynodioecy. *Am. Nat.* **1978**, *112*, 975–997. [CrossRef]
21. Charlesworth, D. Theories of the Evolution of Dioecy. In *Gender and Sexual Dimorphism in Flowering Plants*; Geber, M.A., Dawson, T.E., Delph, L.F., Eds.; Springer: Berlin/Heidelberg, Germany, 1999. [CrossRef]
22. Barrett, S.C.H. The evolution of plant sexual diversity. *Nat. Rev. Genet.* **2002**, *3*, 274–284. [CrossRef]
23. Charlesworth, D. Does sexual dimorphism in plants promote sex chromosome evolution? *Environ. Exp. Bot.* **2018**, *146*, 5–12. [CrossRef]
24. Ming, R.; Bendahmane, A.; Renner, S.S. Sex Chromosomes in Land Plants. *Annu. Rev. Plant Biol.* **2011**, *62*, 485–514. [CrossRef]
25. Charlesworth, D. Plant sex chromosome evolution. *J. Exp. Bot.* **2013**, *64*, 405–420. [CrossRef] [PubMed]
26. Darwin, C.R. *The Different Forms of Flowers on Plants of the Same Species*; John Murray: London, UK, 1877.
27. Munné-Bosch, S. Sex ratios in dioecious plants in the framework of global change. *Environ. Exp. Bot.* **2015**, *109*, 99–102. [CrossRef]
28. Grant, V. Sexual selection in plants: Pros and cons. *Proc. Natl. Acad. Sci. USA* **1995**, *92*, 1247–1250. [CrossRef] [PubMed]
29. Sakai, A.K.; Wagner, W.L.; Ferguson, D.M.; Herbst, D.R. Biogeographical and Ecological Correlates of Dioecy in the Hawaiian Flora. *Ecology* **1995**, *76*, 2530–2543. [CrossRef]
30. Oliveira, P.E. Dioecy in the Cerrado vegetation of Central Brazil. *Flora* **1996**, *191*, 235–243. [CrossRef]
31. Webb, C.J.; Lloyd, D.G.; Delph, L.F. Gender dimorphism in indigenous New Zealand seed plants. *N. Z. J. Bot.* **1999**, *37*, 119–130. [CrossRef]
32. Bawa, K.S.; Opler, P.A. Dioecism in tropical forest trees. *Evolution* **1975**, *29*, 167–179. [CrossRef]
33. Givnish, T.J. Ecological constraints on the evolution of breeding systems in seed plants: Dioecy and dispersal in gymnosperms. *Evolution* **1980**, *34*, 959–972. [CrossRef]
34. Weller, S.G.; Sakai, A.K. Using Phylogenetic Approaches for the Analysis of Plant Breeding System Evolution. *Annu. Rev. Ecol. Syst.* **1999**, *30*, 167–199. [CrossRef]

35. Vamosi, J.C.; Otto, S.; Barrett, S.C.H. Phylogenetic analysis of the ecological correlates of dioecy in angiosperms. *J. Evol. Biol.* **2003**, *16*, 1006–1018. [CrossRef]
36. Wallnöfer, B. The Biology and Systematics of Ebenaceae: A Review. *Nat. Mus. Wien* **2001**, *103*, 485–512.
37. Santos, M.F.; Sano, P.T. Flora de Grão-Mogol, Minas Gerais: Ebenaceae. *Bol. Bot. Univ. São Paulo* **2004**, *22*, 93–95. [CrossRef]
38. Santos, M.F.; Sano, P.T. Flora da Serra do Cipó, Minas Gerais: Ebenaceae. *Bol. Bot. Univ. São Paulo* **2018**, *36*, 23–28. [CrossRef]
39. Duangjai, S.; Wallnöfer, B.; Samuel, R.; Munzinger, J.; Chase, M.W. Generic delimitation and relationships in Ebenaceae sensu lato: Evidence from six plastid DNA regions. *Am. J. Bot.* **2006**, *93*, 1808–1827. [CrossRef]
40. Samuel, R.; Turner, B.; Duangjai, S.; Munzinger, J.; Paun, O.; Barfuss, M.H.J.; Chase, M.W. Systematics and evolution of the Old World Ebenaceae, a review with emphasis on the large genusDiospyrosand its radiation in New Caledonia. *Bot. J. Linn. Soc.* **2019**, *189*, 99–114. [CrossRef]
41. Flora do Brasil. Jardim Botânico do Rio de Janeiro. 2020. Available online: http://floradobrasil.jbrj.gov.br/reflora/floradobrasil/FB7429 (accessed on 15 April 2021).
42. Wallnöfer, B. A revision of neotropical *Diospyros* (Ebenaceae): Part 10. *Nat. Mus. Wien* **2017**, *119*, 183–226.
43. Rebouças, N.C.; Cordeiro, L.S.; Araújo, R.S.; Ribeiro, R.T.M.; Loiola, M.I.B. Flora do Ceará, Brasil: Ebenaceae. *Rodriguésia* **2020**, *71*, e02122018. [CrossRef]
44. Sothers, C. *Diospyros cavalcantei* (Ebenaceae): A New Species from Amazonia. *Kew Bull.* **2000**, *55*, 471. [CrossRef]
45. De Campos, S.S.; Wittmann, M.T.S.; Veit, P.A.; Schwarz, S.F. Biologia Floral e viabilidade de pólen em cultivares de caquizeiro (*Diospyros kaki* L.) e *Diospyros virginiania* L. *Rev. Bras. Frutic.* **2015**, *37*, 685–691. [CrossRef]
46. Lopes, R.C. Ebenaceae Vent. do Estado do Rio de Janeiro. *Rev. Rodriguésia* **1999**, *50*, 85–107. [CrossRef]
47. Bawa, K.S. Patterns of flowering in tropical plants. In *Handbook of Experimental Pollination Biology*; Jones, C.E., Little, R.J., Eds.; Scientific and Academic Editions: New York, NY, USA, 1983; pp. 394–410.
48. Lloyd, D.G.; Webb, C.J. Secondary sex characters in plants. *Bot. Rev.* **1977**, *43*, 177–216. [CrossRef]
49. Forero-Montaña, J.; Zimmerman, J.K. Sexual dimorphism in the timing of flowering in two dioecious trees in a subtropical wet forest, Puerto Rico. *Caribb. J. Sci.* **2010**, *46*, 88–95. [CrossRef]
50. Bullock, S.H.; Bawa, K.S. Sexual dimorphism and the annual flowering pattern in *Jacaratia dolichaula* (D. Smith) Woodson (Caricaceae) in a Costa Rican rain forest. *Ecology* **1981**, *62*, 1494–1504. [CrossRef]
51. Rocca, M.A.; Sazima, M. The dioecious, sphingophilous species *Citharexylum myrianthum* (Verbenaceae): Pollination and visitor diversity. *Flora* **2016**, *201*, 440–450. [CrossRef]
52. Otárola, M.F.; Sazima, M.; Solferini, V.N. Tree size and its relationship with flowering phenology and reproductive output in Wild Nutmeg trees. *Ecol. Evol.* **2013**, *3*, 3536–3544. [CrossRef] [PubMed]
53. Delph, L.F.; Galloway, L.F.; Stanton, M.L. Sexual dimorphism in flower size. *Am. Nat.* **1996**, *148*, 299–320. [CrossRef]
54. Murphy, C.G. Interaction-independent sexual selection and the mechanisms of sexual selection. *Evolution* **1998**, *52*, 8–18. [CrossRef] [PubMed]
55. Foresto, E.B. Levantamento Florístico dos Estratos Arbustivos e Arbóreo de Uma Mata de Galeria em Meio a Compôs Rupestres no Parque Estadual do Rio Preto, São Gonçalo do Rio Preto, MG. Master's Thesis, Universidade de São Paulo, Instituto de Biociências, Departamento de Botânica, São Paulo, Brazil, 2018.
56. Stanton, M.L. Male-male competition during pollination in plant populations. *Am. Nat.* **1994**, *144*, S40–S68. [CrossRef]
57. Kudo, G. Flowering phenologies of animal-pollinated plants: Reproductive strategies and agents of selection. p. 139–158. In *Ecology and Evolution of Flowers*; Harder, L.D., Barrett, S.C.H., Eds.; Oxford University Press: New York, NY, USA, 1996; p. 370.
58. Frankie, G.W.; Baker, H.G.; Opler, P.A. Comparative phonological studies of trees in tropical lowland wet and dry forest sites of Costa Rica. *J. Ecol.* **1947**, *62*, 881–913. [CrossRef]
59. Grison-Pigé, L.; Bessière, J.-M.; Turlings, T.C.J.; Kjellberg, F.; Roy, J.; Hossaert-Mckey, M.M. Limited intersexmimicry of floral odour. *Ficus Carica. Funct. Ecol.* **2001**, *15*, 551–558. [CrossRef]
60. Bawa, K.S. Plant-pollinator interactions in tropical rain forests. *Annu. Rev. Ecol. Syst.* **1990**, *21*, 399–422. [CrossRef]
61. Renner, S.S.; Feil, J.P. Pollinators of Tropical Dioecious Angiosperms. *Am. J. Bot.* **1993**, *80*, 1100. [CrossRef]
62. Baker, H.G. "Mistake" pollination as a reproductive system with special reference to the Caricaceae. *Linn. Soc. Symp. Ser.* **1976**, *2*, 161–169.
63. Willson, M.F.; Ågren, J. Differential Floral Rewards and Pollination by Deceit in Unisexual Flowers. *Oikos* **1989**, *55*, 23. [CrossRef]
64. Otárola, M.F.; Rocca, M.A. Flores no tempo: A floração como uma fase da fenologia reprodutiva. In *Biologia da Polinização*, 1st ed.; Rech, A.R., Agostini, K., Oliveira, P.E.., Machado, I.C., Eds.; Projeto Cultural: Rio de Janeiro, Brazil, 2014; ISBN 978-85-68126-01-1.
65. Barth, F.G. *Insects and Flowers: The Biology of a Partnership*; Princeton University Press: Princeton, NJ, USA, 1985.
66. Schiestl, F.P. On the success of a swindle: Pollination by deception in orchids. *Naturwissenschaften* **2015**, *92*, 255–264. [CrossRef]
67. Knudsen, J.T.; Tollsten, L. Floral scent and intrafloral scent differentiation in *Moneses* and *Pyrola* (*Pyrolaceae*). *Pl. Syst. Evol.* **1991**, *177*, 81–91. [CrossRef]
68. Proctor, M.; Yeo, P.; Lack, A. *The Natural History of Pollination*; Timber Press Inc.: Portland, OR, USA, 1996.
69. Raguso, R.A. Why are some floral nectars scented? *Ecology* **2014**, *85*, 1486–1494. [CrossRef]
70. Piratelli, A.J.; Piña-Rodrigues, F.C.M.; Gandara, F.B.; Santos, E.M.G.; Costa, L.G.S. Biologia da polinização de *Jacaratia spinosa* (AUBL) adc. (caricaceae) em mata residual do sudeste brasileiro. *Rev. Bras. Biol.* **1998**, *58*, 671–679. [CrossRef]

71. Lenza, E.; Oliveira, P.E. Reprodução de Virola sebifera em mata mesofítica de Uberlândia, MG, Brasil. *Rev. Bras. Bot.* **2006**, *29*, 443–451. [CrossRef]
72. Ashman, T.-L. Sniffing out patterns of sexual dimorphism in floral Scent. *Funct. Ecol.* **2009**, *23*, 852–862. [CrossRef]
73. Aguiar, B.I.; Sebbenn, A.M.; Tarazi, R.; Vogado, N.O.; Morellato, L.P.C.; Tambarussi, E.V.; Moreno, M.A.; Pereira, L.C.S.M.; Montibeller, C.; Ferraz, E.M.; et al. Phenology, Seed Germination, and Genetics Explains the Reproductive Strategies of *Diospyros lasiocalyx* (Mart.) B. Wall. *Trop. Plant Biol.* **2020**, *13*, 23–35. [CrossRef]
74. Reekie, E.; Bazzaz, F.A. *Reproductive Allocation in Plants*; Elsevier Academic Press: Amsterdam, The Netherlands, 2005.
75. Chapin, F.S., III; Schulze, E.-D.; Mooney, H.A. The storage and economics of storage in plants. *Annu. Rev. Ecol. Syst.* **1990**, *21*, 423–447. [CrossRef]
76. Escobar, D.E.; Silveira, F.A.O.; Morellato, L.P.C. Timing of seed dispersal and seed dormancy in Brazilian savanna: Two solutions to seasonality. *Ann. Bot.* **2018**, *121*, 1197–1209. [CrossRef]
77. Batalha, M.A.; Martins, F.R. Reproductive phenology of the cerrado plant community in Emas National Park (Central Brazil). *Aust. J. Bot.* **2004**, *52*, 149–161. [CrossRef]
78. Koltunow, A.M.; Grossniklaus, U. Apomixis: A developmental perspective. *Annu. Rev. Plant Biol.* **2003**, *54*, 547–574. [CrossRef]
79. Bicknell, R.A.; Koltunow, A.M. Understanding apomixis: Recent advances and remaining conundrums. *Plant Cell* **2004**, *16*, S228–S245. [CrossRef]
80. Tomlinson, P. Family: Ebenaceae. In *The Botany of Mangroves*; Cambridge University Press: Cambridge, UK, 2016; pp. 242–243. [CrossRef]
81. Ikeda, K.; Sugiura, A. Apomictic seed formation from inter- and intra- specific crosses of diospyros lotus. *Acta Hortic.* **2003**, *601*, 209–211. [CrossRef]
82. Ashman, T.-L. The evolution of separate sexes: A focus on the ecological context. In *Ecology and Evolution of Flowers*; Harder, L.D., Barrett, S.H., Eds.; Oxford University Press: Oxford, UK, 2003; pp. 204–222, 390p.
83. Lloyd, D.G. The distributions of gender in four angiosperm species illustrating two evolutionary pathways to dioecy. *Evolution* **1980**, *34*, 123–134. [CrossRef]
84. Chagas, F.P.; Cabral, K.S.; Araújo, M.; Rocha, S.S. *Proposta de ampliação do Parque Estadual Serra Nova e do Talhado*; Instituto Estadual de Florestas: Belo Horizonte, Brazil, 1978.
85. Wallnöfer, B.; Ebenaceae in Lista de Espécies da Flora do Brasil. Jardim Botânico do Rio de Janeiro. 2015. Available online: http://floradobrasil.jbrj.gov.br/jabot/floradobrasil/FB17314 (accessed on 15 April 2021).
86. Marimon, B.S.; Felfili, J.M. Ethnobotanical comparison of "Pau Brasil" (*Brosimum rubescens* Taub.) forests in a xavante Indian and a non-xavante community in eastern Mato Grosso state, Brazil. *Econ. Bot.* **2001**, *55*, 555–569. [CrossRef]
87. Funch, L.S.; Harley, R.; Funch, R.; Giulietti, A.M.; Melo, E. *Plantas Úteis—Chapada Diamantina—Useful Plants*; RiMa Editora: Sao Paulo, Brazil, 2004; p. 187.
88. Forti, G.; Tambarussi, E.; Kageyama, P.; Moreno, M.; Ferraz, E.; Ibañes, B.; Vencovsky, R.; Mori, G.; Sebbenn, A. Low genetic diversity and intrapopulation spatial genetic structure of the Atlantic Forest tree, Esenbeckia leiocarpa Engl. (Rutaceae). *Ann. For. Res.* **2014**, *57*, 165–174. [CrossRef]
89. Silvério, D.V.; Lenza, E. Phenology of woody species in a typical cerrado in the Bacaba Municipal Park, Nova Xavantina, Mato Grosso, Brazil. *Biol. Neotrop.* **2010**, *10*, 205–218. [CrossRef]
90. Fournier, L.A. Un método cuantitativo para la medición de características fenológicas en árboles. *Turrialba* **1974**, *24*, 422–423.
91. Morellato, L.P.C.; Leitão-Filho, H.F.; Rodrigues, R.R.; Joly, C.A. Estratégias fenológicas de espécies arbóreas em floresta de altitude na Serra do Japi, Jundiaí, São Paulo. *Rev. Bras. Biol.* **1990**, *50*, 149–162.
92. Sarmiento, G.; Monasterio, M. Life forms and phenology. In *Ecosystems of the World: Tropical Savannas*; Bouliere, F., Ed.; Elsevier: Amsterdam, The Netherlands, 1983; pp. 79–108.
93. Cesário, L.F.; Gaglianone, M.C. Biologia floral e fenologia reprodutiva de *Schinus terebinthifolius* Raddi (Anacardiaceae) em Restinga do Norte Fluminense. *Acta Bot. Bras.* **2008**, *22*, 828–833. [CrossRef]
94. Kearns, C.A.; Inouye, D.W. *Techniques for Pollination Biologists*; University Press of Colorado: Niwot, CO, USA, 1993; p. 583.
95. Instituto Nacional de Meteorologia—INMET. Available online: http://www.inmet.gov.br/portal/index.php?r=home2/index (accessed on 20 September 2016).
96. Ohya, I.; Nanami, S.; Itoh, A. Dioecious plants are more precocious than cosexual plants: A comparative study of relative sizes at the onset of sexual reproduction in woody species. *Ecol. Evol.* **2017**, *7*, 5660–5668. [CrossRef] [PubMed]
97. Queenborough, S.A.; Mazer, S.J.; Vamosi, S.M.; Garwood, N.C.; Valencia, R.; Freckleton, R.P. Seed mass, abundance and breeding system among tropical forest species: Do dioecious species exhibit compensatory reproduction or abundances? *J. Ecol.* **2009**, *97*, 555–566. [CrossRef]

plants MDPI

Article

A Foundational Population Genetics Investigation of the Sexual Systems of *Solanum* (Solanaceae) in the Australian Monsoon Tropics Suggests Dioecious Taxa May Benefit from Increased Genetic Admixture via Obligate Outcrossing

Jason T. Cantley [1,2,*], Ingrid E. Jordon-Thaden [2,3], Morgan D. Roche [2,4], Daniel Hayes [2], Stephanie Kate [1] and Christopher T. Martine [2]

1 Department of Biology, San Francisco State University, San Francisco, CA 94132, USA
2 Biology Department, Bucknell University, Lewisburg, PA 17837, USA
3 Department of Botany, University of Wisconsin Madison, Madison, WI 53706, USA
4 Department of Ecology and Evolutionary Biology, University of Tennessee, Knoxville, TN 37996, USA
* Correspondence: cantley@sfsu.edu

Abstract: *Solanum* section *Leptostemonum* is an ideal lineage to test the theoretical framework regarding proposed evolutionary benefits of outcrossing sexual systems in comparison to cosexuality. Theoretically, non-cosexual taxa should support more genetic diversity within populations, experience less inbreeding, and have less genetic structure due to a restricted ability to self-fertilize. However, many confounding factors present challenges for a confident inference that inherent differences in sexual systems influence observed genetic patterns among populations. This study provides a foundational baseline of the population genetics of several species of different sexual systems with the aim of generating hypotheses of any factor—including sexual system—that influences genetic patterns. Importantly, results indicate that dioecious *S. asymmetriphyllum* maintains less genetic structure and greater admixture among populations than cosexual *S. raphiotes* at the same three locations where they co-occur. This suggests that when certain conditions are met, the evolution of dioecy may have proceeded as a means to avoid genetic consequences of self-compatibility and may support hypotheses of benefits gained through differential resource allocation partitioned across sexes. Arguably, the most significant finding of this study is that all taxa are strongly inbred, possibly reflective of a shared response to recent climate shifts, such as the increased frequency and intensity of the region's fire regime.

Keywords: dioecy; fire; population genetics; sexual system; Solanaceae

Citation: Cantley, J.T.; Jordon-Thaden, I.E.; Roche, M.D.; Hayes, D.; Kate, S.; Martine, C.T. A Foundational Population Genetics Investigation of the Sexual Systems of *Solanum* (Solanaceae) in the Australian Monsoon Tropics Suggests Dioecious Taxa May Benefit from Increased Genetic Admixture via Obligate Outcrossing. *Plants* **2023**, *12*, 2200. https://doi.org/10.3390/plants12112200

Academic Editors: Brenda Molano-Flores and James Cohen

Received: 23 February 2023
Revised: 21 May 2023
Accepted: 26 May 2023
Published: 2 June 2023

1. Introduction

During the 19th Century, Charles Darwin commented on challenges in understanding the genetic, demographic, and ecological implications of dioecy in angiosperms. In his seminal work *'Forms and Flowers'*, Darwin wrote *"There is much difficulty in understanding why hermaphrodite plants should ever have been rendered dioecious."* [1]. Dioecy is a sexual system wherein plants of a population are unisexual, with staminate and carpellate flowers on separate individuals. In *'Effects of Cross and Self Fertilization in the Vegetable Kingdom'*, Darwin postulated that the selective mechanism driving the evolution of dioecy was a benefit gained through the separation of male and female sexual function thereby alleviating resource allocation costs of both sexes on the same individual, particularly under stressful environmental conditions [2]. Within this framework, Darwin argued against the idea that any advantages gained via obligate outcrossing aiding in inbreeding avoidance were not as significant as improved sexual function due to better resource allocation partitioned across separate plants. This determination was made with the assumption that hermaphroditism (i.e., a sexual system in which individuals of a population exhibit only flowers with both male and female functionality, henceforth referred to in this paper as "cosexuality") had evolved prior to dioecy in angiosperms. Darwin was only partially correct. Since those initial hypotheses, subsequent studies have shown that both the costs of resource allocation between sexes and outbreeding advantages are important factors shaping the evolution of sexual systems, including dioecy, across different lineages of angiosperms [3].

Despite being a sexual system that can overcome some effects of inbreeding, dioecy is often discussed in terms of being an evolutionary 'dead-end' for sexual system evolution because reversions to a sexual system in which male and female function occurs on one individual seemed unlikely, particularly in animals [4]. For angiosperms, the dead-end hypothesis has been supported by the observation that many clades of dioecious taxa are less species rich than their sister taxa that are capable of self-fertilization, suggesting that dioecy might lead to increased extinction rates [5,6] Theoretically, an increased extinction rate or the genetic disadvantages of maintaining a dioecious sexual system could result from a necessity to maintain smaller spatial distributions of individuals in populations since (a) only females contribute to seed production and dispersal of progeny and (b) unisexuality requires the presence of a nearby partner of the opposite sex. Therefore, intrasexual competition for local resources among individuals could, by limiting geographic distributions and local abundance of individuals due to finite resources, have genetic consequences. This would not be the case for cosexual taxa since cosexual flowers exist on all individuals and resources are not in competition between sexes partitioned across separate individuals in different locations. To circumvent a loss of genetic diversity potentially inherent to dioecious sexual systems, dioecious taxa are associated with a number of correlated life history traits whose evolution may be the result of the reallocation of energy that would usually be used by the missing sex into other beneficial traits. Some examples include the evolution of larger fruits with many additional ovules [7] and shifts to wind pollination to bypass issues linked to limited local pollinator density [8,9]. Similarly, woodiness and the often-associated increased longevity of individuals are frequently correlated [7,10,11]. Increased longevity lengthens the duration that optimal combinations of genetic diversity persist in a population, which thereby statistically increases the chances of admixture and introgression of these genes into the future. Similarly, woodiness increases the chances of genetic mixing among populations as it can physically support both larger fruits with more ovules and an increased total number of fruits.

The focal lineage of this study, the "*S. dioicum* + *S. echinatum* Group" [12], is a set of woody *Solanum* taxa of the Australian monsoon tropics (Figure 1) that variously exhibit one of three different sexual systems: cosexuality, andromonoecy (i.e., individual plants with cosexual and staminate flowers borne in each inflorescence), or functional dioecy. Andromonoecious *Solanum* taxa have long been recognized and are most prevalent in the large subgenus *Leptostemonum* [13] of ~550 species [14]. Dioecy was first suggested for the Caribbean species *S. polygamum* Vahl in the 1700s [15], but it was not until the 1970s that additional dioecious *Solanum* taxa were reported [16,17]. As currently understood, dioecy is rare within *Solanum*. Roughly 1% of species, or ≈21 of the ca. 1400 currently described species [18], are dioecious [19–23]. There are an estimated four to six distinct evolutionary events that have led to dioecy in *Solanum* [21,24–27]. Of these, the largest radiation is from the Australian monsoon tropics, where 13 currently described dioecious species occur [12,23]. Australian taxa may represent either one or two transitions to dioecy and are closely related to a radiation of ca. 15 andromonoecious taxa [28,29] and a clade of ca. 15 cosexual taxa, including the widespread and highly variable *Solanum echinatum* R. Brown [12,29,30] from which *S. raphiotes* is a recent segregate taxon [31]. As is the case with all instances of dioecy in *Solanum* (first formally described by for the Mexican species *S. appendiculatum* Dunal [32]), the condition among the Australian taxa is best described as "functional" dioecy, although these species appear morphologically to be androdioecious [16,17]. The cryptic nature of this system among these Australian taxa manifests itself as female inflorescences which are reduced to a solitary and morphologically appearing cosexual flower bearing anthers producing pollen. However, pollen developed in these anthers is inaperturate and ingerminable [33]. Meanwhile, male plants bear inflorescences that present as a simple monochasial helicoid type cyme consisting of a few to dozens of staminate flowers, all producing porate and germinable pollen [16,17,34]. Additionally, functionally female flowers have larger corollas than those of their male counterparts [33] but produce pollen of lower nutritional quality than the porate pollen produced by staminate flowers [35].

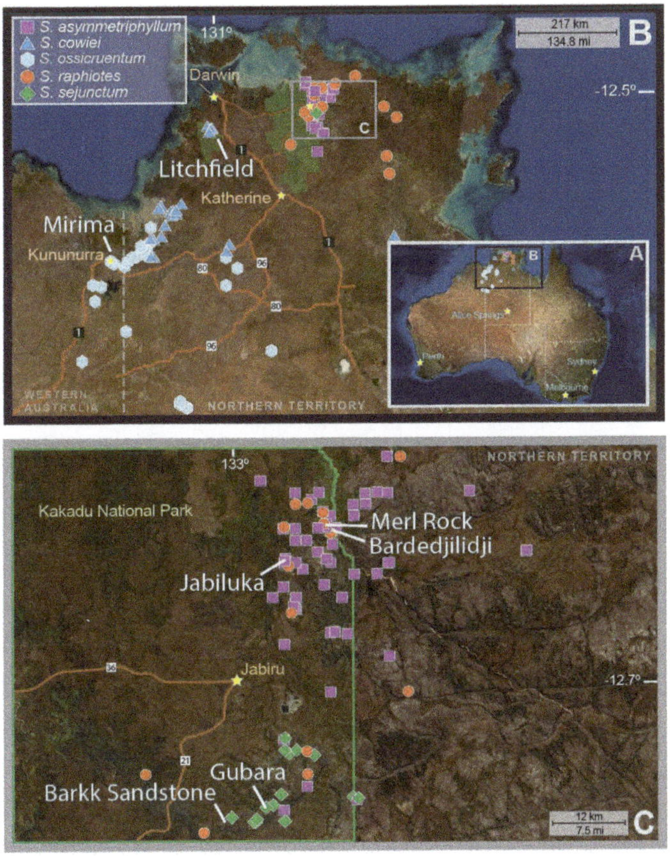

Figure 1. Map indicating known herbarium collection localities of each of the five species for which population genetics data were analyzed in this study. A view of collections within (**A**) Australia, (**B**) the northwest region, and (**C**) a closer view of Kakadu National Park and the Arnhemland Plateau. White lines and text indicate population collection localities used in this study (note that Merl Rock and Bardedjilidji populations are discrete and separated by approximately 2.5 km despite overlapping collection icons). Stars indicate human population centers along roads (denoted by orange lines). Green polygons in (**B**) indicate Litchfield National Park and Kakadu National Park.

It remains equally likely that dioecy among Australian *Solanum* taxa has arisen from andromonoecious or cosexual ancestry [12,29,33]. Given the rarity of dioecy in the genus *Solanum*, the theoretical work of Martine and Anderson [28] plus Anderson and Symon [33] details necessary transitional steps among *Solanum* sexual systems and the formulation of hypotheses regarding the evolution of dioecy: (a) populations must be small and widely separated from one another, (b) population sizes must be effectively limited due to individual 'plants' of a population being the result from only a few ramets of a few surviving genets, and (c) the pollinator fauna must be small and have relatively local ranges such that pollination between widespread populations is infrequent to non-existent. All three of these factors theoretically increase the probability of inbreeding among closely related individuals through self-fertilization. Therefore, these three factors could serve as the selective pressures promoting a physical separation of sexes in order to compensate for the effects of inbreeding and resource allocation. In turn, inducing a mechanism of even partial female or male sterility in a population could play a role in mitigating the effects of inbreeding by allowing for an increase in genetic mixing among andromonoecious or cosexual individuals. Moreover, if outcrossing and sex-related resource reallocation are indeed important to populations of individuals, then dioecy could hypothetically be more efficient in achieving increased genetic mixing among individuals compared to taxa that cannot self-fertilize [36].

No studies have investigated the population genetics of sympatrically occurring Australian *Solanum* taxa directly in the wild. Because sympatric taxa in our study group often have different sexual systems, the group is ideal for testing hypotheses related to the theoretical implications of dioecy in comparison to taxa in which self-fertilization is possible. For dioecy (and perhaps andromonoecy), the aforementioned hypotheses suggest these taxa at some point in time should have quantifiable genetic advantages. Specifically, we hypothesize that patterns observed in population genetics analyses of dioecious species should appear as a greater amount of genetic diversity in populations, exhibit less inbreeding, and be less genetically structured. However, a suite of factors, which are difficult to isolate in situ, present significant challenges for precise inferences on the role sexual systems play in shaping the genetic landscape of taxa. These include microhabitat niche preferences, the duration of site occupation, the degree of initial genetic diversity at the time of population founding events, the typical life spans of each taxon, and the potential for hybridization. Therefore, the primary aim of this investigatory study was to explore population-level genetics of sympatrically occurring *Solanum* taxa occupying the extreme ends of the sexual system variation found in the "*S. dioicum* + *S. echinatum* Group": dioecy and cosexuality. In doing so, we hope to establish a foundational understanding of how these taxa, operating in the wild, may (or may not) fit into existing theoretical frameworks of sexual system evolution, as well as contribute to the broader understanding of why transitions to dioecy, though rare in number, are still a widespread occurrence among angiosperms.

2. Results

2.1. Sequencing and Ipyrad Filtering

Across the non-monophyletic assemblage of the five taxa sequenced, a total of 308 variant bi-allelic loci across 193 individuals, representing 10 total populations, were retained from the original 360 original samples sent for sequencing. The removal of data was necessary to produce a 'hard-filtered' dataset in which individual samples and loci were retained only when passing the following described strict filtering steps. First, raw sequences were demultiplexed in ipyrad [37], wherein seven samples were removed as they had <200,000 raw sequencing reads. Next, loci were removed if they were not shared across at least 100 of the 358 samples. This resulted in a total of 3288 loci retained. Individuals with >60% missing data were then removed, reducing the number of individuals to 193 and 1458 loci. Following this, 763 loci were removed since they were <10 bp apart from each other. Removal of 50 invariant loci was performed through the application of a 1% Minor Allele Frequency (MAF) threshold. An additional 166 loci were removed as they could be associated with linkage disequilibrium. When filtering for loci that were within 1 kb base pairs of each other and those with an R_2 value > 0.8, a final 239 linked loci were removed.

2.2. Summary Statistics

Summary statistics, geographic locations of sampled populations, and voucher information of each of the five species and all 10 populations are given in Table 1. For species represented by more than one population, species-level statistics are provided as well as statistics for each population. In all cases, regardless of hierarchical level, H_o was determined to be significantly less than H_e. Bartlett's tests corroborate the statistical significance suggesting that heterozygosity was significantly less than expected at Hardy–Weinberg equilibrium. The F_{IS} values for all five species and 10 populations were all >0.9, indicating that homozygosity is acutely high and suggesting significant levels of inbreeding for all taxa (Table 1).

Table 1. The sexual system, geographic location, reference vouchers, and summary statistics of the five species and 10 populations of this study.

Species	Sexual System	Population	*n*	Latitude	Longitude	Reference Voucher (Herbarium)	H_o	H_e	F_{IS}
S. ossicruentum	Dioecious	Mirima	9	15.76378	128.75175	CTM 4011 (BUPL)	0.0014	0.0924	0.9849
S. cowiei	Dioecious	Florence Falls	9	13.21958	130.73645	CTM 1751 (BUPL)	0.0004	0.1101	0.9968
S. sejunctum	Dioecious		43				0.0000	0.1215	1.0000
		Gubara Pools	24	12.82928	132.8756	CTM 1739 (BUPL)	0.0000	0.1288	1.0000
		Barkk Sandstone	19	12.85907	132.81788	CTM 1729 (BUPL)	0.0000	0.0953	1.0000
S. asymmetri-phyllum	Dioecious		49				0.0004	0.1582	0.9973
		Merl Rock	32	12.42622	132.96022	CTM 1702 (BUPL)	0.0001	0.1333	0.9993
		Bardedjilidji	16	12.43727	132.96803	CTM 1721 (BUPL)	0.0006	0.1825	0.9967
		Jabaluka	12	12.47702	132.90275	CTM 1700 (BUPL)	0.0011	0.1286	0.9920
S. raphiotes	Cosexual		83				0.0009	0.1778	0.9951
		Merl Rock	47	12.42622	132.96022	CTM 1709 (BUPL)	0.0004	0.1735	0.9977
		Bardedjilidji	24	12.43727	132.96803	CTM 1714 (BUPL)	0.0004	0.1439	0.9974
		Jabaluka	12	12.47702	132.90275	CTM 737 (CONN)	0.0038	0.1043	0.9635

2.3. *Pairwise-F_{ST} and AMOVA*

Heatmaps of Weir–Cockerham adjusted [38] Pairwise-F_{ST} values were generated at the two hierarchical levels (species, populations) to visualize genetic relationships (Figure 2). When considering species, all Pairwise-F_{ST} values were >0.8, suggesting that the five species are strongly differentiated from each other. At the population level, the three species represented by more than one population (*S. asymmetriphyllum*, *S. raphiotes*, and *S. sejunctum*) each separately have lower Pairwise-F_{ST} values among their own populations, pointing to a moderate to high degree of intraspecific genetic similarity. However, some genetic structure is noted at the population level. Importantly, at the three locations where more than one species was sampled (i.e., Jabiluka, Merl Rock, and Bardedjilidji), Pairwise-F_{ST} values were >0.8, indicating strong genetic structure separates the sympatrically occurring populations of different species with different sexual systems.

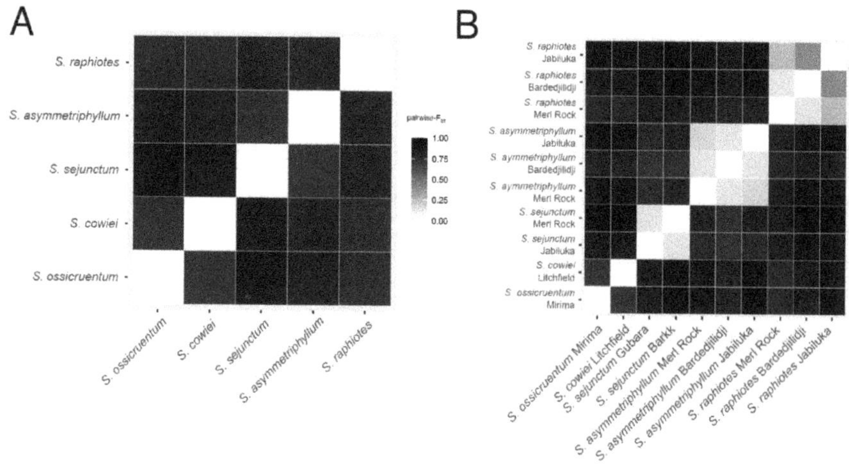

Figure 2. Heatmap of Weir–Cockerham-adjusted Pairwise-F_{ST} values among (**A**) the five Australian *Solanum* taxa of this study and (**B**) among all 10 sampled populations from 193 total individuals and 308 bi-allelic single-nucleotide polymorphisms.

Echoing the findings of the Pairwise-F_{ST} values, the results of AMOVA analyses confirm a pattern of genetic diversity with strong species-level structuring and less structure among the populations of species represented by more than one population. Genetic diversity was most pronounced at the species level, where 92.5% of the variance among samples occurs. More variation was found within populations (5.62%) than between populations of a species (1.61%), indicating little genetic diversity among different populations of a species.

2.4. PCA and DAPC Analyses

The optimal number of *k*-means clusters useful for describing the data ranged between 8 and 10 as determined by a series of the smallest obtained BIC values. This range parallels the findings of both the AMOVA and Pairwise-F_{ST} statistical analyses. More precisely, the upper limit of *k*-means clusters (10) matches the total number of a priori defined populations, while the lower limit of 8 reflects less genetic structuring among populations of the three species represented by more than one population. The optimal number of principal components (PCs) identified using an *a*-score spline interpolation approach followed by cross-validation was nine. With nine PCs and three discriminant analysis eigenvalues retained, 90% of the variance was represented in this optimally parameterized DAPC. A PCA scatter plot and a compoplot (i.e., a structure-like bar plot) were generated to visualize each individual's proportional assignment to the 10 genetic clusters (Figure 3). We chose to visualize 10 *k*-means clusters to show the finest scale pattern observed among the populations.

The DAPC structure-like bar plot visualizations (Figure 4) reveal that all five species occupy mutually exclusive *k*-means clusters without a signal of admixture among species, even at the three localities where two species with different sexual systems occur sympatrically. Individuals of two species—*S. ossicruentum* and *S. cowiei*—each represent a unique *k*-means cluster, while the other three species—*S. asymmetriphyllum*, *S. sejunctum*, and *S. raphiotes*—are each represented by two or three *k*-means clusters, reflecting migration among these species' populations and pointing to a strong shared ancestry. Both *S. asymmetriphyllum* and *S. raphiotes* share a pattern across the three sites where they occur sympatrically. For these two species, Jabiluka reflects a slightly higher level of genetic distinctiveness than Merl Rock and Bardedjilidji localities, which are similar to each other in their proportional assignments of the three *k*-means clusters assigned to each species.

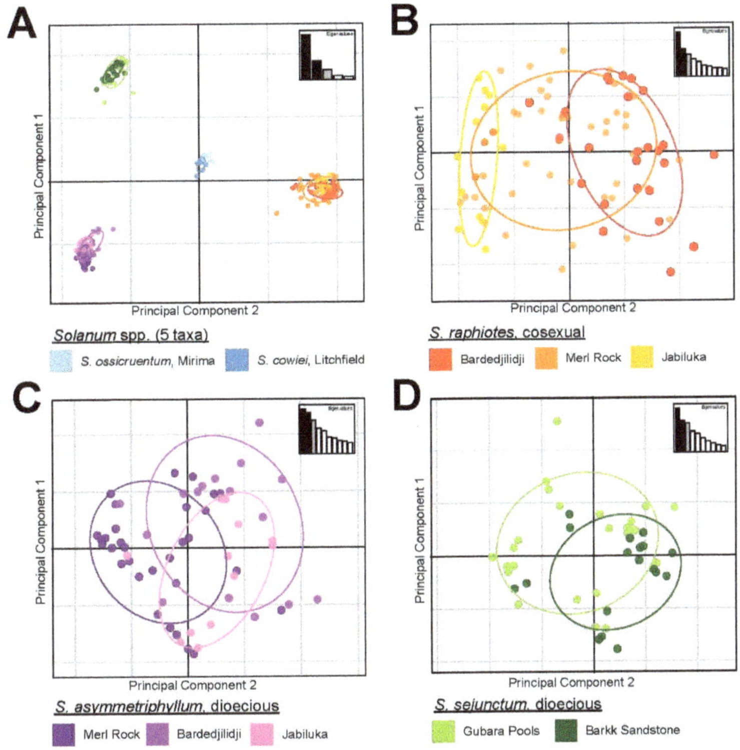

Figure 3. Multivariate analysis visualizations of all five species representing all 10 genetically determined populations of this study from 212 total individuals and 308 bi-allelic single-nucleotide polymorphisms. (**A**) A PCA scatterplot comparison of PC1 and PC2 for all five taxa in this study. Colored points and their corresponding labels indicate species and population. (**B**) PCA for *S. raphiotes*. (**C**) PCA for *S. asymmetriphyllum*. (**D**) PCA for *S. sejunctum*.

When investigating the genetic structure of populations on a more granular scale using sub-sampled datasets obtained using the repool function of adegenet, a number of additional insights were possible from PCA and DAPC analyses. First, three sub-sampled datasets investigated the within-species relationships of populations. For dioecious *S. asymmetriphyllum* and cosexual *S. raphiotes*, which sympatrically occur at the same three geographic population locations, these finer scale datasets corroborate that both species occupy three *k*-means clusters, with each cluster loosely representing each population locality. However, some patterns of relationships are different between the two species. For *S. raphiotes*, the PCA scatterplot indicates that most variation captured in PC1 represents differences among populations. The distribution of individuals from populations along PC2 is similar, but Merl Rock individuals occupy a wider distribution (Figure 3B). Additionally, Merl Rock has the widest span of the x-axis, suggesting it maintains the most variation of the three populations. *Solanum raphiotes* individuals of Jabiluka and Bardedjilidji do not overlap, but the wider variation of Merl Rock overlaps with both. The DAPC bar plot visualization suggests a signal of gene flow from Jabiluka into Merl Rock and Bardedjilidji populations because of the lack of shared genotypes among the individuals between the two populations (Figure 4B). However, the reciprocal signal of Jabiluka as a sink for gene flow is less prominent as only three individuals are admixed with less than a 0.03 proportional *k*-means cluster assignment from either Merl Rock or Bardedjilidji. *Solanum asymmetriphyllum* differs from *S. raphiotes* in these same locations in several ways. First, the PCA suggests that populations of *S. asymmetriphyllum* are less genetically distinct than they are for *S. raphiotes*, as evidenced by a larger number of admixed individuals (Figure 4C), especially in the population from Bardedjilidji. Unlike the strong genetic structure of the *S. raphiotes* population at Jabiluka, the DAPC of *S. asymmetriphyllum* indicates that Jabiluka is both a stronger source and sink of gene flow. There is also greater exchange between Merl Rock and Bardedjilidji. Overall, (i) the total number of admixed individuals, (ii) an overall greater proportional assignment of the Jabiluka *k*-means cluster to individuals of *S. asymmetriphyllum* at Merl Rock and Bardedjilidji, and (iii) the observation that Jabiluka is both a more significant source and sink of genetic diversity suggests that migration between populations of *S. asymmetriphyllum* is more readily achieved and subsequently maintained than it is for *S. raphiotes*.

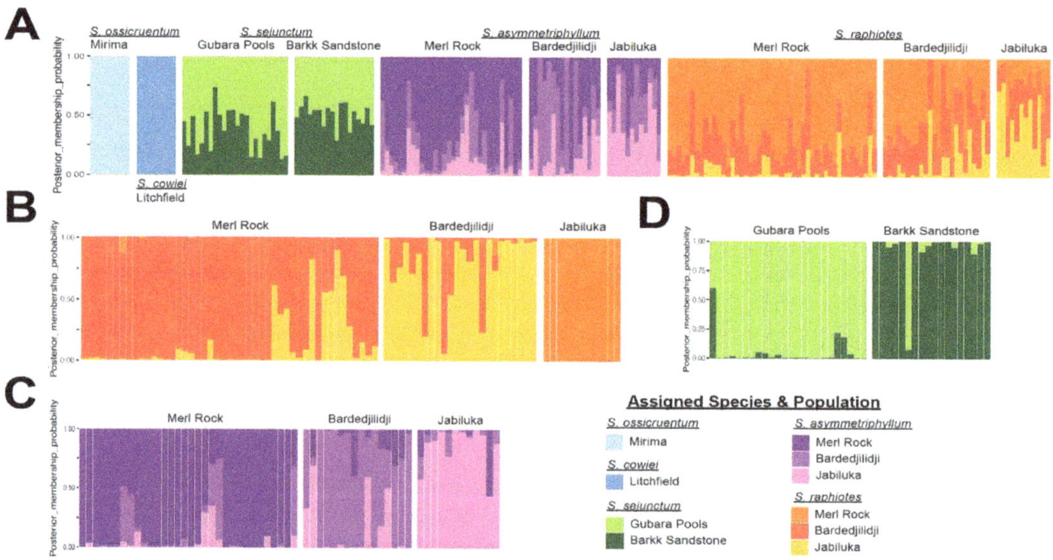

Figure 4. Results of multivariate DAPC analyses for all species (**A**) and sub-sampled analyses of the three individual species that are represented by more than one population (**B–D**). Each panel visualizes a *k*-means DAPC structure-like bar plot. This visualization of the DAPC retains the first three eigenvalues and the optimal nine PCs. Vertical bars represent the proportional assignment of each individual to one or several of the eleven *k*-means clusters, as indicated by colors in the legend. Species and populations are as indicated along the x-axis.

Solanum sejunctum is represented by only two populations, Gubara Pools and Barkk Sandstone. In the DAPC analysis, each population is proportionally represented, predominantly by their own unique *k*-means cluster; however, each maintains a weak signal of admixture from the other population (Figure 4C). Nearly half of the individuals from each population are admixed. One individual in each population has a greater than 0.50 proportional assignment of the other population's *k*-means cluster, but all other admixed individuals have proportional assignments of the other population's *k*-means cluster at levels less than 0.25. When assessing the placement of individuals in the scatterplot of PC1 vs. PC2, individuals of populations overlap partially; however, there is obvious clustering by a priori population assignment. Generally speaking, *S. sejunctum* exhibits a similar pattern to *S. asymmetriphyllum* and *S. raphiotes* in terms of the number of admixed individuals (Figure 4D).

Side-by-side comparisons of species pairs that occur sympatrically in PCA and DAPC analyses yield interesting insights (Figure 5). First, no admixture is observed between either *S. asymmetriphyllum* or *S. raphiotes*. In both repooled datasets, PCA analyses separate species along the x-axis of PC1, indicating this is where the majority of variation is located (between species). Variation along the y-axis of PC2 indicates the variation within individuals, and populations of the species echo the patterns observed in the species-specific repooled datasets. Overall, cosexual *S. raphiotes* occupies more variation than dioecious *S. asymmetriphyllum*.

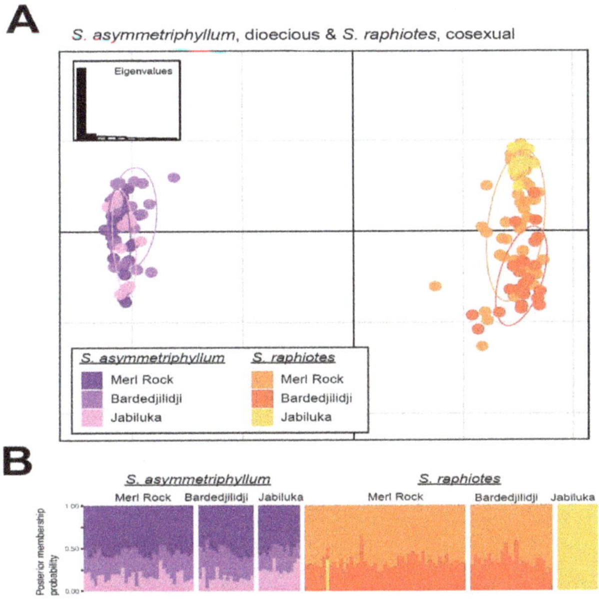

Figure 5. Visualizations of multivariate analyses of the three locations with sympatrically occurring *Solanum* taxa with different sexual systems. The top panel (**A**) is a PCA scatterplot and below (**B**) is the DAPC structure-like bar plot for dioecious *S. asymmetriphyllum* and cosexual *S. raphiotes*, each from Merl Rock, Bardedjilidji, and Jabiluka.

2.5. Isolation by Distance

Mantel tests were performed for *S. asymmetriphyllum* and *S. raphiotes* to test for an isolation by distance (IBD) model. Mantel tests for both taxa indicated that the genetic structure of these populations does not follow a classic IBD model. Simulated *p*-values of 0.3334967 and 0.5000145 were recovered for *S. asymmetriphyllum* and *S. raphiotes*, respectively. Local density measured using a 2-dimensional kernel density estimation reveals that Jabiluka represents a 'distant patch' rather than clinal variation, as would be expected in a classic IBD model. One difference is noted between the populations of the two species. The variation in *S. asymmetriphyllum* at Bardedjilidji and Merl Rock appears to be one genetic cluster that is clinal, but *S. raphiotes* retains greater genetic separation among these localities.

3. Discussion

3.1. Genetic Diversity and Structure in Dioecious Australian Solanum

Our findings offer support for both the avoidance of inbreeding and resource allocation hypotheses for the selective pressures promoting the evolution of dioecy in Australian *Solanum*. However, the patterns of population structure and diversity recorded for four different dioecious taxa suggest (i) a pattern for *S. asymmetriphyllum* with reduced variation compared to that of the sympatrically occurring cosexual taxon *S. raphiotes* (though with widespread admixture) and (ii) two dioecious taxa, *S. ossicruentum* and *S. cowiei*, with limited variation. Patterns inferred for *S. asymmetriphyllum* and potentially for *S. sejunctum* leave open the possibility that dioecy may have evolved as a mechanism to avoid the genetic cost of inbreeding in dioecious Australian *Solanum* taxa. A direct comparison of the genetic structure of dioecious *S. asymmetriphyllum* and cosexual *S. raphiotes* at three co-occurring locations (Jabiluka, Merl Rock, and Bardedjilidji) reveals that migration between populations of *S. asymmetriphyllum* is more readily achieved (thus genetic diversity is decreased) and subsequently maintained than it is currently for *S. raphiotes* (which shows higher genetic diversity). All analyses conveyed a similar finding; across three localities, *S. asymmetriphyllum* and *S. raphiotes* were similar in overall variance and levels of inbreeding, but *S. asymmetriphyllum* importantly maintains more admixture across all populations. This pattern suggests that symmetrical migration among populations of *S. asymmetriphyllum* has occurred and was subsequently maintained. We can therefore conclude that the gene pools of this dioecious taxon are continuously being mixed, resulting in decreasing overall genetic diversity. Comparatively, *S. raphiotes* maintains less admixture at the same population locations, and the pattern of migration is asymmetrical. The gene pools of this cosexual taxon indicate that the presence of balanced male-to-female floral parts requires less migration between populations to maintain the species. This, therefore, results in higher genetic diversity between populations as fewer genes are shared over time. Specifically, the *S. raphiotes* population at Jabiluka appears to currently be only a source of introgressive gene flow to Merl Rock and Bardedjilidji. Given that H_o is significantly lower than expected and F_{IS} values are elevated to similar levels across both taxa, it is possible that the admixture observed among *S. asymmetriphyllum* populations represents a beneficial mechanism of dioecy that helps avoid genetic consequences of self-fertilization through introgression via obligate outcrossing and maintenance of interspecific hybrids from nearby populations. Interestingly, the populations of both *S. asymmetriphyllum* and *S. raphiotes* found at Bardedjilidji seem to have the most diverse genetic makeup; perhaps this is a function of sharing the same local suite of pollinators and/or seed dispersers, but we cannot be certain. However, an important caveat to the conclusions drawn above is that one must assume demographic history at the site is similar for both species to allow for these comparisons. Demographic differences, such as time of site occupation (i.e., founder events) or recent bottleneck events, and even ecological niche preferences could be influential factors driving the different patterns observed between these two species.

While fundamental data regarding certain life history traits and site demographics remain unknown and beyond the scope of this study, novel insights help sharpen the evolutionary understanding of selective pressures that may have shaped the genetic landscape of dioecious vs. cosexual taxa of the Australian monsoon tropics. Several known correlates of dioecy are differentially expressed in these taxa, including pronounced secondary growth resulting in larger and taller individuals, larger fruits each with hundreds of seeds, larger female corollas, and more conspicuous male floral displays [34,39]. Additional factors shaping population structure and gene flow among populations—while not known to be necessarily correlated with dioecy—may also be important, such as inflorescences that are elevated higher above the ground, greater fecundity via an increased number of fruits per individual per season, and the potential exclusion of smaller vertebrate dispersal vectors resulting from larger and heavier fruit and seed sizes. There are also abiotic factors, such as soil chemistry, fire, and precipitation regimes (i.e., abiotic environmental stochasticity), spatiotemporal patch dynamics, and the finite availability of suitable habitat. However, our clearest inference detailing the evolutionary strategy for dioecy in the Australian monsoon tropics is for three Kakadu dioecious *Solanum* (KDS): *S. asymmetriphyllum*, *S. sejunctum*, and the southern undescribed taxon *S.* 'sp. Deaf Adder' [40]. KDS taxa are each other's closest relatives [12,30] and occupy very similar ecological niches. All have the intriguing ability to produce clonal genets arising from underground rhizomes emerging through cracks and fissures of dissected sandstone [34,41]. Underground rhizomes of KDS taxa are not unique for Australian *Solanum* [42,43], but it seems KDS and closely related dioecious taxa are more strongly reliant upon this ability to regenerate post fire, hinting at a strategy for long-term site occupancy in the finite locations of suitable dissected sandstone habitat. Although the average longevity of KDS genets is not known, in situ observations and greenhouse experiments suggest they likely persist many generations longer than most cosexual Australian *Solanum* taxa of the focal lineage [34]. An additional factor of consideration regarding increased longevity as part of the evolutionary strategy of KDS taxa must also include the formation of a genetically diverse soil seed bank. Fruit-producing KDS plants annually produce thousands, if not tens of thousands, of seeds that lay dormant in the soil until optimal conditions are met for germination [33,34]. The genetic structure of this seed bank theoretically represents half maternal genetics, and when the mother plant eventually senesces, perhaps following fire, competition among seedlings should allow for its replacement by the most optimal combination of beneficial alleles stored in the bank. Consideration of these three factors—increased longevity, prolific post-fire resprouting, and comparatively enormous seed banks—may have evolved as a result of newly gained flexibility in resource allocation correlated with the evolution of dioecy aiding in increased reproductive assurance and may explain the pattern of maintained admixture observed in our data.

In contrast to the wider variation observed for dioecious *S. asymmetriphyllum* and *S. sejunctum*, populations of dioecious *S. cowiei* and *S. ossicruentum* appear to be less genetically diverse (despite the lack of direct comparisons with sympatric cosexual taxa), though similar patterns of low heterozygosity and high F_{IS} values were found. Several factors could explain this contrasting pattern. Both taxa may still fundamentally represent the same evolutionary strategy as KDS taxa as they are noted to have many of the same adaptations, such as large fruits with hundreds of seeds, underground rhizomes allowing for post-fire resprouting, and large seed banks. Some notable differences among these taxa are simply the locations that they inhabit—both *S. cowiei* and *S. ossicruentum* occur as many fragmented populations across a much wider and more western distribution [20,22], which may have had different stochastic environmental conditions. In particular, both the precipitation and fire regimes of these areas could have differentially caused bottlenecks or led to recent founding events different from those of KDS taxa. Furthermore, *S. cowiei* and *S. ossicruentum* may be descendants from an independent evolution of dioecy due to differential life history traits compared to the KDS sandstone endemic taxa, thus allowing them to evolve along divergent paths and ultimately resulting in the reduced variance observed.

From this study, it seems that the genetic status of dioecious Australian *Solanum* could support both the avoidance of inbreeding and resource allocation hypotheses for the selective pressures promoting the evolution of dioecy; however, more research is needed. These taxa are an ideal system in which to investigate how variance among correlated dioecious life history traits variously contributes to resilience to environmental stochasticity. For example, some traits of *S. cowiei* and *S. ossicruentum* are not as pronounced and may have had less of an evolutionary benefit for maintaining genetic diversity when compared to KDS taxa. Obvious differences include smaller, dry, bony fruits with tiny seeds in *S. ossicruentum* [22] and the growth habit of *S. cowiei* that includes the production of rhizomes directly in sand at the base of sandstone boulders, which may offer less, or at least differential, protection from fire and herbivory [20]. In short, it may be that differences across a suite of life history traits, demographics, and environmental stochasticity of a heterogeneous landscape may have led to the wider variation (and assumedly greater evolutionary fitness) observed for KDS taxa.

3.2. Insights from a Shared Background of Elevated Inbreeding

Some of the most striking results of this study are the ubiquitously elevated levels of homozygosity and the high values of the inbreeding co-efficient (F_{IS}) across all taxa. This indicates that inbreeding is an important selective pressure for all taxa. It is suggested that heterozygotes should be selected for over homozygotes on account of 'hybrid-vigor', which should benefit an individual's evolutionary fitness in comparison to homozygotes. Moreover, because homozygotes are thought to perpetuate deleterious alleles in a population, homozygotes are subject to the effects of inbreeding depression [44]. Evidence for the selective favoring of heterozygotes is supported in several studies through the demonstration of differential heterozygosity among plants of different size classes or generations [44,45], though no such pattern is observed here for *Solanum*. An obligately outcrossing sexual system of dioecy should theoretically mitigate the effects of inbreeding (at least at the time of its evolution), but mitigation can only persist in large populations of many individuals that allow for outcrossing to occur among more distantly related individuals rather than outcrossing with siblings, which would subsequently increase homozygosity in the population.

Our results raise the question as to why inbreeding and homozygosity are so pronounced in our data. Our data only provide a snapshot in time of the life histories of these *Solanum* taxa. These species differ from each other in several ways (sexual system, longevity, fecundity, size, specific ecological niche preference, etc.), but all have evolved and adapted to the climatic processes of the Australian monsoon tropics [46]. The AMT is a dynamic biome that has had a fluctuating climate for many millions of years. With the Pliocene closure of the Isthmus of Panama 3.0–2.5 Ma [47,48], the modern circulation patterns of Earth's oceans were established and several cycles of glaciation–interglaciation followed. During warmer interglacial periods, Australia became more arid overall, though in the north the monsoon cycles strengthened, bringing more precipitation to the region. This would have resulted in the production of more biofuel, effectively changing the intensity and frequency of fires. One product of this was the emergence of fires as a principal ecological process across the AMT [49,50]. However, the once much more contiguous rainforests of the region contracted as the overall aridification of the continent commenced, resulting in intensified drought periods following periodic monsoonal precipitation events. Aridity in the Kimberley region was more pronounced, and it is suggested that the lower biodiversity observed there today is in part a result of a higher extinction rate driven by these intensified droughts in comparison to the sandstone flora of the Top End in the Northern Territory [46,51]. During periods of glaciation, global temperatures were cooler, resulting in weakened monsoon cycles with less precipitation and less accumulation of organic material to fuel fires. As a result, fires may not have been a significant abiotic pressure shaping the ecosystems during these glacial periods [46].

AMT fires have been characterized as likely to have been infrequent and of higher intensity prior to human arrival [49,52,53]. These landscapes were host to far fewer fire-adapted species, a protracted savanna of fire-promoting trees and grasses, widespread evergreen dry forests, and more contiguous rainforest habitat. All *Solanum* taxa of this study are associated with patchily distributed sandstone escarpment habitats existing throughout a widespread and fire-prone savanna of open eucalypt woodland on plains of oligotrophic sandy soils [22,34]. This heterogenous matrix of habitats influenced by stochastic processes of fire provides context into the finding that populations of *S. asymmetriphyllum* and *S. raphiotes* were not found to be isolated by distance. Many taxa of the sandstone flora have narrow ranges of specific habitat restricted to deep gorges, steep escarpments, and cliffs [54]. Today, sandstone habitats harboring *Solanum* taxa range from *Allosyncarpia* (Myrtaceae)-dominated rainforest relic patches to xeric heathlands—habitats that are partially protected from fires due to the topographic complexity of their terrain. As with many endemic sandstone taxa, AMT *Solanum* are thought to have evolved strategies for survival with fire. For example, although KDS taxa can resprout following fire via underground rhizomes or germinate post fire from large seedbanks, they also are largely restricted to these partially fire-protected refugia.

The arrival of the first humans to the AMT ushered in a regime of frequent, small, and low-intensity fires that likely contributed to an expansion of fire-tolerant and fire-adapted species coincident with the retraction of fire-sensitive taxa to habitat refugia [55–57]. Much more recently, human alterations to the environment have initiated a stark shift in the regional fire regime [58], which is likely a major and alarming factor in the abundant homozygosity and inbreeding we have documented in this study. Beginning with the arrival of Europeans less than 200 years ago, an unambiguously different fire regime now dominates the Top End, characterized by fires that are more extensive, frequent, and intense than occurred previously under the traditional management of indigenous peoples [54–57]. Consequently, the AMT has seen a recent expansion of grassy understories. In turn, this produces more flammable biomass, promoting more intense and frequent fires [59]. Throughout this rapid shift in regional fire regime (<200 years), fire-tolerant *Solanum* taxa sensitive to higher fire temperatures at longer durations would become more and more restricted to many fragmented refugial habitats and forced to primarily exchange genes with small numbers of individuals in close proximity, increasing the possibility of newly elevated levels of inbreeding.

The extreme inbreeding and homozygosity observed in these Australian *Solanum* taxa fits with previous research indicating that anthropogenically driven fire regime changes over the last two centuries are directly responsible for significant biodiversity loss and a contraction of suitable habitat for many taxa across sandstone environments [54,60–62]. The two habitats with the highest endemism—heathlands and monsoon rainforests—are documented as having ongoing decline in the abundance of fire-sensitive taxa [56,63–65]. Research indicates that over 40% of sandstone vegetation in Kakadu National Park alone had been burnt over a 14-year period from 1980 to 1994 at frequencies of at least one year in every three years [54]. This work found the risk of extirpation due to fire was particularly acute for obligately reseeding species such as *Solanum*, which comprise over half of the heathland flora sampled, thus highlighting these taxa as a conservation management priority.

4. Materials and Methods

4.1. Taxon Sampling and Field-Observed Life History

Initial collections for this study included samples from 360 individuals representing five *Solanum* species across three sexual systems from 10 locations of the Top End region of the Northern Territory, Australia, during three collecting expeditions from 2009 to 2014 (Table 1, Figure 1). Ultimately, the only samples retained following strict filtering steps required for downstream analyses (see below) were those of taxa representing the sexual systems of primary interest in this study: cosexuality and dioecy.

All taxa in this study share a common Australian ancestor [30,66], estimated at approximately 5 million years old [67]; chosen taxa represent a polyphyletic assemblage. Four taxa have a dioecious sexual system (*S. asymmetriphyllum* R.L. Specht, *S. cowiei* Martine, *S. ossicruentum* Martine and J. Cantley, and *S. sejunctum* K. Brennan, Martine, Symon), and *S. raphiotes* A.R. Bean is cosexual [31]. The knowledge of relationships among this group of Australian *Solanum* taxa is still in flux, but of the four dioecious taxa, *S. asymmetriphyllum* and *S. sejunctum* are sister taxa (i.e. Kakadu Clade) [66] and are in turn sister to a clade of cosexual taxa, which includes *S. raphiotes* [12]. These two clades appear to be reciprocally sister to a larger dioecious clade of 13 taxa (i.e., Kimberley Dioecious Clade) [23,66], which include dioecious *S. cowiei* and *S. ossicruentum*. These dioecious and cosexual taxa are in turn sister to a clade of ca. 15 andromonoecious taxa (i.e., Andromonoecious Bush Tomato Clade) [66]. Evidence of hybridization among any of the taxa with different sexual systems has not been observed in the field via intermediate morphological characters, but attempts to form hybrids of taxa with different sexual systems in a greenhouse setting have proven occasionally successful [68] and are currently under study [69].

The sampling strategy for the collection of populations included three sites where two *Solanum* species with different sexual systems were present. We were able to collect three population pairs of dioecious *S. asymmetriphyllum* and cosexual *S. raphiotes* at three locations in Kakadu National Park where they sympatrically occur: at Merl Rock, Bardedjilidji, and Jabiluka. During collection, populations were discerned by geographic limitations of localities where collections could be safely made. Notably, individuals of all taxa, except for *S. raphiotes*, tended to occur in higher densities around sandstone escarpment and boulders. This differential in density may play a role in the genetic patterns observed in our data. However, in an effort to mitigate this potential effect, we collected individuals of both cosexual and dioecious taxa occurring as evenly as possible across the same accessible geographic area.

4.2. DNA Library Preparation and Sequencing

Genomic DNA was extracted from dried leaf material using a modified CTAB method [70]. Leaf tissue was pulverized using a GenoGrinder (SPEX Sample Prep, Metuchen, NJ, USA) and steel beads in 2 mL microfuge tubes. Leaf material was hydrated in lysis buffer and incubated for 30 min at 37 °C and centrifuged, and the supernatant was then extracted with a chloroform:isoamyl alcohol (24:1) solution before being precipitated at −20 °C in ice-cold 100% isopropanol for 20 min. DNA pellets were cleaned with two consecutive washes of ethanol, 75% and 95%, respectively, resuspended in ddH_2O, and then incubated at 37 °C with RNase A (Sigma-Aldrich, St. Louis, MO, USA) for one hour. For restriction enzyme digestion, we found that samples benefited from desalting, which resulted in more complete and clean digestion in preparation for double digest restriction enzyme site-associated DNA sequencing (ddRADseq) library preparation following a slightly modified protocol [71,72]. Desalting was conducted by repeating the above isopropanol precipitation step and pellet washes with ethanol. DNA concentration was tested with a Qubit 2.0 (Invitrogen, Waltham, MA, USA) using the High Sensitivity Kit, and the samples were then normalized in concentration to have ≈300 ng of DNA per sample in a 96-well plate with a concentration of 60 ng/µL, adding one well of a purchased commercial control gDNA of *Solanum tuberosum* (BioChain, Newark, CA, USA) in each plate. EcoRI HF and SphI HF restriction enzymes were chosen bioinformatically for gDNA digestion as they have comparable cut site frequency for the focal *Solanum* taxa, as determined in silico using *Solanum tuberosum* as a closely related reference genome using Geneious (San Diego, CA, USA). All adapters, primers, and barcodes were ordered from Eurofins Operon (Louisville, KY, USA). All enzymes used within (T4 DNA Ligase, Phusion PCR kit, EcoRI HF, SphI HF) were ordered from NE Biolabs (Ipswich, MA, USA).

Using 300 ng of gDNA template at 60 ng/µL (5 µL volume), 0.75 µL of SphI-HF and 0.75 µL of EcoRI-HF (15 units each per reaction), 2.5 µL of 10× CutSmart Buffer, and 16 µL of water (making a total of 25 µL of reaction volume per well), the DNA was digested for 3 h at 37 °C in an incubator, kept at 4 °C, and then cleaned using AMPure XP Beads as suggested by the manufacturer (Beckman Coulter). This and the following procedural tests were performed to determine the optimum number of digestion enzyme units (5 to 25 units) with a sample pool of 48 individuals that were not the experimental samples but instead from extra spiny *Solanum* leaf tissue collected from greenhouse grown plants. This digested and cleaned DNA was quantified with a Qubit BR Kit and normalized from 0.1 to 0.2 µg (100 to 200 ng). The annealed/barcoded adapters were ligated to the digested and cleaned DNA samples using a simple mixture of adapters, 2 µL of each primer, 2 µL of T4 DNA Ligase, and 4 µL 10 X ligase buffer per well added to 30 µL of digested DNA, resulting in a total ligase reaction volume of 40 µL per sample. The selected barcodes and adapters were ligated in a thermocycler according to the following routine: heating to 37 °C for 30 min, heating to 65 °C for 10 min, cooling slowly by 2 °C every 90 sec until 21 °C was reached, holding the temperature at 21 °C for 10 min, holding the temperature at 4 °C, and then cleaning using AMPure XP Beads. Digested DNA fragments with ligated adapters and barcodes were then size-selected using a 10 mm thick 2% agarose maxi-gel and prestained with 17.5 µL of GelRed (Phenix Research, Swedesboro, NJ, USA), with 2 combs with a thickness of 1.5 mm and 20 wells; the pools of 48 barcoded samples each were loaded into the wells after being mixed with 10 µL of 10× GoTaq Green buffer (undiluted) (Promega, Madison, WI, USA). Using the UV lamp on the gel visualization box (Major Science, San Jose, CA, USA), the gel regions between 200 and 500 bp were cut from the gel with a razor blade and frozen at −20 °C before then being placed in a 2 mL cellulose filter tube (Spin X Column, Costar, Washington, DC, USA) and centrifuged at room temperature for 30 min at 13,000 rpm. To ensure all DNA was removed from the agarose, 100 µL of water was added, and the columns were spun again for 10 min at 15,000 rpm. Each pool was cleaned with the AMPure XP beads again with 1.5 X beads to pool volume to reach a final volume of ~30 µL and quantified with either the Qubit BR or HS Kit to ensure 100 to 300 ng of DNA per pool. For the final library PCR amplification step, each pool was normalized to 100 ng of DNA. The PCR reaction used the Phusion High Fidelity PCR Kit (200 reactions per kit) and included 8.5 µL of water, 4 µL of 5 X HF Buffer, 0.4 µL of 10 µM dNTPs, 0.6 µL of DMSO, 2.0 µL of 10 µM PCR1 primer, 2.0 µL of 10 µM PCR2-index primer (1 to 12 indices), 20–25 ng of DNA/water in 10 µL, and 1.0 µL of DNA Phusion Polymerase per reaction.

The thermocycler settings were as follows: step 1: 98 °C for 30 s; step 2: 98 °C for 10 s; step 3: 79 °C for 30 s; step 4: 72 °C for 30 s; step 5: go to 2, 14 times; step 6: 72 °C for 10 min; step 7: 4 °C forever. Tests were performed to determine the best number of cycle repetitions (from 5 to 30) for optimum library amplification. These tests were conducted on a sample pool separate from the experimental sample set. To determine the success and quantity of the pooled library amplifications, the concentration of the libraries before and after PCR, as well as before and after magnetic bead cleaning of the libraries, were determined with the Qubit HS Kit. Using this method, we could determine that optimum and/or successful amplification was on average 50 times the original product. Gels of amplified libraries were not used as it was found that too much material was lost and visualization was inconsistent. Each final amplified library was quantified with a Bioanalyzer (Agilent, Santa Clara, CA, USA) fragment analyzer at iBEST (University of Idaho), which averaged ~37 ng/µL per pool, ranging from 1000 to 6000 ng of total amplified DNA. In some cases, the pools were size-selected again with PippenPrep at iBEST to remove small molecular weight contaminants before sequencing. Each pool was sequenced at the QB3 Vincent J. Coates Genomics Sequencing Laboratory at the University of California Berkeley on an Illumina MiSeq in a 150-paired-end read rapid run. Four pools of 48 barcoded and indexed samples were combined into one lane, resulting in 192 individual samples per lane (384 individuals were sequenced in two lanes).

4.3. Sequenced Data Processing and 'Hard-Filtering'

Raw paired-end Illumina sequencing reads were demultiplexed, followed by the removal of their adaptors and de novo assembly into loci using the program 3.0.63 ipyrad [37]. All Fastq sequence files are accessible from GenBank's National Center for Biotechnology Information Short Read Archive database (SUB4712600/PRJNA498556). Two samples were removed prior to assembly, which had low numbers of sequencing reads. Loci were removed if they were not shared across at least 100 of the 358 samples. Using the output from the ipyrad run, vcftools [73] was used to identify individuals with >60% missing data, which were then systematically removed. Loci with >50% missingness and loci that were very close to each other (<10 bp) were then removed. Output files were converted to formats compatible with plink [74], where additional filtering of minor alleles with frequencies of >1% plus linked loci within 1 kb of each other and with R_2 values > 0.8 to account for linkage disequilibrium were removed. The resulting 'hard-filtered' dataset reduced the total number of individuals to 193, representing 1458 loci. A full detailing of all parameters used in ipyrad assembly followed by hard filtering performed in vcftools and plink are available in Supplemental File S1 and S2, respectively.

All following analyses were conducted using R Studio (ver 1.4.1717, Rstudio, Boston, MA, USA) and associated packages [75]. Descriptive statistics such as F_{ST}, F_{IS}, and observed and expected heterozygosity (H_O, H_E) were calculated using the packages hierfstat [76], pegas [77], and adegenet [78]. Pairwise-F_{ST} values were calculated at two different hierarchical levels: among the 10 populations and separately for the five species. Both calculations implemented the Weir–Cockerham correction method of Pairwise-F_{ST} estimation to account for differences in the number of individuals of each species or population [38]. Bartlett's tests were conducted at both hierarchical levels to investigate the statistical significance of variance among calculated values of H_O and H_E. To further understand the variance at the different hierarchical levels, an AMOVA was performed in the package poppr [79].

The genetic structure of the populations and species was estimated and visualized using two multivariate methods: Principal Components Analysis (PCA) and Discriminant Analysis of Principle Components (DAPC) on the hard-filtered dataset in the adegenet program. Prior to running PCAs, we first used *k*-means clustering to identify the appropriate number of genetic clusters as indicated by the lowest Bayesian information criterion value using the find.clusters function. Principle Component (PC) data were then transformed by discriminant analysis. DAPC requires users to define the number of PCs retained in the analysis to help mitigate the generation of questionable results since too few or too many retained PCs can affect the balance between overfitting the data and the power of the discriminant analysis. We analyzed between 3 and 250 PCs and then used an *a*-score spline interpolation approach (i.e., the proportion of successful reassignments of group membership of all individuals in the dataset corrected for the number of retained PCs) followed by implementation of the xvalDAPC function for cross-validation to determine that 9 PCs was the optimal number to retain before discriminant analysis. To visualize the DAPC, we generated a 'compoplot' (i.e., a structure-like bar plot) in which each individual was proportionally assigned to one or many *k*-means clusters to illustrate population and species membership probability and to identify admixed individuals.

To understand different axes of variation in our dataset for species and populations at a more nuanced level, the hard-filtered dataset was sub-sampled in several ways using the repool function before additional PCA and DAPCs were performed. The four sub-sampled datasets were (i) the two populations of *S. sejunctum*, (ii) three populations of *S. asymmetriphyllum*, (iii) three populations of *S. raphiotes* and, (iv) the combination of dioecious *S. asymmetriphyllum* and cosexual *S. raphiotes* used to compare the variation in sympatrically occurring species with different sexual systems at three different geographic locations (Jabiluka, Merl Rock, and Bardedjilidji).

For *S. asymmetriphyllum* and *S. raphiotes*, which each had three populations, we used Mantel tests to estimate whether isolation by distance influenced the observed distribution of genetic diversity of these taxa. Additionally, as isolation may not be the only correlating factor of geographic distance to genetic distance, we explored the subtle difference of whether populations of each species conformed more to predictions of clinal genetic differentiation (i.e., a classic IBD scenario) or if they functioned more similarly to a model of distant patches in which each population is genetically differentiated and distantly located but not behaving in a cline-like fashion. This was explored by plotting an IBD plot using the 2-dimensional kernel density estimation function kde2d in adegenet.

5. Conclusions

Arguably, our most significant finding is that all five study species of Australian *Solanum* of the Australian monsoon tropics are highly inbred. Secondarily, and more directly addressing the main hypotheses tested in this research, no pattern of genetic structure or diversity can be definitively attributed as resulting from the presence of a sexual system alone. However, in the portion of the complete dataset that we were able to compile, we can confidently say that the dioecious taxa have decreased genetic diversity and more admixture between populations in comparison to the cosexual taxon. Several traits present in the focal taxa—which may or may not be correlated to dioecy—likely shape the genetic landscape of each taxon, but conclusions could not be made given the limited populations that ended up in the final cleaned sequence dataset. When all factors are taken into consideration, our data suggest differential combinations of sexual system, correlated life history traits, and demographic history of populations better explain the observed patterns than sexual system alone. Furthermore, when certain conditions are met, obligately outcrossing dioecious taxa may be more capable of maintaining a greater degree of admixture among populations than cosexual taxa. These insights complement the theoretical framework hypothesizing that the evolution of dioecy in Australian *Solanum* may have proceeded as a means to avoid the genetic consequences of self-fertilization and may even support hypotheses of benefits gained through differential resource allocation partitioned across male and female individuals. Emerging from our data are new hypotheses of a testable multi-factorial framework in which benefits of the evolution of dioecy can be teased apart for a more nuanced understanding in future research.

An alarming finding of this research was elevated homozygosity across all taxa, regardless of the sexual system. Although it appears that *Solanum* taxa have some evolved strategies to live in the presence of fire, many of these taxa are presumably fire-sensitive, and the recent human-mediated shift in regional fire regime looms large as a significant factor fueling reduced genetic diversity across many taxa of the Australian monsoon tropics.

Frequent fires are noted as occurring more often than the time necessary for obligate reseeding taxa such as *Solanum* to reach full reproductive maturity. From a standpoint of genetic diversity and inbreeding, these selective forces could certainly promote a high occurrence of inbreeding for taxa such as *Solanum* that are tolerant of historically infrequent low-temperature fires but inadequately adapted for recent and future fire intensities. In fact, future studies should be designed to assess the hypothesis that high levels of inbreeding and elevated homozygosity may be a common feature for many fire-sensitive sandstone taxa of the Top End, and perhaps across all of the Australian monsoon tropics. Resultantly, this work highlights a pressing need to include fire-sensitive taxa in the AMT as an important conservation priority for conservation managers.

Supplementary Materials: The following supporting information can be downloaded at: https://www.mdpi.com/article/10.3390/plants12112200/s1, Figure S1: Isolation by distance plots for (A) *S. asymmetriphyllum*, (B) *S. raphiotes*.

Author Contributions: Conceptualization, J.T.C., I.E.J.-T. and C.T.M.; methodology: J.T.C., I.E.J.-T. and C.T.M.; software, J.T.C. and I.E.J.-T.; validation, C.T.M.; formal analysis, J.T.C.; investigation, J.T.C., I.E.J.-T., M.D.R. and D.H.; resources, C.T.M.; data curation, C.T.M.; writing—original draft preparation, J.T.C.; writing—review and editing, all authors; visualization, J.T.C. and S.K.; supervision, C.T.M.; project administration, J.T.C. and C.T.M.; funding acquisition, J.T.C. and C.T.M. All authors have read and agreed to the published version of the manuscript.

Funding: This work was supported by the David Burpee Endowment at Bucknell University, with additional support from a Botanical Society of American Undergraduate Student Research Award (to MR), San Francisco State University College of Science and Engineering, and a BioLuminary Fellowship supporting undergraduate research awarded through the SF State Department of Biology.

Data Availability Statement: The data presented in this study are openly available on GitHub at https://github.com/cantley (accessed on 31 May 2023) and NCBI's Sequence Read Archive (SRA) at https://www.ncbi.nlm.nih.gov/sra/docs/ (accessed on 31 May 2023).

Acknowledgments: The authors appreciate the generous assistance from the staff of the Northern Territory Herbarium (Palmerston (DNA) and Alice Springs (NT)) and the George Safford Torrey Herbarium, University of Connecticut (CONN), as well as assistance with collecting permits from Parks Australia, the Northern Territory Parks and Wildlife Commission, and the Western Australia Department of Parks and Wildlife. We gratefully acknowledge the Traditional Owners of the lands where many of our collections were made. Numerous individuals assisted with collection of specimens in the field during trips between 2009 and 2014, including R. Martine, I. Martine, J. Martine, E. Sullivan, E. Capaldi, D. Vogt, B. Figley, G. Lionheart, and B. Lavoie. This work used the Vincent J. Coates Genomics Sequencing Laboratory at UC Berkeley, supported by NIH S10 Instrumentation Grants S10RR029668 and S10RR027303.

Conflicts of Interest: The authors report no conflict of interest.

References

1. Darwin, C. *The Different Forms of Flowers on Plants of the Same Species*; John Murray: London, UK, 1877.
2. Darwin, C. *The Effects of Cross and Self Fertilization in the Vegetable Kingdom*; Ams Press Inc.: London, UK, 1876.
3. Barrett, S.C. Understanding plant reproductive diversity. *Philos. Trans. R. Soc. B Biol. Sci.* **2010**, *365*, 99–109. [CrossRef] [PubMed]
4. Bull, J.J.; Charnov, E.L. On irreversible evolution. *Evolution* **1985**, *39*, 1149–1155. [CrossRef] [PubMed]
5. Heilbuth, J. Lower species richness in dioecious clades. *Am. Nat.* **2000**, *156*, 221–241. [CrossRef]
6. Barrett, S.C. The evolution of plant reproductive systems: How often are transitions irreversible? *Proc. R. Soc. B Biol. Sci.* **2013**, *280*, 20130913. [CrossRef]
7. Muenchow, G.E. Is dioecy associated with fleshy fruit? *Am. J. Bot.* **1987**, *74*, 287–293. [CrossRef]
8. Vamosi, J.C.; Vamosi, S.M. The role of diversification in causing the correlates of dioecy. *Evolution* **2004**, *58*, 723–731.
9. Muyle, A.; Martin, H.; Zemp, N.; Mollion, M.; Gallina, S.; Tavares, R.; Silva, A.; Bataillon, T.; Widmer, A.; Glémin, S.; et al. Dioecy is associated with high genetic diversity and adaptation rates in the plant genus Silene. *Mol. Biol. Evol.* **2021**, *38*, 805–818. [CrossRef]
10. Renner, S.S.; Ricklefs, R.E. Dioecy and its correlates in the flowering plants. *Am. J. Bot.* **1995**, *82*, 595–606. [CrossRef]
11. Sakai, A.K.; Wagner, W.L.; Ferguson, D.M.; Herbst, D.R. Origins of dioecy in the Hawaiian flora. *Ecology* **1995**, *76*, 2517–2529. [CrossRef]

12. Martine, C.T.; Jordon-Thaden, I.E.; McDonnell, A.J.; Cantley, J.T.; Hayes, D.; Roche, M.; Frawley, E.S.; Gilman, I.S.; Tank, D. Phylogeny of the Australian *Solanum dioicum* group using seven nuclear genes: Testing Symon's fruit and seed dispersal hypotheses. *PLoS ONE* **2019**, *14*, e0207564. [CrossRef]
13. Whalen, M.D.; Costich, D.E. Andromonoecy in *Solanum*. In *Solanaceae: Biology and Systematics*; Columbia University Press: New York, NY, USA, 1986; pp. 284–302.
14. Knapp, S.; Sagona, E.; Carbonell, A.; Chiarini, F. A revision of the *Solanum elaeagnifolium* clade (Elaeagnifolium clade; subgenus Leptostemonum, Solanaceae). *PhytoKeys* **2017**, *84*, 1–104. [CrossRef]
15. Knapp, S. Solanaceae Source. PBI *Solanum*: A Worldwide Treatment. 2014. Available online: http://www.solanaceaesource.org/ (accessed on 31 May 2023).
16. Symon, D. Dioecious solanums. *Taxon* **1970**, *19*, 909–910. [CrossRef]
17. Symon, D.E. Sex forms in *Solanum* (Solanaceae) and the role of pollen collecting insects. In *The Biology and Taxonomy of the Solanaceae*; Hawkes, J., Lester, G., Skelding, N.A., Eds.; Academic Press for the Linnean Society: London, UK, 1979; pp. 385–397.
18. Gagnon, E.; Hilgenhof, R.; Orejuela, A.; McDonnell, A.J.; Sablok, G.; Aubriot, X.; Giacomin, L.; Gouvêa, L.; Bohs, L.; Dodsworth, S.; et al. Phylogenomic data reveal hard polytomies across the backbone of the large genus *Solanum* (Solanaceae). *Am. J. Bot.* **2022**, *109*, 1–22.
19. Barrett, R.L. *Solanum zoeae* (Solanaceae), a new species of bush tomato from the North Kimberley, Western Australia. *Nuytsia* **2013**, *23*, 5–21. [CrossRef]
20. Martine, C.T.; Symon, D.E.; Evans, E.C. A new cryptically dioecious species of bush tomato (*Solanum*) from the Northern Territory, Australia. *PhytoKeys* **2013**, *30*, 23–32. [CrossRef]
21. Anderson, G.J.; Anderson, M.K.J.; Patel, N. The ecology, evolution, and biogeography of dioecy in the genus *Solanum*: With paradigms from the strong dioecy in *Solanum polygamum*, to the unsuspected and cryptic dioecy in *Solanum conocarpum*. *Am. J. Bot.* **2015**, *102*, 471–486. [CrossRef] [PubMed]
22. Martine, C.T.; Cantley, J.T.; Frawley, E.S.; Butler, A.R.; Jordon-Thaden, I.E. New functionally dioecious bush tomato from northwestern Australia, *Solanum ossicruentum*, may utilize "trample burr" dispersal. *PhytoKeys* **2016**, *63*, 19–29. [CrossRef]
23. Williams, T.M.; Hayes, J.; Cantley, J.T.; McDonnell, A.J.; Jobson, P.; Martine, C.T. *Solanum scalarium* (Solanaceae), a newly-described dioecious bush tomato from Judbarra/Gregory National Park, Northern Territory, Australia. *PhytoKeys* **2022**, *216*, 103–116. [CrossRef]
24. D'Arcy, W.G. Solanaceae studies II. Typification of the subdivisions of *Solanum*. *Ann. Mo. Bot. Gard.* **1972**, *59*, 262–278. [CrossRef]
25. Knapp, S. A revision of the *Solanum sessile* species group (section *Geminata* pro parte: Solanaceae). *Bot. J. Linn. Soc.* **1991**, *105*, 179–210. [CrossRef]
26. Knapp, S.; Persson, V.; Blackmore, S. Pollen morphology and functional dioecy in *Solanum* (Solanaceae). *Plant Syst. Evol.* **1998**, *210*, 113–139. [CrossRef]
27. Whalen, M.D. Conspectus of species groups in *Solanum* subgenus *Leptostemonum*. *Gentes Herb.* **1984**, *12*, 179–282.
28. Martine, C.T.; Anderson, G.J. Dioecy, pollination, and seed dispersal in Australian spiny *Solanum*. In Proceedings of the VIth International Solanaceae Conference: Acta Horticulturae, Madison, WI, USA, 23–27 July 2006; ISHS: Leuven, Belgium, 2007; Volume 745, pp. 269–283.
29. McDonnell, A.J.; Wetreich, H.B.; Cantley, J.T.; Jobson, P.C.; Martine, C.T. *Solanum plastisexum*, an enigmatic new bush tomato from the Australian Monsoon Tropics exhibiting breeding system fluidity. *PhytoKeys* **2019**, *124*, 39–55. [CrossRef] [PubMed]
30. Martine, C.T.; Anderson, G.J.; Les, D.H. Gender-bending aubergines: Molecular phylogenetics of cryptically dioecious *Solanum* in Australia. *Aust. Syst. Bot.* **2009**, *22*, 107–120. [CrossRef]
31. Bean, A.R. A taxonomic revision of the *Solanum echinatum* group (Solanaceae). *Phytotaxa* **2012**, *57*, 33–50. [CrossRef]
32. Anderson, G.J. Dioecious *Solanum* of hermaphroditic origin is an example of a broad convergence. *Nature* **1979**, *282*, 836–838. [CrossRef]
33. Anderson, G.J.; Symon, D.E. Functional dioecy and andromonoecy in *Solanum*. *Evolution* **1989**, *43*, 204–219. [CrossRef]
34. Symon, D.E. A revision of the genus *Solanum* in Australia. *J. Adel. Bot. Gard.* **1981**, *4*, 1–367.
35. Ndem-Galbert, J.R.; Hall, J.; McDonnell, A.J.; Martine, C.T. Differential reward in "male" versus "female" pollen of functionally dioecious *Solanum* (Solanaceae). *Am. J. Bot.* **2021**, *108*, 2282–2293. [CrossRef]
36. Miller, J.S.; Diggle, P.K. Correlated evolution of fruit size and sexual expression in andromonoecious *Solanum* sections *Acanthophora* and *Lasiocarpa* (Solanaceae). *Am. J.* **2007**, *94*, 1706–1715.
37. Eaton, D.A.; Overcast, I. ipyrad: Interactive assembly and analysis of RADseq datasets. *Bioinformatics* **2020**, *36*, 2592–2594. [CrossRef]
38. Weir, B.S.; Cockerham, C.C. Estimating F-statistics for the analysis of population structure. *Evolution* **1984**, *38*, 1358–1370.
39. Hilgenhof, R.; Gagnon, E.; Knapp, S.; Aubriot, X.; Tepe, E.J.; Bohs, L.; Giacomin, L.L.; Gouvêa, Y.F.; Stehmann, J.R.; Martine, C.T.; et al. Morphological trait evolution in *Solanum* (Solanaceae): Evolutionary lability of key taxonomic characters. *bioRxiv* **2023**. bioRxiv:2023.02. 24.529849.
40. Marino, C. Cleaning the Variable Mess: A Population Genomics Approach to Understanding the Evolutionary History of a Complicated Plant Group (2023). Honors Thesis 645, Bucknell University, Lewisburg, PA, USA, 2023. Available online: https://digitalcommons.bucknell.edu/honors_theses/645 (accessed on 31 May 2023).

41. Brennan, K.; Martine, C.T.; Symon, D. *Solanum sejunctum* (Solanaceae), a new functionally dioecious species from Kakadu National Park, Northern Territory, Australia. *Beagle Rec. Mus. Art Gall. North. Territ.* **2006**, *22*, 1–7. [CrossRef]
42. Bean, A.R. 2012 Onwards. *Solanum* species of eastern and northern Australia. Version: 25th January 2023. Available online: http://www.delta-intkey.com (accessed on 31 May 2023).
43. Martine, C.T.; Lavoie, E.; Tippery, N.L.; Vogt, F.D.; Les, D.H. *Solanum* from Litchfield National Park is a relative of *S. dioicum*. *North. Territ. Nat.* **2011**, *23*, 29–38.
44. Ennos, R.A. Detection and measurement of selection: Genetic and ecological approaches. In *Plant Population Genetics, Breeding and Genetic Resources*; Brown, A.H.D., Clegg, M.T., Kahler, A.L., Wier, B.S., Eds.; Sinauer Associates: Sunderland, MA, USA, 1989; pp. 200–214.
45. Farris, M.A.; Mitton, J.B. Population density, outcrossing rate, and heterozygote superiority in ponderosa pine. *Evolution* **1984**, *38*, 1151–1154. [CrossRef]
46. Bowman, D.M.; Brown, G.K.; Braby, M.F.; Brown, J.R.; Cook, L.G.; Crisp, M.D.; Ford, F.; Haberle, S.; Hughes, J.; Isagi, Y.; et al. Biogeography of the Australian monsoon tropics. *J. Biogeogr.* **2010**, *37*, 201–216. [CrossRef]
47. Haug, G.H.; Tiedemann, R. Effect of the formation of the Isthmus of Panama on Atlantic Ocean thermohaline circulation. *Nature* **1998**, *393*, 673–676. [CrossRef]
48. O'Dea, A.; Lessios, H.A.; Coates, A.G.; Eytan, R.I.; Restrepo-Moreno, S.A.; Cione, A.L.; Collins, L.S.; De Queiroz, A.; Farris, D.W.; Norris, R.D.; et al. Formation of the Isthmus of Panama. *Sci. Adv.* **2016**, *2*, e1600883. [CrossRef]
49. Bowman, D.M.J.S. The Australian summer monsoon: A biogeographic perspective. *Aust. Geogr. Stud.* **2002**, *40*, 261–277. [CrossRef]
50. Felderhof, L.; Gillieson, D. Comparison of fire patterns and fire frequency in two tropical savanna bioregions. *Austral Ecol.* **2006**, *31*, 736–746. [CrossRef]
51. Byrne, M.; Yeates, D.K.; Joseph, L.; Kearne, M.; Bowler, J.; Williams, M.A.J.; Cooper, S.; Donnellan, S.C.; Keogh, J.S.; Leys, R.; et al. Birth of a biome: Insights into the assembly and maintenance of the Australian arid zone biota. *Mol. Ecol.* **2008**, *17*, 4398–4417. [CrossRef] [PubMed]
52. Vigilante, T. Analysis of explorer's records of Aboriginal landscape burning in the Kimberley region of Western Australia. *Aust. Geogr. Stud.* **2001**, *39*, 135–155. [CrossRef]
53. Russell-Smith, J.; Yates, C.; Edwards, A.; Allan, G.E.; Cook, G.D.; Cooke, P.; Craig, R.; Heath, B.; Smith, R. Contemporary fire regimes of northern Australia, 1997–2001: Change since Aboriginal occupancy, challenges for sustainable management. *Int. J. Wildland Fire* **2003**, *12*, 283–297. [CrossRef]
54. Russell-Smith, J.; Ryan, P.G.; Klessa, D.; Waight, G.; Harwood, R. Fire regimes, fire-sensitive vegetation and fire management of the sandstone Arnhem Plateau, monsoonal northern Australia. *J. Appl. Ecol.* **1998**, *35*, 829–846. [CrossRef]
55. Bowman, D.M.J.S. The impact of Aboriginal landscape burning on the Australian biota. *New Phytol.* **1998**, *140*, 385–410. [CrossRef]
56. Bowman, D.M.J.S.; Price, O.; Whitehead, P.J.; Walsh, A. The 'wilderness effect' and the decline of *Callitris intratropica* on the Arnhem Land Plateau, northern Australia. *Aust. J. Bot.* **2001**, *49*, 665–672. [CrossRef]
57. Bowman, D.M.J.S.; Walsh, A.; Prior, L.D. Landscape analysis of Aboriginal fire management in central Arnhem Land, north Australia. *J. Biogeogr.* **2004**, *31*, 207–223. [CrossRef]
58. Kelly, L.T.; Giljohann, K.M.; Duane, A.; Aquilué, N.; Archibald, S.; Batllori, E.; Bennett, A.F.; Buckland, S.T.; Canelles, Q.; Clarke, M.F.; et al. Fire and biodiversity in the Anthropocene. *Science* **2020**, *370*, eabb0355. [CrossRef] [PubMed]
59. Bowman, D.M.J.S.; Franklin, D.C.; Price, O.F.; Brook, B.W. Land management affects grass biomass in the *Eucalyptus tetrodonta* savannas of monsoonal Australia. *Austral Ecol.* **2007**, *32*, 446–452. [CrossRef]
60. Russell-Smith, J.; Bowman, D.M.J.S. Conservation of monsoon rainforest isolates in the Northern Territory, Australia. *Biol. Conserv.* **1992**, *59*, 51–63. [CrossRef]
61. Russell-Smith, J.; Lucas, D.E.; Brock, J.; Bowman, D.M.J.S. *Allosyncarpia*-dominated rain forest in monsoon northern Australia. *J. Veg. Sci.* **1993**, *4*, 67–82. [CrossRef]
62. Russell-Smith, J.; Craig, R.; Gill, A.M.; Smith, R.; Williams, J.E. *Australian Fire Regimes: Contemporary Patterns (April 1998–March 2000) and Changes since European Settlement*; Department of the Environment and Heritage: Canberra, Australia, 2002.
63. Bowman, D.M.J.S.; Panton, W.J. Decline of *Callitris intratropica* in the Northern Territory: Implications for pre- and post-colonisation fire regimes. *J. Biogeogr.* **1993**, *20*, 373–381. [CrossRef]
64. Bowman, D.M.J.S. Preliminary observations on the mortality of *Allosyncarpia ternata* stems on the Arnhemland plateau, northern Australia. *Aust. For.* **1994**, *57*, 62–64. [CrossRef]
65. Price, O.; Bowman, D.M.J.S. Fire-stick forestry: A matrix model in support of skillful fire management of *Callitris intratropica* R.T. Baker by north Australian Aborigines. *J. Biogeogr.* **1994**, *21*, 573–580. [CrossRef]
66. Martine, C.T.; Vanderpool, D.; Anderson, G.J.; Les, D.H. Phylogenetic relationships of andromonoecious and dioecious Australian species of *Solanum* subgenus *Leptostemonum* section *Melongena*: Inferences from ITS sequence data. *Syst. Bot.* **2006**, *31*, 410–420. [CrossRef]
67. Särkinen, T.; Bohs, L.; Olmstead, R.G.; Knapp, S. A phylogenetic framework for evolutionary study of the nightshades (Solanaceae): A dated 1000-tip tree. *BMC Evol. Biol.* **2013**, *13*, 214. [CrossRef]

68. Hayes, D.S. Ex Situ Interspecies Crossing Rates Infer Importance of Geographic Barriers in Speciation among Closely Related *Solanum* Species of the Australian Monsoon Tropics. Honors Thesis, Bucknell University, Lewisburg, PA, USA, 2018. p. 447. Available online: https://digitalcommons.bucknell.edu/honors_theses/447 (accessed on 31 May 2023).
69. Zizis, D.; Williams, T.M.; Martine, C.T. Heading for a Breakdown: Assessing Evolution through the Hybridization of Two Sexual Systems. Honors Thesis 650, Bucknell University, Lewisburg, PA, USA, 2023. Available online: https://digitalcommons.bucknell.edu/honors_theses/650 (accessed on 31 May 2023).
70. Doyle, J.J.; Doyle, J.L. Isolation of Plant DNA from Fresh Tissue. *Am. J. Plant Sci.* **1990**, *12*, 13–15.
71. Peterson, B.K.; Weber, J.N.; Kay, E.H.; Fisher, H.S.; Hoekstra, H.E. Double digest RADseq: An inexpensive method for de novo SNP discovery and genotyping in model and non-model species. *PLoS ONE* **2012**, *7*, e37135. [CrossRef] [PubMed]
72. Jordon-Thaden, I.E.; Beck, J.; Rushworth, C.; Windham, M.; Diaz, N.; Cantley, J.T.; Martine, C.T.; Rothfels, C. A basic ddRADseq two-enzyme protocol performs well with herbarium and silica-dried tissues across four genera. *Appl. Plant Sci.* **2020**, *8*, e11344. [CrossRef]
73. Danecek, P.; Auton, A.; Abecasis, G.; Albers, C.A.; Banks, E.; DePristo, M.A.; Handsaker, R.E.; Lunter, G.; Marth, G.T.; Sherry, S.T.; et al. The variant call format and VCFtools. *Bioinformatics* **2011**, *27*, 2156–2158. [CrossRef]
74. Purcell, S.; Neale, B.; Todd-Brown, K.; Thomas, L.; Ferreira, M.A.; Bender, D.; Maller, J.; Sklar, P.; De Bakker, P.I.; Daly, M.J.; et al. PLINK: A tool set for whole-genome association and population-based linkage analyses. *Am. J. Hum. Genet.* **2007**, *81*, 559–575. [CrossRef] [PubMed]
75. R Core Team. *R: A Language and Environment for Statistical Computing*; R Foundation for Statistical Computing: Vienna, Austria, 2021; Available online: https://www.R-project.org/ (accessed on 31 May 2023).
76. Goudet, J. Hierfstat, a package for R to compute and test hierarchical F-statistics. *Mol. Ecol. Notes* **2005**, *5*, 184–186. [CrossRef]
77. Paradis, E. Pegas: An R package for population genetics with an integrated–modular approach. *Bioinformatics* **2010**, *26*, 419–420. [CrossRef] [PubMed]
78. Jombart, T.; Ahmed, I. adegenet 1.3-1: New tools for the analysis of genome-wide SNP data. *Bioinformatics* **2011**, *27*, 3070–3071. [CrossRef]
79. Kamvar, Z.N.; Tabima, J.F.; Grünwald, N.J. Poppr: An R package for genetic analysis of populations with clonal, partially clonal, and/or sexual reproduction. *PeerJ* **2014**, *2*, e281. [CrossRef]

MDPI AG
Grosspeteranlage 5
4052 Basel
Switzerland
Tel.: +41 61 683 77 34

Plants Editorial Office
E-mail: plants@mdpi.com
www.mdpi.com/journal/plants

www.ingramcontent.com/pod-product-compliance
Lightning Source LLC
Chambersburg PA
CBHW040759250926
51850CB00007BA/1035
* 9 7 8 3 7 2 5 8 1 9 5 7 7 *